Quantitative Analysis in Financial Markets

Collected papers of the New York University
Mathematical Finance Seminar, Volume II

QUANTITATIVE ANALYSIS IN FINANCIAL MARKETS:
Collected Papers of the New York University Mathematical Finance Seminar

Editor: Marco Avellaneda (New York University)

Published

Vol. 1: ISBN 981-02-3788-X
ISBN 981-02-3789-8 (pbk)

Quantitative Analysis in Financial Markets

Collected papers of the New York University
Mathematical Finance Seminar, Volume II

Editor

Marco Avellaneda

Professor of Mathematics
Director, Division of Quantitative Finance
Courant Institute
New York University

Published by

World Scientific Publishing Co. Pte. Ltd.

P O Box 128, Farrer Road, Singapore 912805

USA office: Suite 1B, 1060 Main Street, River Edge, NJ 07661

UK office: 57 Shelton Street, Covent Garden, London WC2H 9HE

British Library Cataloguing-in-Publication Data
A catalogue record for this book is available from the British Library.

QUANTITATIVE ANALYSIS IN FINANCIAL MARKETS:
Collected Papers of the New York University Mathematical Finance Seminar, Volume II

ISBN 981-02-4225-5
ISBN 981-02-4226-3 (pbk)

Printed in Singapore by Fulsland Offset Printing

INTRODUCTION

It is a pleasure to edit the second volume of papers presented at the Mathematical Finance Seminar of New York University. These articles, written by some of the leading experts in financial modeling cover a variety of topics in this field. The volume is divided into three parts: (I) Estimation and Data-Driven Models, (II) Model Calibration and Option Volatility and (III) Pricing and Hedging.

The papers in the section on "Estimation and Data-Driven Models" develop new econometric techniques for finance and, in some cases, apply them to derivatives. They showcase several ways in which mathematical models can interact with data. Andrew Lo and his collaborators study the problem of dynamic hedging of contingent claims in incomplete markets. They explore techniques of minimum-variance hedging and apply them to real data, taking into account transaction costs and discrete portfolio rebalancing. These dynamic hedging techniques are called "epsilon-arbitrage" strategies. The contribution of Yacine Ait-Sahalia describes the estimation of stochastic processes for financial time-series in the presence of missing data. Andreas Weigend and Shanming Shi describe recent advances in non-parametric estimation based on Neural Networks. They propose new techniques for characterizing time-series in terms of Hidden Markov Experts. In their contribution on the statistics of prices, Geman, Madan and Yor argue that asset price processes arising from market clearing conditions should be modeled as pure jump processes, with no continuous martingale component. However, they show that continuity and normality can always be obtained after a time change. Kaushik Ronnie Sircar studies dynamic hedging in markets with stochastic volatility. He presents a set of strategies that are robust with respect to the specification of the volatility process. The paper tests his theoretical results on market data.

The second section deals with the calibration of asset-pricing models. The authors develop different approaches to model the so-called "volatility skew" or "volatility smile" observed in most option markets. In many cases, the techniques can be applied to fitting prices of more general instruments. Peter Carr and Dilip Madan develop a model for pricing options based on the observation of the implied volatilities of a series of options with the same expiration date. Using their model, they obtain closed-form solutions for pricing plain-vanilla and exotic options in markets with a volatility skew. Thomas Coleman and collaborators attack the problem of the volatility smile in a different way. Their method combines the use of numerical optimization, spline approximations, and automatic differentiation. They illustrate the effectiveness of their approach on both synthetic and real data for option pricing and hedging. Leisen and Laurent consider a discrete model for option pricing based on Markov chains. Their approach is based on finding a probability measure on the Markov chain which satisfies an optimality criterion. Avellaneda, Buff, Friedman, Kruk and Newman develop a methodology for calibrating Monte Carlo models. They show how their method can be used to calibrate models to the prices of traded options in equity and FX markets and to calibrate models of the term-structure of interest rates.

In the section entitled "Pricing and Risk-Management". Alexander Levin discusses a lattice-based methodology for pricing mortgage-backed securities. Peter Carr and Guang Yang consider the problem of pricing Bermudan-style interest rate options using Monte Carlo simulation. Alexander Lipton studies the symmetries and scaling relations that exist in the Black–Scholes equation and applies them to the valuation of path-dependent options. Cardenas and Picron, from Summit Systems, describe accelerated methods for computing the Value-at-Risk of large portfolios using Monte Carlo simulation. The closing paper, by Katherine Wyatt, discusses algorithms for portfolio optimization under structural requirements, such as trade amount limits, restrictions on industry sector, or regulatory requirements. Under such restrictions, the optimization problem often leads to a "disjunctive program". An example of a disjunctive program is the problem to select a portfolio that optimally tracks a benchmark, subject to trading amount requirements.

I hope that you will find this collection of papers interesting and intellectually stimulating, as I did.

Marco Avellaneda
New York, October 1999

ACKNOWLEDGEMENTS

The Mathematical Finance Seminar was supported by the New York University Board of Trustees and by a grant from the Belibtreu Foundation. It is a pleasure to thank these individuals and organizations for their support. We are also grateful to the editorial staff of World Scientific Publishing Co., and especially to Ms. Yubing Zhai.

THE CONTRIBUTORS

Yacine Ait-Sahalia is Professor of Economics and Finance and Director of the Bendheim Center for Finance at Princeton University. He was previously an Assistant Professor (1993–1996), Associate Professor (1996–1998) and Professor of Finance (1998) at the University of Chicago's Graduate School of Business, where he has been teaching MBA, executive MBA and Ph.D. courses in investments and financial engineering. He received the University of Chicago's GSB award for excellence in teaching and has been consistently ranked as one of the best instructors. He was named an outstanding faculty by *Business Week*'s 1997 *Guide to the Best Business Schools*. Outside the GSB, Professor Ait-Sahalia has conducted seminars in finance for investment bankers and corporate managers, both in Europe and the United States. He has also consulted for financial firms and derivatives exchanges in Europe, Asia and the United States. His research concentrates on investments, fixed-income and derivative securities, and has been published in leading academic journals. Professor Ait-Sahalia is a Sloan Foundation Research Fellow and has received grants from the National Science Foundation. He is also an associate editor for a number of academic finance journals, and a Research Associate for the National Bureau of Economic Research. He received his Ph.D. in Economics from the Massachusetts Institute of Technology in 1993 and is a graduate of France's Ecole Polytechnique.

Marco Avellaneda is Professor of Mathematics and Director of the Division of Financial Mathematics at the Courant Institute of Mathematical Sciences of New York University. He earned his Ph.D. in 1985 from the University of Minnesota. His research interests center around pricing derivative securities and in quantitative trading strategies. He has also published extensively in applied mathematics, most notably in the fields of partial differential equations, the design of composite materials and hydrodynamic turbulence. He was consultant for Banque Indosuez, New York, where he established a quantitative modeling group in FX options in 1996. Subsequently, he moved to Morgan Stanley & Co., as Vice-President in the Fixed-Income Division's Derivatives Products Group, where he remained until 1998, prior to returning to New York University. He is the managing editor of the *International Journal of Theoretical and Applied Finance*, and an associate editor of *Communications in Pure and Applied Mathematics*. He has published approximately 80 research papers, written a textbook entitled "*Quantitative Modeling of Derivative Securities: From Theory to Practice*" and edited the previous volume of the NYU Mathematical Finance Seminar series.

Robert Buff earned his Ph.D. in the Computer Science Department of the Courant Institute of Mathematical Sciences at New York University. He enjoys building interactive computational finance applications with intranet and internet technology. He implemented several online pricing and calibration tools for the Courant Finance webserver. Currently, he works in credit derivatives research at J. P. Morgan.

Juan D. Cardenas is Manager of Market and Credit Risk in the Financial Technology Group at Summit Systems, Inc. in New York. He joined Summit as Financial Engineer in 1993, previously working as a Financial Analyst at Banco de Occidente — Credencial in Bogotá, Colombia, 1986–1987. He was also an instructor in Mathematics at Universidad de Los Andes in Bogotá, Colombia, 1987. His education includes B.S. in Mathematics from Stanford University in 1985, and Ph.D. in Mathematics from Courant Institute of New York University in 1993. Publications: "VAR: One Step Beyond" (co-author) *RISK Magazine*, October 1997.

Peter Carr has been a Principal at Banc of America Securities LLC since January of 1999. He is the head of equity derivatives research and is also a visiting assistant professor at Columbia University. Prior to his current position, he spent three years in equity derivatives research at Morgan Stanley and eight years as a professor of finance at Cornell University. Since receiving his Ph.D. in Finance from UCLA in 1989, he has published articles in numerous finance journals. He is currently an associate editor for six academic journals and is the practitioner director for the Financial Management Association. His research interests are primarily in the field of derivative securities, especially American-style and exotic derivatives. He has consulted for several firms and has given numerous talks at both practitioner and academic conferences.

Thomas F. Coleman is Professor of Computer Science and Applied Mathematics at Cornell University and Director of a major Cornell research center: The Cornell Theory Center (a supercomputer center). He is the Chair of the SIAM Activity Group on Optimization (1998–2001) and is on the editorial board of several journals. Professor Coleman is the author of two books on computational mathematics. He is also the editor of four proceedings and has published over 50 journal articles. Coleman is a Mathworks, Inc. consultant. He established and now directs the Financial Industry Solutions Center (FISC), a computational finance joint venture with SGI located at 55 Broad Street in New York.

Craig A. Friedman is a Vice-President in the Fixed Income Division of Morgan Stanley (Global High Yield Group), working on quantitative trading strategies, pricing, and asset allocation problems. He received his Ph.D. from the Courant Institute of Mathematical Sciences at New York University.

Emmanuel Fruchard now in charge of the Front Office and Risk Management product line for continental Europe, has previously been leading the Financial Engineering group of Summit for three years. This group is in charge of the design of advanced valuation models and market & credit risk calculation methods. Before joining Summit in 1995, Mr. Fruchard was the head of Fixed Income & FX Research at Credit Lyonnais in Paris. He holds a BA degree in Economics and M.S. degrees in Mathematics and Computer Science.

Hélyette Geman is Professor of Finance at the University Paris IX Dauphine and at ESSEC Graduate Business School. She is a graduate from Ecole Normale Supérieure, holds a master's degree in Theoretical Physics and a Ph.D. in Mathematics from the University Paris VI Pierre et Marie Curie and a Ph.D. in Finance from the University Paris I Panthéon Sorbonne. Dr Geman is also a member of honor of the French Society of Actuaries. Previously a Director at Caisse des Dépôts in charge of Research and Development, she is currently a scientific adviser for major financial institutions and industrial firms. Dr Geman has extensively published in international journals and received in 1993 the first prize of the Merrill Lynch awards for her work on exotic options and in 1995 the first AFIR (Actuarial Approach for Financial Risk) International prize for her pioneering research on catastrophe and extreme events derivatives. She is the co-founder and editor of *European Finance Review*, associate editor of the journals *Mathematical Finance*, *Geneva Papers on Insurance*, and the *Journal of Risk* and the author of the book "Insurance and Weather Derivatives".

Lukasz Kruk is currently a Postdoctoral Associate at the Department of Mathematics, Carnegie Mellon University. He earned his Ph.D. in 1999 at the Courant Institute of New York University. His research interests include limit theorems in probability theory, stochastic control, queuing theory and mathematical finance.

Dietmar P.J. Leisen is a Postdoctoral Fellow in Economics at Stanford University's Hoover Institution. He earned his Ph.D. in 1998 from the University of Bonn. His research interests include pricing and hedging of futures and options, risk management, financial engineering, portfolio management, financial innovation; publications on financial engineering appeared in the journals Applied Mathematical Finance and the Journal of Economic Dynamics and Control. He worked as a Consultant for The Boston Consulting Group, Frankfurt, on shareholder value management in banking and with the Capital Markets Division of Societe Generale (SG), Paris, on the efficiency of pricing methods for derivatives.

Alexander Levin is a Vice President and Treasury R&D Manager of The Dime Bancorp., Inc. He holds Soviet equivalents of a M.S. in Applied Mathematics from University of Naval Engineering, and a Ph.D. in Control and Dynamic Systems from Leningrad State University (St. Petersburg). His career began in the field of control system engineering. His results on stability of interconnected systems and differential equations, aimed for the design of automated multi-machine power plants, were published in the USSR, USA and Europe. He taught at the City College of New York and worked as a quantitative system developer at Ryan Labs, Inc., a fixed income research and money management company, before joining The Dime Bancorp. His current interests include developing efficient numerical and analytical tools for pricing complex term-structure-contingent, dynamic assets, risk measurement and management, and modeling mortgages and deposits. He has recently published a number of papers in this field and is the author of *Mortgage Solutions*, *Deposit Solutions*, and *Option Solutions*, proprietary computer pricing systems at The Dime.

Yuying Li received her Ph.D. from the Computer Science Department at University of Waterloo, Canada, in 1988. She is the recipient of the 1993 Leslie Fox Prize in numerical analysis. Yuying Li is a senior research associate in computer science and a member of the Cornell/SGI Financial Industrial Solution Center (FISC). She has been working at Cornell since 1988. Her main research interests include scientific computing, computational optimization and computational finance.

Alex Lipton is a Vice President at the Deutsche Bank Forex Product Development Group and an Adjunct Professor of Mathematics at the University of Illinois. Alex earned his Ph.D. in pure mathematics from Moscow State University. At Deutsche Bank, he is responsible for modeling exotic multi-currency options with a particular emphasis on stochastic volatility and calibration aspects. Prior to joining Deutsche Bank, he worked at Bankers Trust where his responsibilities included research on foreign exchange, equity and fixed income derivatives and risk management. Alex worked for the Russian Academy of Sciences, MIT, the University of Massachusetts and the University of Illinois where he was a Full Professor of Applied Mathematics; in addition, for several years he was a Consultant at Los Alamos National Laboratory. Alex conducted research and taught numerous courses on analytical and numerical methods for fluid and plasma dynamics, astrophysics, space physics, and mathematical finance. He is the author of one book and more than 75 research papers. His latest book *Mathematical Methods for Foreign Exchange* will be published shortly by World Scientific Publishing Co. In January 2000, Alex became the first recipient of the prestigious "Quant of the Year" award by Risk Magazine for his work on a range of new derivative products.

Andrew W. Lo is the Harris & Harris Group Professor of Finance at MIT's Sloan School of Management and the director of MIT's Laboratory for Financial

Engineering. He received his Ph.D. in Economics from Harvard University in 1984, and taught at the University of Pennsylvania's Wharton School as the W.P. Carey Assistant Professor of Finance from 1984 to 1987, and as the W.P. Carey Associate Professor of Finance from 1987 to 1988. His research interests include the empirical validation and implementation of financial asset pricing models; the pricing of options and other derivative securities; financial engineering and risk management; trading technology and market microstructure; statistical methods and stochastic processes; computer algorithms and numerical methods; financial visualization; nonlinear models of stock and bond returns; and, most recently, evolutionary and neurobiological models of individual risk preferences. He has published numerous articles in finance and economics journals, and is a co-author of *The Econometrics of Financial Markets* and *A Non-Random Walk Down Wall Street*. He is currently an associate editor of the *Financial Analysis Journal*, the *Journal of Portfolio Management*, the *Journal of Computational Finance*, and the *Review of Economics and Statistics*. His recent awards include the Alfred P. Sloan Foundation Fellowship, the Paul A. Samuelson Award, the American Association for Individual Investors Award, and awards for teaching excellence from both Wharton and MIT.

Dilip B. Madan obtained Ph.D. degrees in Economics (1971) and Mathematics (1975) from the University of Maryland and then taught econometrics and operations research at the University of Sydney. His research interests developed in the area of applying the theory of stochastic processes to the problems of risk management. In 1988 he joined the Robert H. Smith School of Business where he now specializes in mathematical finance. His work is dedicated to improving the quality of financial valuation models, enhancing the performance of investment strategies, and advancing the understanding and operation of efficient risk allocation in modern economies. Of particular note are contributions to the field of option pricing and the pricing of default risk. He is a founding member and treasurer of the Bachelier Finance Society and Associate Editor for Mathematical Finance. Recent contributions have appeared in *European Finance Review, Finance and Stochastics, Journal of Computational Finance, Journal of Financial Economics, Journal of Financial and Quantitative Analysis, Mathematical Finance,* and *Review of Derivatives Research.*

Jean-Francois Picron is a Senior Consultant in Arthur Andersen's Financial and Commodity Risk Consulting practice, where he is responsible for internal systems development and works with major financial institutions on risk model reviews, derivatives pricing and systems implementation. Before joining Arthur Andersen, he was a Financial Engineer at Summit Systems, where he helped design and implement the market and credit risk modules. He holds an M. Eng. in Applied Mathematics from the Universite Catholique de Louvain and an MBA in Finance from Cornell University.

Shanming Shi works in the quantitative trading group of proprietary trading at J. P. Morgan. He earned his Ph.D. of Systems Engineering in 1994 from the

Tianjin University. He then earned his Ph.D. of Computer Science in 1998 from the University of Colorado at Boulder. His interests focus on mathematical modeling of financial markets. He has published in the fields of hidden Markov models, neural networks, combination of forecasts, task scheduling of parallel systems, and mathematical finance.

Ronnie Sircar is an Assistant Professor in the Mathematics Department at the University of Michigan in Ann Arbor. His Ph.D. is from Stanford University (1997). His research interests are applied and computational mathematics, particularly stochastic volatility modeling in financial applications.

Kristen Walters is a Director of Product Management at Measurisk.com, a Web-based risk measurement company serving the buy-side market. Kristen has 13 years of experience in capital markets and risk management. Prior to joining Measurisk, she consulted to major trading banks and end-users of derivatives at both KPMG and Arthur Andersen LLP. She was also responsible for market and credit risk management product development at Summit Systems, Inc. She has a BBA in Accounting from the University of Massachusetts at Amherst and an MBA in Finance from Babson College.

Katherine Wyatt received her Ph.D. in Mathematics in 1997 from the Graduate Center of the City University of New York. Her research interests include applications of mathematical programming in finance, in particular using disjunctive programming in modeling accounting regulations and in problems in risk management. She has worked as a financial services consultant at KPMG and is presently Assistant Director of Banking Research and Statistics at the New York State Banking Department.

Guang Yang is a quantitative analyst for the commercial team and research and development team at NumeriX. Guang has a Ph.D. in Aerospace Engineering from Cornell University, and also held a post-doctoral position at Cornell researching the direct simulation of turbulent flows on parallel computers and on mathematical finance. Prior to joining NumeriX, he worked at Open Link Financial as a Vice President, leading research and development on derivatives modeling.

Jean-Paul Laurent is Professor of Mathematics and Finance at ISFA Actuarial School at University of Lyon, Research Fellow at CREST and Scientific Advisor to Paribas. He has previously been Research Professor at CREST and Head of the quantitative finance team at Compagnie Bancaire in Paris. He holds a Ph.D. degree from University of Paris-I. His interests center on quantitative modeling for financial risks and the pricing of derivatives. He has published in the fields of hedging in incomplete markets, financial econometrics and the modeling of default risk.

Weiming Yang is senior application developer of Summit System Incorporation. He earned his Ph.D. in 1991 from the Chinese Academy of Science. He has published in the fields of nonlinear dynamics, controlling chaos, stochastic processes, recognition process and mathematical finance.

Andreas Weigend is the Chief Scientist of ShockMarket Corporation. From 1993 to 2000, he worked concurrently as full-time faculty and as independent consultant to financial firms (Goldman Sachs, Morgan Stanley, J. P. Morgan, Nikko Securities, UBS). He has published more than 100 scientific articles, some cited more than 250 times, and co-authored six books including *Computational Finance* (MIT Press, 2000), *Decision Technologies for Financial Engineering* (World Scientific, 1997), and *Time Series Prediction* (Addison-Wesley, 1994). His research integrates concepts and analytical tools from data mining, pattern recognition, modern statistics, and computational intelligence. Before joining ShockMarket Corporation, Andreas Weigend was an Associate Professor of Information Systems at New York University's Stern School of Business. He received an IBM Partnership Award for his work on discovering trading styles, as well as a 1999 NYU Curricular Development Challenge Grant for his innovative course Data Mining in Finance. He also organized the sixth international conference *Computational Finance CF99* that brought together decision-makers and strategists from the financial industries with academics from finance, economics, computer science and other disciplines. Prior to NYU, he was an Assistant Professor of Computer Science and Cognitive Science at the University of Colorado at Boulder. His research was supported by the National Science Foundation and the Air Force Office of Scientific Research. He received his Ph.D. from Stanford in Physics, and was a postdoc at Xerox PARC (Palo Alto Research Center).

CONTENTS

Reprinted from J. Finance **LIV**(4) (1999) 1361–1395

TRANSITION DENSITIES FOR INTEREST RATE
AND OTHER NONLINEAR DIFFUSIONS

YACINE AÏT-SAHALIA*

*Department of Economics, Princeton University,
Princeton, NJ 08544-1021, USA
E-mail: yacine@princeton.edu*

This paper applies to interest rate models the theoretical method developed in
Aït-Sahalia (1998) to generate accurate closed form approximations to the transition
function of an arbitrary diffusion. While the main focus of this paper is on the maximum-
likelihood estimation of interest rate models with otherwise unknown transition functions, applications to the valuation of derivative securities are also briefly discussed.

Continuous-time modeling in finance, though introduced by Louis Bachelier's
1900 thesis on the theory of speculation, really started with Merton's seminal
work in the 1970s. Since then, the continuous-time paradigm has proved to be an
immensely useful tool in finance and more generally economics. Continuous-time
models are widely used to study issues that include the decision to optimally con-
sume, save, and invest, portfolio choice under a variety of constraints, contingent
claim pricing, capital accumulation, resource extraction, game theory, and more
recently contract theory. Many refinements and extensions are possible, the basic
dynamic model for the variable(s) of interest X_t is a stochastic differential equation,

$$dX_t = \mu(X_t; \theta)dt + \sigma(X_t; \theta)dW_t , \qquad (1)$$

where W_t a standard Brownian motion, the drift μ and diffusion σ^2 are known
functions except for an unknown parameter[a] vector θ in a bounded set $\Theta \subset R^d$.

One major impediment to both theoretical modeling and empirical work with
continuous-time models of this type is the fact that in most cases little can be
said about the implications of the dynamics in Eq. (1) for longer time intervals.
Though Eq. (1) fully describes the evolution of the variable X over each infinitesimal

*_Mathematica_ code to implement this method can be found at http://www.princeton.edu/ yacine.
I am grateful to David Bates, René Carmona, Freddy Delbaen, Ron Gallant, Lars Hansen, Per
Mykland, Peter C. B. Phillips, Peter Robinson, Angel Serrat, Suresh Sundaresan and George
Tauchen for helpful comments. Robert Kimmel provided excellent research assistance. This research was conducted during the author's tenure as an Alfred P. Sloan Research Fellow. Financial
support from the NSF (Grant SBR-9996023) is gratefully acknowledged.
[a]Non- and semiparametric approaches, which do not constrain the functional form of the functions
μ and/or σ^2 to be within a parametric class, have been developed (see Aït-Sahalia, 1996a, 1996b
and Stanton, 1997).

instant, one cannot in general characterize in closed-form an object as simple (and fundamental for everything from prediction to estimation and derivative pricing) as the conditional density of $X_{t+\Delta}$ given the current value X_t. For a list of the rare exceptions, see Wong (1964). In finance, the well-known models of Black and Scholes (1973), Vasicek (1977) and Cox, Ingersoll and Ross (1985) rely on these existing closed-form expressions. In this paper, I will describe and implement empirically a method developed in a companion paper (Aït-Sahalia, 1998) which produces very accurate approximations *in closed-form* to the unknown transition function $p_X(\Delta, x|x_0; \theta)$, the conditional density of $X_{t+\Delta} = x$ given $X_t = x_0$ implied by the model in Eq. (1).

These closed-form expressions can be useful for at least two purposes. First, they let us estimate the parameter vector θ by maximum-likelihood.[b] In most cases, we observe the process at dates $\{t = i\Delta | i = 0,\ldots,n\}$, where $\Delta > 0$ is generally small, but fixed as n increases. For instance, the series could be weekly or monthly. Collecting more observations means lengthening the time period over which data are recorded, not shortening the time interval between successive existing observations.[c] Because a continuous-time diffusion is a Markov process, and that property carries over to any discrete subsample from the continuous-time path, the log-likelihood function has the simple form

$$\ell_n(\theta) \equiv n^{-1} \sum_{i=1}^{n} \ln\{p_X(\Delta, X_{i\Delta}|X_{(i-1)\Delta}; \theta)\}. \tag{2}$$

With a given Δ, two methods are available in the literature to compute p_X numerically. They involve either solving numerically the Kolmogorov partial differential equation known to be satisfied by p_X (see, e.g., Lo, 1988), or simulating a large number of sample paths along which the process is sampled very finely (see Pedersen, 1995; Honoré, 1997 and Santa-Clara, 1995). Neither method however produces a closed-form expression to be maximized over θ, and the calculations for all the pairs (x, x_0) must be repeated separately every time the value of θ changes. By contrast, the closed-form expressions in this paper make it possible to maximize the expression in Eq. (2) with p_X replaced by its closed-form approximation.

---

[b]A large number of new approaches have been developed in recent years. Some theoretical estimation methods are based on the generalized method of moments (Hansen and Scheinkman, 1995, Bibby and Sørensen, 1995) and on nonparametric density-matching (Aït-Sahalia, 1996a, 1996b), others on nonparametric approximate moments (Stanton, 1997), simulations (Duffie and Singleton, 1993; Gouriéroux, Monfort and Renault, 1993; Gallant and Tauchen, 1998, Pedersen, 1995), the spectral decomposition of the infinitesimal generator (Hansen, Scheinkman and Touzi, 1998; and Florens, Renault and Touzi, 1995), random sampling of the process to generate moment conditions (Duffie and Glynn, 1997), or finally Bayesian approaches (Eraker, 1997; Jones, 1997 and Elerian, Chib and Shephard, 1998).

[c]Discrete approximations to the stochastic differential Eq. (1) could be employed (see Kloeden and Platen, 1992): see Chan *et al.* (1992) for an example. As discussed by Merton (1980), Lo (1988), and Melino (1994), ignoring the difference generally results in inconsistent estimators, unless the discretization happens to be an exact one, which is tantamount to saying that p_X would have to be known in closed-form.

Derivative pricing provides a second natural outlet for applications of this methodology. Suppose that we are interested in pricing at date zero a derivative security written on an asset with price process $\{X_t | t \geq 0\}$, and with payoff function $\Psi(X_\Delta)$ at some future date Δ. For simplicity, assume that the underlying asset is traded, so that its risk-neutral dynamics have the form

$$dX_t/X_t = \{r - \delta\}dt + \sigma(X_t; \theta)dW_t\,, \tag{3}$$

where r is the riskfree rate and δ the dividend rate paid by the asset — both constant again for simplicity.

It is well-known that when markets are dynamically complete, the only price of the derivative security that is compatible with the absence of arbitrage opportunities is

$$P_0 = e^{-r\Delta}E[\Psi(X_\Delta)|X_0 = x_0] = e^{-r\Delta}\int_0^{+\infty}\Psi(x)p_X(\Delta, x|x_0; \theta)\,dx\,, \tag{4}$$

where p_X is the transition function (or risk-neutral density, or state-price density) induced by the dynamics in Eq. (3).

The Black–Scholes option pricing formula is the prime example of Eq. (4), when $\sigma(X_t; \theta) = \sigma$ is constant. The corresponding p_x is known in closed-form (as a lognormal density) and so the integral in Eq. (4) can be evaluated explicitly for specific payoff functions (see also Cox and Ross, 1976). In general, of course, no known expression for p_X is available and one must rely on numerical methods such as solving numerically the PDE satisfied by the derivative price, or Monte Carlo integration of Eq. (3). These methods are the exact parallels to the two existing approaches to maximum-likelihood estimation that I described earlier.

Here, given the sequence $\{\tilde{p}_X^{(K)} | K \geq 0\}$ of approximations to p_X, the valuation of the derivative security would be based on the explicit formula

$$P_0^{(K)} = e^{-r\Delta}\int_0^{+\infty}\Psi(x)\tilde{p}_X^{(K)}(\Delta, x|x_0; \theta)\,dx\,. \tag{5}$$

Formulas of the type given in Eq. (4) where the unknown p_X is replaced by another density have been proposed in the finance literature (see, e.g., Jarrow and Rudd, 1982). There is an important difference, however, between what I propose and the existing formulae: the latter are based on calculating the integral in Eq. (4) with an ad hoc density $\tilde{p}_X$ — typically adding free skewness and kurtosis parameters to the lognormal density, so as to allow for departures from the Black–Scholes formula. In doing so, these formulas ignore the underlying dynamic model specified in Eq. (3) for the asset price, whereas my method gives in closed-form the option pricing formula (of order of precision corresponding to that of the approximation used) which corresponds to the given dynamic model in Eq. (3). Then one can, for instance, explore how changes in the specification of the volatility function $\sigma(x; \theta)$ affect the derivative price, which is obviously impossible when the specification of the density $\hat{p}_X$ to be used in Eq. (4) in lieu of p_X is unrelated to Eq. (3).

The paper is organized as follows. In Section 1, I briefly describe the approach used in Aït-Sahalia (1998) to derive a closed-form sequence of approximations to p_X, give the expressions for the approximation and describe its properties. I then study in Section 2 a number of interest rate models, some with unknown transition functions, and give the closed-form expressions of the corresponding approximations. Section 3 reports maximum-likelihood estimates for these models, using the Federal Funds rate, sampled monthly between 1963 and 1998. Section 4 concludes, while a statement of the technical assumptions is in the appendix.

1. Closed-Form Approximations to the Transition Function

1.1. *Tail standardization via transformation to unit diffusion*

The first step towards constructing the sequence of approximations to p_X consists in standardizing the diffusion function of X — that is, transforming X into another diffusion Y defined as

$$Y_t \equiv \gamma(X_t; \theta) = \int^{X_t} du/\sigma(u; \theta), \qquad (6)$$

where any primitive of the function $1/\sigma$ may be selected.

Let $D_X = (\underline{x}, \bar{x})$ denote the domain of the diffusion X. I will consider two cases where $D_X = (-\infty, +\infty)$ or $D_X = (0, +\infty)$. The latter case is often relevant in finance, when considering models for asset prices or nominal interest rates. Moreover, the function σ is often specified in financial models in such a way that $\sigma(0; \theta) = 0$ and μ and/or σ violate the linear growth conditions near the boundaries. The assumptions in the appendix allow for this behavior.

Because $\sigma > 0$ on the interior of the domain D_X, the function γ in Eq. (6) is increasing and thus invertible. It maps D_X into $D_X = (\underline{y}, \bar{y})$, the domain of Y. For a given model under consideration, I will assume that the parameter space Θ is restricted in such a way that D_X is independent of θ in Θ. This restriction on Θ is inessential, but it helps keep the notation simple. Again, in finance, most, if not all cases, will have D_X and D_Y be either the whole real line $(-\infty, +\infty)$ or the half line $(0, +\infty)$.

By applying Itô's Lemma, Y has unit diffusion as desired:

$$dY_t = \mu_Y(Y_t; \theta)dt + dW_t, \qquad (7)$$

where

$$\mu_Y(y; \theta) = \frac{\mu(\gamma^{-1}(y; \theta); \theta)}{\sigma(\gamma^{-1}(y; \theta); \theta)} - \frac{1}{2}\frac{\partial \sigma}{\partial x}(\gamma^{-1}(y; \theta); \theta). \qquad (8)$$

Finally, note that it can be convenient to define Y_t instead as minus the integral in Eq. (6) if that makes $Y_t > 0$, for instance if $\sigma(x; \theta) = x^\rho$ and $\rho > 1$. For example, if $D_X = (0, +\infty)$ and $\sigma(x; \theta) = x^\rho$, then $Y_t = (1-\rho)X_t^{1-\rho}$ if $0 < \rho < 1$ (so $D_Y = (0, +\infty)$, $Y_t = \ln(X_t)$ if $\rho = 1$ (so $D_Y = (-\infty, +\infty)$), and $Y_t = (\rho-1)X_t^{-(\rho-1)}$ if $\rho > 1$ (so $D_Y = (0, +\infty)$ again). In all cases, Y has unit diffusion; that is,

$\sigma_Y^2(y;\theta) = 1$. When the transformation $Y_t \equiv \gamma(X_t;\theta) = -\int^{X_t} du/\sigma(u;\theta)$ is used, the drift $\mu_Y(y;\theta)$ in $dY_t = \mu_Y(Y_t;\theta)dt - dW_t$ is, instead of Eq. (8),

$$\mu_Y(y;\theta) = -\frac{\mu(\gamma^{-1}(y;\theta);\theta)}{\sigma(\gamma^{-1}(y;\theta);\theta)} + \frac{1}{2}\frac{\partial\sigma}{\partial x}(\gamma^{-1}(y;\theta);\theta). \tag{9}$$

The point of making the transformation from X to Y is that it is possible to construct an expansion for the transition density of Y. Of course, this would be of little interest since we only observe X, not the artificially introduced Y, and the transformation depends upon the unknown parameter vector θ. However, the transformation is useful because one can obtain the transition density p_X from p_Y through the Jacobian formula

$$
\begin{aligned}
p_X(\Delta, x|x_0;\theta) &= \frac{\partial}{\partial x}\mathrm{Prob}(X_{t+\Delta} \le x|X_t = x_0;\theta) \\[2mm]
&= \frac{\partial}{\partial x}\mathrm{Prob}(Y_{t+\Delta} \le \gamma(x;\theta)|Y_t = \gamma(x_0;\theta);\theta) \\[2mm]
&= \frac{\partial}{\partial x}\left[\int_{\underline{y}}^{\gamma(x;\theta)} p_Y(\Delta, y|\gamma(y_0;\theta);\theta)dy\right] \\[2mm]
&= \frac{p_Y(\Delta, \gamma(x;\theta)|\gamma(x_0;\theta);\theta)}{\sigma(\gamma(x;\theta);\theta)}.
\end{aligned}
\tag{10}
$$

Therefore, there is never any need to actually transform the data $\{X_{i\Delta}, i = 0,\ldots,n\}$ into observations on Y (which depends on θ anyway). Instead, the transformation from X to Y is simply a device to obtain an approximation for p_X from the approximation of p_Y. Practically speaking, once the approximation for p_X has been derived once and for all as the Jacobian transform of that of Y, the process Y no longer plays any role.

1.2. *Explicit expressions for the approximation*

As shown in Aït-Sahalia (1998), one can derive an explicit expansion for the transition density of the variable Y based on a Hermite expansion of its density $y \mapsto p_Y(\Delta, y|y_0;\theta)$ around a Normal density function. The analytic part of the expansion of p_Y up to order K is given by

$$\tilde{p}_Y^{(K)}(\Delta, y|y_0;\theta) = \Delta^{-1/2}\phi\left(\frac{y - y_0}{\Delta^{1/2}}\right)\exp\left(\int_{y_0}^{y}\mu_Y(w;\theta)dw\right)$$

$$\times \sum_{k=0}^{K} c_k(y|y_0;\theta)\frac{\Delta^k}{k!}, \tag{11}$$

where $\phi(z) \equiv e^{-z^2/2}/\sqrt{2\pi}$ denotes the $N(0,1)$ density function, $c_0(y|y_0;\theta) = 1$ and for all $j \ge 1$,

$$c_j(y|y_0;\theta) = j(y - y_0)^{-j} \int_{y_0}^{y} (w - y_0)^{j-1}$$

$$\times \{\lambda_Y(w)c_{j-1}(w|y_0;\theta) + (\partial^2 c_{j-1}(w|y_0;\theta)/\partial w^2)/2\}\, dw, \qquad (12)$$

where $\lambda_Y(y;\theta) \equiv -(\mu_Y^2(y;\theta) + \partial\mu_Y(y;\theta)/\partial y)/2$.

Tables 1 through 5 give the explicit expression of these coefficients for popular models in finance, which I discuss in detail in Section 2. Before turning over to these examples, a few general remarks are in order. The general structure of the expansion in Eq. (11) is as follows: the leading term in the expansion is Gaussian, $\Delta^{-1/2}\phi(y - y_0)/\Delta^{1/2})$, followed by a correction for the presence of the drift, $\exp(\int_{y_0}^{y} \mu_Y(w;\theta)dw)$, and then additional correction terms which depend upon the specification of the function $\lambda_Y(y;\theta)$ and its successive derivatives. These correction terms play two roles: first, they account for the nonnormality of p_Y and second they correct for the discretization bias implicit in starting the expansion with a Gaussian term with no mean adjustment and variance Δ (instead of $\text{Var}[Y_{t+\Delta}|Y_t]$, which is equal to Δ only in the first order).

In general, the function p_Y is not analytic in time. Therefore Eq. (11) must be interpreted strictly as the analytic part, or Taylor, series. In particular, for given

Table 1. Explicit sequence for the Vasicek model. This table contains the coefficients of the density approximation for p_Y corresponding to the Vasicek model in Example 1, $dX_t = \kappa(\alpha - X_t)dt + \sigma dW_t$. The terms in the expansion are evaluated by applying the formulas in Eq. (12). From Eq. (11), the $K = 0$ term in this expansion is $\tilde{p}_Y^{(0)}(\Delta, y|y_0;\theta)$, the $K = 1$ term is

$$\tilde{p}_Y^{(1)}(\Delta, y|y_0;\theta) = \tilde{p}_Y^{(0)}(\Delta, y|y_0;\theta)\{1 + c_1(y|y_0;\theta)\Delta\},$$

and the $K = 2$ term is

$$\tilde{p}_Y^{(2)}(\Delta, y|y_0;\theta) = \tilde{p}_Y^{(0)}(\Delta, y|y_0;\theta)\{1 + c_1(y|y_0;\theta)\Delta + c_2(y|y_0;\theta)\Delta^2/2\}.$$

Additional terms can be obtained in the same manner by applying Eq. (12) further. These computations and those of Tables 2 to 5 were all carried out in *Mathematica*.

$$\tilde{p}_Y^{(0)}(\Delta, y|y_0, \theta) = \frac{1}{\sqrt{\Delta}\sqrt{2\pi}} \exp\left[-\frac{(y - y_0)^2}{2\Delta} - \frac{y^2\kappa}{2} + \frac{y_0^2\kappa}{2} + \frac{y\alpha\kappa}{\sigma} - \frac{y_0\alpha\kappa}{\sigma}\right].$$

$$c_1(y|y_0, \theta) = -\frac{1}{6\sigma^2}(\kappa(3\alpha^2\kappa - 3(y + y_0)\alpha\kappa\sigma + (-3 + y^2\kappa + yy_0\kappa + y_0^2\kappa)\sigma^2)).$$

$$c_2(y|y_0, \theta) = \frac{1}{36\sigma^4}(\kappa^2(9\alpha^4\kappa^2 - 18y\alpha^3\kappa^2\sigma + 3\alpha^2\kappa(-6 + 5y^2\kappa)\sigma^2$$

$$- 6y\alpha\kappa(-3 + y^2\kappa)\sigma^3 + (3 - 6y^2\kappa + y^4\kappa^2)\sigma^4$$

$$+ 2\kappa\sigma(-3\alpha + y\sigma)(3\alpha^2\kappa - 3y\alpha\kappa\sigma) + (-3 + y^4\kappa)\sigma^2)y_0$$

$$+ 3\kappa\sigma^2(5\alpha^2\kappa - 4y\alpha\kappa\sigma + (-2 + y^2\kappa)y_0^2 + 2\kappa^2\sigma^3(-3\alpha + y\sigma)y_0^3$$

$$+ \kappa^2\sigma^4 y_0^4)).$$

(y, y_0, θ), it will generally have a finite convergence radius in Δ. As we will see below, however, the series in Eq. (11) with $K = 1$ or 2 at most is very accurate for the values of Δ that one encounters in empirical work in finance.

The sequence of explicit functions $\tilde{p}_Y^{(K)}$ in Eq. (11) is designed to approximate p_Y. As discussed above, one can then approximate p_X (the object of interest) by using the Jacobian formula for the inverted change of variable $Y \to X$:

$$\tilde{p}_X^{(K)}(\Delta, x|x_0; \theta) \equiv \sigma(x; \theta)^{-1} \tilde{p}_Y^{(K)}(\Delta, \gamma(x; \theta)|\gamma(x_0; \theta); \theta). \tag{13}$$

The main objective of the transformation $X \to Y$ was to provide a method of controlling the size of the tails of the transition density. As shown in Aït-Sahalia (1998), the fact that Y has unit diffusion makes the tails of the density p_Y, in the limit where Δ goes to zero, similar in magnitude to those of a Gaussian variable. That is, the tails of p_Y behave like $\exp[-y^2/2\Delta]$ as is apparent from Eq. (11). However, the tails of the density p_X are proportional to $\exp[-\gamma(x; \theta)^2/2\Delta]$. So for instance, if $\sigma(x; \theta) = 2\sqrt{x}$ then $\gamma(x; \theta) = \sqrt{x}$ and the right tail of p_X becomes proportional to $\exp[-x^2/2\Delta]$; this is verified by Eq. (13). Not surprisingly, this is the tail behavior for Feller's transition density in the Cox, Ingersoll and Ross model. If now $\sigma(x; \theta) = x$, then $\gamma(x; \theta) = \ln(x)$ and the tails of p_X are proportional to $\exp[-\ln(x)^2/2\Delta]$: this is what happens in the log-Normal case (see the Black–Scholes model). In other words, while the leading term of the expansion in Eq. (11) for p_Y is Gaussian, the expansion for p_X will start with a deformed or "stretched" Gaussian term, with the specific form of the deformation given by the function $\gamma(x; \theta)$.

The sequence of functions in Eq. (11) solves the forward and backward Kolmogorov equations up to order Δ^K; that is,

$$\begin{cases} \dfrac{\partial \tilde{p}_Y^{(K)}}{\partial \Delta} + \dfrac{\partial}{\partial y}\{\mu_Y(y; \theta)\tilde{p}_Y^{(K)}\} - \dfrac{1}{2}\dfrac{\partial^2 \tilde{p}_Y^{(K)}}{\partial y^2} = O(\Delta^K) \\[3mm] \dfrac{\partial \tilde{p}_Y^{(K)}}{\partial \Delta} - \mu_Y(y_0; \theta)\dfrac{\partial \tilde{p}_Y^{(K)}}{\partial y_0} - \dfrac{1}{2}\dfrac{\partial^2 \tilde{p}_Y^{(K)}}{\partial y_0^2} = O(\Delta^K) \end{cases} \tag{14}$$

The boundary behavior of the transition density $\tilde{p}_Y^{(K)}$ is similar to that of p_Y; under the assumptions made, $\lim_{y \to \underline{y} \text{ or } \bar{y}} p_Y = 0$. The expansion is designed to deliver an approximation of the density function $y \mapsto p_Y(\Delta, y|y_0; \theta)$ for a fixed value of conditioning variable y_0. Therefore, except in the limit where Δ becomes infinitely small, it is not designed to reproduce the limiting behavior of p_Y in the limit where y_0 tends to the boundaries.

Finally, note that the form of the expansion is compatible with the expression that arises out of Girsanov's Theorem in the following sense. Under the assumptions made, the process Y can be transformed by Girsanov's Theorem into a Brownian motion if $D_Y = (-\infty, +\infty)$, or into a Bessel process in dimension 3 if $D_Y = (0, +\infty)$. This gives rise to a formulation of p_Y in a form that involves the conditional expectation of the exponential of the integral of function of a Brownian Bridge (see Gihman and Skorohod, 1972, chapter 3) for the case where $D_Y = (-\infty, +\infty)$, or a Bessel Bridge if $D_Y = (0, +\infty)$. This conditional expectation term can either be expressed in terms of the conditional densities of the Brownian Bridge when $D_Y = (-\infty, +\infty)$ (see Dacunha-Castelle and Florens-Zmirou, 1986), or integrated by Monte Carlo simulation. Further discussion of these and other theoretical properties of the expansion is contained in Aït-Sahalia (1998).

2. Examples

2.1. *Comparison of the approximation to the closed-form densities for specific models*

In this section, I study the size of the approximation made when replacing p_X by $\tilde{p}_X^{(K)}$, in the case of typical examples in finance where p_x is known in closed-form and sampling is at the monthly frequency. Since the performance of the approximation improves as Δ gets smaller, monthly sampling is taken to represent a worst-case scenario as the upper bound to the sampling interval relevant for finance. In practice, most continuous-time models in finance are estimated with monthly, weekly, daily or higher frequency observations. The examples studied below reveal that including the term $c_2(y, y_0; \theta)$ generally provides an approximation to p_X which is better by a factor of at least ten than what one obtains when only the term $c_1(y, y_0; \theta)$ is included. Further calculations show that each additional order produces additional improvements by an additional factor of at least ten.

I will often compare the expansion in this paper to the Euler approximation; the latter corresponds to a simple discretization of the continuous-time stochastic differential equation, where the differential Eq. (1) is replaced by the difference equation

$$X_{t+\Delta} - X_t = \mu(X_t; \theta)\Delta + \sigma(X_t; \theta)\sqrt{\Delta}\epsilon_{t+\Delta} \tag{15}$$

with $\epsilon_{t+\Delta} \sim N(0, 1)$, so that

$$p_X^{\text{Euler}}(\Delta, x|x_0; \theta) = (2\pi\Delta\sigma^2(x_0; \theta))^{-1/2}$$

$$\times \exp\{-(x - x_0 - \mu(x_0; \theta)\Delta)^2/2\Delta\sigma^2(x_0; \theta)\}. \tag{16}$$

Example 1 (Vasicek's Model). Consider the Ornstein–Uhlenbeck specification proposed by Vasicek (1977) for the short term interest rate:

$$dX_t = \kappa(\alpha - X_t)dt + \sigma dW_t. \tag{17}$$

X is distributed on $D_X = (-\infty, +\infty)$ and has the Gaussian transition density

$$p_X(\Delta, x|x_0; \theta) = (\pi\gamma^2/\kappa)^{-1/2}\exp\{-(x - \alpha - (x_0 - \alpha)e^{-\kappa\Delta})^2\kappa/\gamma^2\}, \tag{18}$$

where $\theta \equiv (\alpha, \kappa, \sigma)$ and $\gamma^2 \equiv (1 - e^{-2\kappa\Delta})$. In this case, we have that $Y_t = \gamma(X_t; \theta) = \sigma^{-1}X_t$ and $\mu_Y(y; \theta) = \kappa\alpha\sigma^{-1} - \kappa y$, so that $\lambda_Y(y; \theta) = \kappa/2 - \kappa^2(\alpha - \sigma y)^2/2\sigma^2$.

Table 1 reports the first two terms in the expansion for this model, obtained from applying the general formula in Eq. (11). More terms can be calculated in Eq. (12) one after the other: once $c_2(y|y_0;\theta)$ has been obtained, calculate $c_3(y|y_0;\theta)$, etc. Starting from the closed-form expression, one can show directly that these expressions indeed represent a Taylor series expansion for the closed-form density $p_X(\Delta, x|x_0;\theta)$.

Figure 1(a) plots the density p_X as a function of the interest rate value x for a monthly sampling frequency ($\Delta = 1/12$), evaluated at $x_0 = 0.10$ and for the parameter values corresponding to the maximum-likelihood estimator from the Federal Funds data (see Table 4 in Section 4). Figure 1(b) plots the uniform approximation error $|p_X - \tilde{p}_X^{(K)}|$ for $K = 1$, 2 and 3, *in log-scale*. The error is calculated as the maximum absolute deviation between p_X and $\tilde{p}_X^{(K)}$ over the range ± 4 standard deviations around the mean of the density, and is also compared to the uniform error for the Euler approximation. The striking feature of the results is the speed of convergence to zero of the approximation error as K goes from one to two and from two to three. In effect, one can approximate p_X (which is of order 10^{+1}) within 10^{-3} with the first term alone ($K = 1$) and within 10^{-7} with $K = 3$, even though the interest rate process is only sampled once a month. Similar calculations for a weekly sampling frequency ($\Delta = 1/52$) reveal that the approximation error gets smaller even faster for this lower value of Δ.

In other words, small values of K already produce extremely precise approximations to the true density, p_X, and the approximation is even more precise if Δ is smaller. Of course, the exact density being Gaussian, in this case the expansion, whose leading term is Gaussian, has fairly little "work" to do to approximate the true density. In the Ornstein–Uhlenbeck case, the expansion involves no correction for nonnormality, which is normally achieved through the change of variable X to Y; it reduces here to a linear transformation and therefore does not change the nature of the leading term in the expansion. Comparing the performance of the expansion to that of the Euler approximation in this model (where both have the correct Gaussian form for the density) reveals that the expansion is capable of correcting for the discretization bias involved in a discrete approximation, whereas the Euler approximation is limited to a first order bias correction. In this case, the Euler approximation can be refined by increasing the precision of the conditional mean and variance approximations (see Huggins, 1997). Of course, discrete approximations to Eq. (1) of an order higher than Eq. (15) are available, but they do not lead to explicit density approximations since, compared to the Euler Eq. (15), they involve combinations of multiple powers of $\epsilon_{t+\Delta}$ (see e.g., Kloeden and Platen, 1992).

Example 2 (The CIR Model). Consider Feller's (1952) square-root specification

$$dX_t = \kappa(\alpha - X_t)dt + \sigma\sqrt{X_t}dW_t, \tag{19}$$

proposed as a model for the short term interest rate by Cox *et al.* (1985). X is distributed on $D_X = (0, +\infty)$ provided that $q \equiv 2\kappa\alpha/\sigma^2 - 1 \geq 0$. Its transition density is given by:

$$p_X(\Delta, x|x_0; \theta) = ce^{-u-v}(v/u)^{q/2}I_q(2(uv)^{1/2}), \tag{20}$$

with $\theta \equiv (\alpha, \kappa, \sigma)$ all positive, $c \equiv 2\kappa/(\sigma^2\{1 - e^{-\kappa\Delta}\})$, $u \equiv cx_0e^{-\kappa\Delta}$, $v \equiv cx$, and I_q is the modified Bessel function of the first kind of order q. Here $Y_t = \gamma(X_t; \theta) = 2\sqrt{X_t}/\sigma$ and $\mu_Y(y; \theta) = (q + 1/2)/y - \kappa y/2$.

The first two terms in the explicit expansion are given in Table 2. When evaluated at the maximum-likelihood estimates from Federal Funds data, the results reported in Fig. 2 are very similar to those of Fig. 1, again with an extremely fast convergence even for a monthly sampling frequency. The uniform approximation error is reduced to 10^{-5} with the first two terms, and 10^{-8} with the first three terms included.

Table 2. Explicit sequence for the Cox–Ingersoll–Ross model. This table contains the coefficients of the density approximation for p_Y corresponding to the Cox, Ingersoll and Ross model in Example 2, $dX_t = \kappa(\alpha - X_t)dt + \sigma\sqrt{X_t}dW_t$. The expansion for p_Y in this table applies also to the model proposed by Ahn and Gao (1998) (see Example 3). The terms in the expansion are evaluated by applying the formulae in Eq. (12). From Eq. (11), the $K = 0$ term in this expansion is $\tilde{p}_Y^{(0)}(\Delta, y|y; \theta)$, the $K = 1$ term is

$$\tilde{p}_Y^{(1)}(\Delta, y|y_0; \theta) = \tilde{p}_Y^{(0)}(\Delta, y|y_0; \theta)\{1 + c_1(y|y_0; \theta)\Delta\},$$

and the $K = 2$ term is

$$\tilde{p}_Y^{(2)}(\Delta, y|y_0; \theta) = \tilde{p}_Y^{(0)}(\Delta, y|y_0; \theta)\{1 + c_1(y|y_0; \theta)\Delta + c_2(y|y_0; \theta)\Delta^2/2\}.$$

Additional terms can be obtained in the same manner by applying Eq. (12) further.

$$\tilde{p}_Y^{(0)}(\Delta, y|y_0, \theta) = \frac{1}{\sqrt{\Delta}\sqrt{2\pi}}\exp\left[-\frac{(y - y_0)^2}{2\Delta} - \frac{y^2\kappa}{4} - \frac{\kappa y_0^2}{4}\right]y^{-(1/2)+(2\alpha\kappa/\sigma^2)}y_0^{(1/2)-(2\alpha\kappa/\sigma^2)}.$$

$$c_1(y|y_0, \theta) = \frac{1}{24yy_0\sigma^4}(48\alpha^2\kappa^2 - 48\alpha\kappa\sigma^2 + 9\sigma^4 + y\kappa^2\sigma^2(-24\alpha + y^2\sigma^2)y_0$$

$$+ y^2\kappa^2\sigma^4y_0^2 + y\kappa^2\sigma^4y_0^3).$$

$$c_2(y|y_0, \theta) = \frac{1}{576y^2y_0^2\sigma^8}(9(256\alpha^4\kappa^4 - 512\alpha^3\kappa^3\sigma^2 + 224\alpha^2\kappa^2\sigma^4 + 32\alpha\kappa\sigma^6 - 15\sigma^8)$$

$$+ 6y\kappa^2\sigma^2(-24\alpha + y^2\sigma^2)(16\alpha^2\kappa^2 - 16\alpha\kappa\sigma^2 + 3\sigma^4)y_0$$

$$+ y^2\kappa^2\sigma^4(672\alpha^2\kappa^2 - 48\alpha\kappa(2 + y^2\kappa)\sigma^2 + (-6 + y^4\kappa^2)\sigma^4)y_0^2$$

$$+ 2y\kappa^2\sigma^4(48\alpha^2\kappa^2 - 24\alpha\kappa(2 + y^2\kappa)\sigma^2 + (9 + y^4\kappa^2)\sigma^4)y_0^3$$

$$+ 3y^2\kappa^4\sigma^6(-16\alpha + y^2\sigma^2)y_0^4 + 2y^3\kappa^4\sigma^8y_0^5 + y^2\kappa^4\sigma^8y_0^6).$$

Panel A

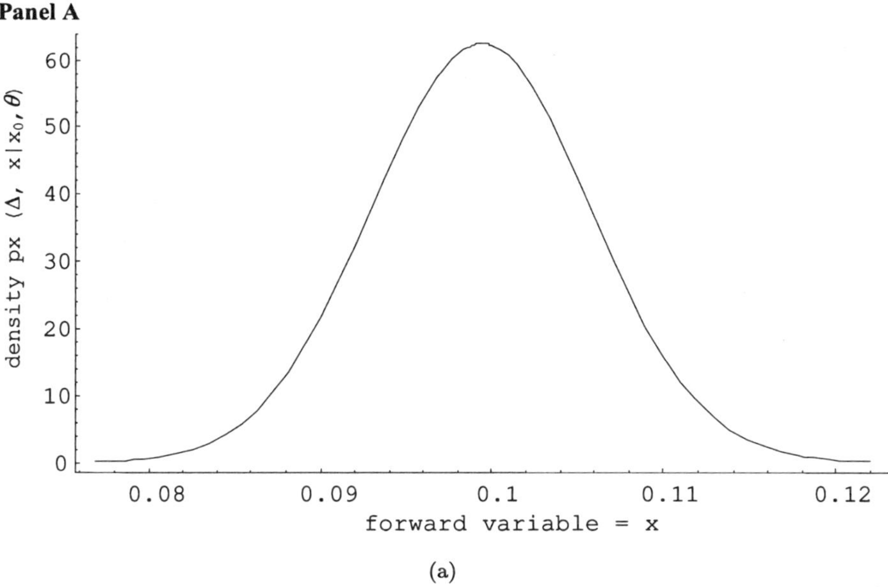

Panel B

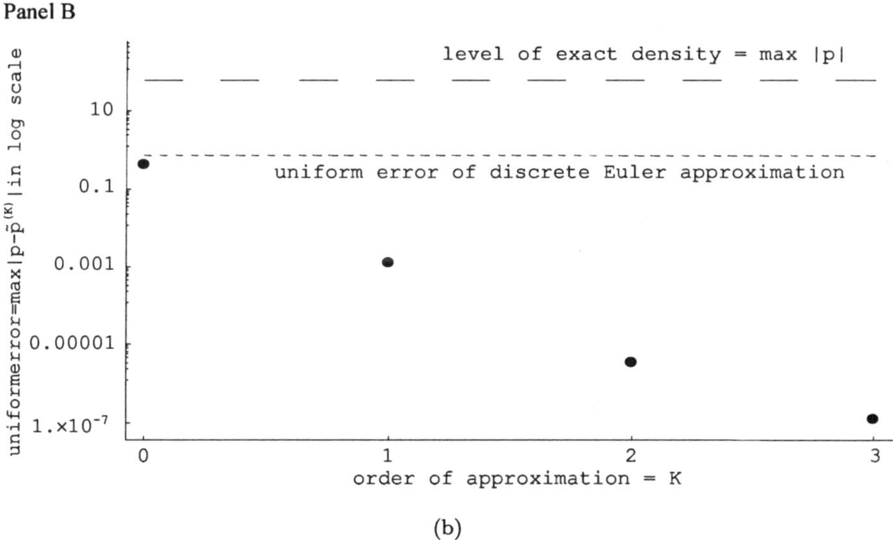

Fig. 1. Exact conditional density and approximation errors for the Vasicek model. Figure 1(a) plots for the Vasicek (1997) model (see Example 1 and Table 1) the closed-form conditional density $x \mapsto p_X(\Delta, x|x_0, \theta)$ as a function of x, with $x_0 = 10\%$, monthly sampling ($\Delta = 1/12$) and θ replaced by the MLE reported in Table 6. Figure 1(b) plots the uniform approximation errors $|p_X - \tilde{p}_X^{(K)}|$ for $K = 1$, 2, and 3, in log-scale, so that each unit on the y-axis corresponds to a reduction of the error by a multiplicative factor of ten. The error is calculated as the maximum absolute deviation between p_X and $\tilde{p}_X^{(K)}|$ over the range ± 4 standard deviations around the mean of the density. Both the value of the exact conditional density at its peak and the uniform error for the Euler approximation p_X^{Euler} are also reported for comparison purposes. This figure illustrates the speed of convergence of the approximation. A lower sampling interval than monthly would provide an even faster convergence of the density approximation sequence.

Panel A

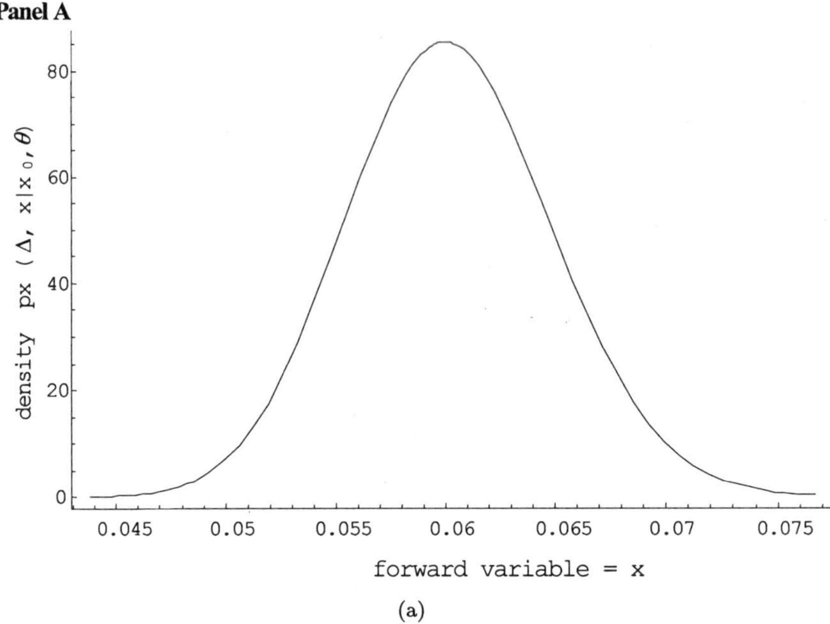

(a)

Panel B

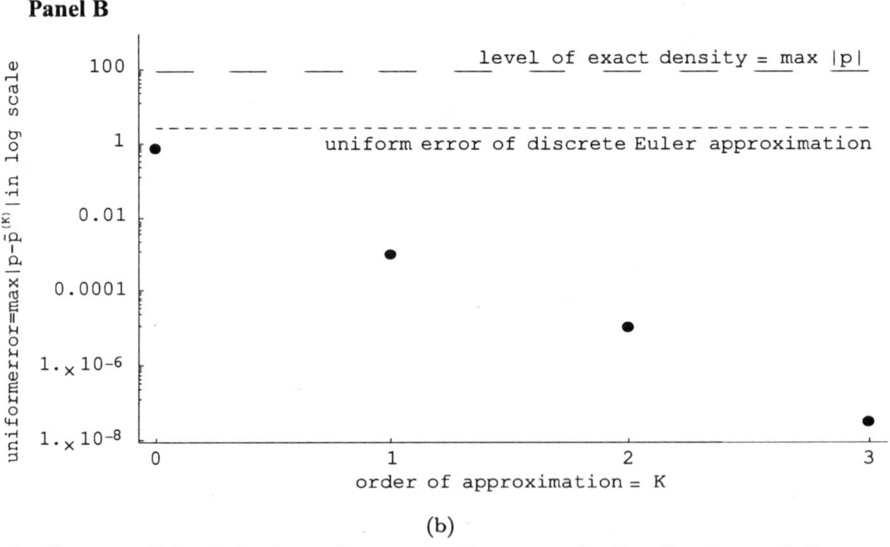

(b)

Fig. 2. Exact conditional density and approximation errors for the Cox–Ingersoll–Ross model. Figure 2(a) plots for the CIR (1985) model (see Example 2 and Table 2) the closed-form conditional density $x \mapsto p_X(\Delta, x|x_0, \theta)$ as a function of x, with $x_0 = 6\%$, monthly sampling ($\Delta = 1/12$) and θ replaced by the MLE reported in Table 6. Figure 2(b) plots the uniform approximation error $|p_X - \tilde{p}_X^{(K)}|$ for $K = 1$, 2, and 3, in log-scale, so that each unit on the y-axis corresponds to a reduction of the error by a multiplicative factor of ten. The error is calculated as the maximum absolute deviation between p_X and $\tilde{p}_X^{(K)}|$ over the range ± 4 standard deviations around the mean of the density. Both the value of the exact conditional density at its peak and the uniform error for the Euler approximation p_X^{Euler} are also reported for comparison purposes. This figure illustrates the speed of convergence of the approximation.

Example 3 (Inverse of Feller's Square Root Model). In this example, I generate densities for Ahn and Gao's (1998) specification of the interest rate process as one over an auxiliary process which follows a Cox–Ingersoll–Ross specification. As a result of Itô's Lemma, the model's specification is

$$dX_t = X_t(\kappa - (\sigma^2 - \kappa\alpha)X_t)dt + \sigma X_t^{3/2}dW_t\,, \tag{21}$$

with closed-form transition density given by

$$p_X(\Delta, x|x_0; \theta) = (1/x^2)p_X^{\mathrm{CIR}}(\Delta, 1/x|1/x_0; \theta)\,, \tag{22}$$

where p_X^{CIR} is the density function given in Eq. (20). The expansion in Eq. (11) for p_Y is identical to that for the CIR model given in Table 2 (because the Y process is the same with the same transformed drift μ_Y and unit diffusion). To get back to an expansion for X, the change of variable $Y \to X$ however is different, and is now given by $Y_t = \gamma(X_t; \theta) = 2/(\sigma\sqrt{X_t})$; hence the expansion for p_X will naturally be different than that of the CIR model (it will now approximate the left-hand side of Eq. (22) rather than Eq. (20)).

Figure 3(a) reports the drift for this model, evaluated at the maximum-likelihood estimates from Table 6 below. This model generates, in an environment where closed-form solutions are available, some of the effects documented empirically by Aït-Sahalia (1996b): almost no drift while the interest rate is in the middle of its range, strong mean-reversion when the interest rate gets large. Figure 3(b) plots the unconditional or marginal density, which is also the stationary density $\pi(x, \theta)$ for this process when the initial data point X_0 has π as its distribution. π is given by

$$\pi(y; \theta) \equiv \exp\left\{2\int^y \mu_Y(u; \theta)\, du\right\} \bigg/ \int_{\underline{y}}^{\bar{y}} \exp\left\{2\int^v \mu_Y(u; \theta)\, du\right\} dv\,. \tag{23}$$

Figure 3(c) compares the exact conditional density in Eq. (22), its Euler approximation and the expansion with $K = 1$ for the conditioning interest rate $x_0 = 0.10$. It is apparent from the figure that including the first term alone is sufficient to make the exact and approximate densities fall on top of one another, whereas the Euler approximation is distinct. Finally, Fig. 3(d) reports the uniform approximation error between the Euler approximation and the exact density on the one hand, and between the first three terms in the expansion and the exact density on the other. As can be seen from these figures, the expansion in Eq. (11) provides again a very accurate approximation to the exact density.

2.2. *Density approximation for models with no closed-form density*

Of course, the usefulness of the method introduced in Aït-Sahalia (1998) lies largely in its ability to deliver explicit density approximations for models which do not have closed-form transition densities. The next two examples correspond

Panel A

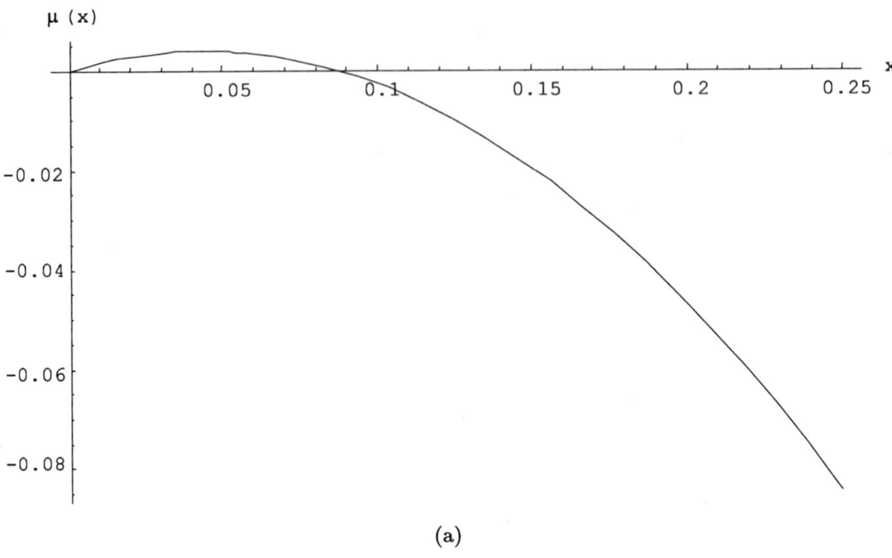

(a)

Panel B

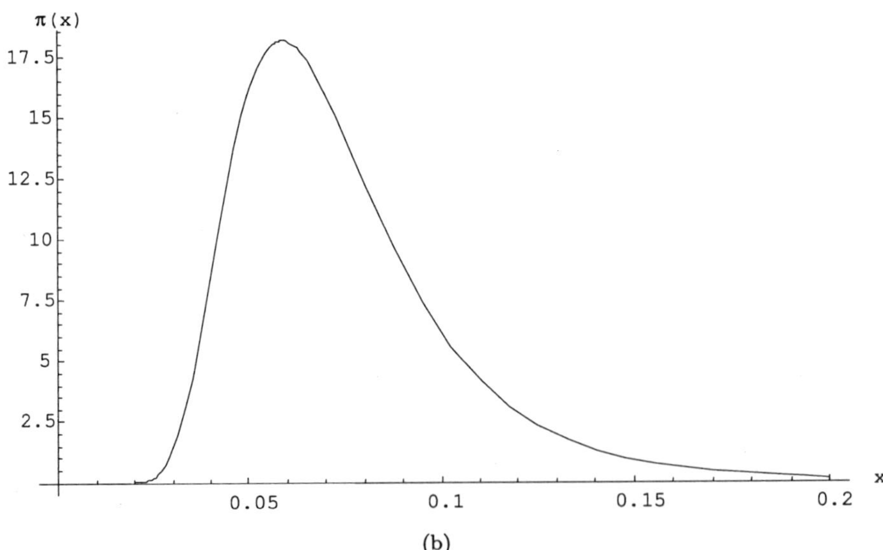

(b)

Fig. 3. Drift, density and approximation errors for the inverse of Feller's process. Results for the model proposed by Ahn and Gao (1998) model (see Example 3 and Table 2) are reported: the drift $\mu(X_t, \theta) = X_t(\kappa - (\sigma^2 - \kappa\alpha)X_t)$ in Fig. 3(a), the marginal density $\pi(X_t, \theta)$ in Fig. 3(b), the exact and conditional density approximation, p_X, p_X^{Euler} and $\tilde{p}_X^{(1)}$ as functions of the forward variable x, for $x_0 = 0.10$ in Fig. 3(c). The sampling frequency is monthly ($\Delta = 1/12$) and the parameter vector θ is evaluated at the the MLE reported in Table 6. Figure 3(d) reports the uniform approximation error $|p_X - \tilde{p}_X^{(K)}|$ for $K = 1$, 2, and 3, in log-scale, as in Figs. 1(b) and 2(b).

Panel C

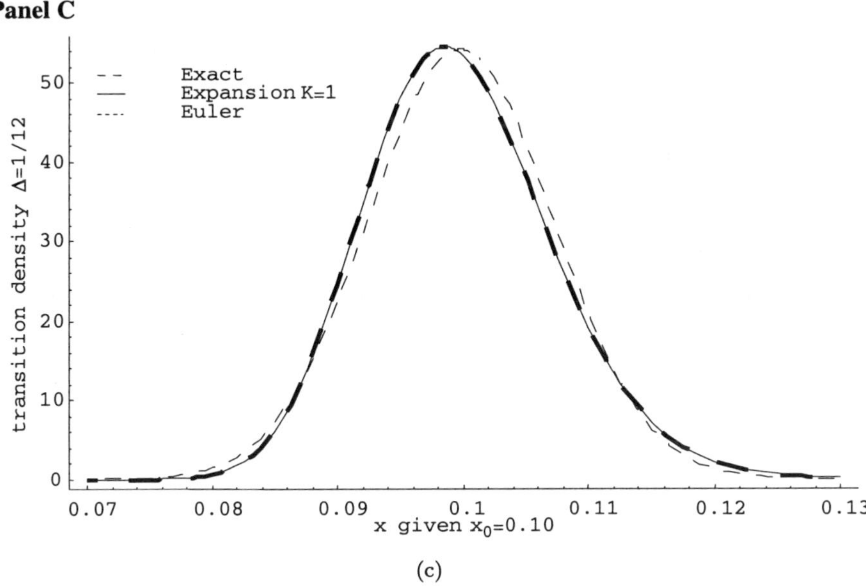

(c)

Panel D

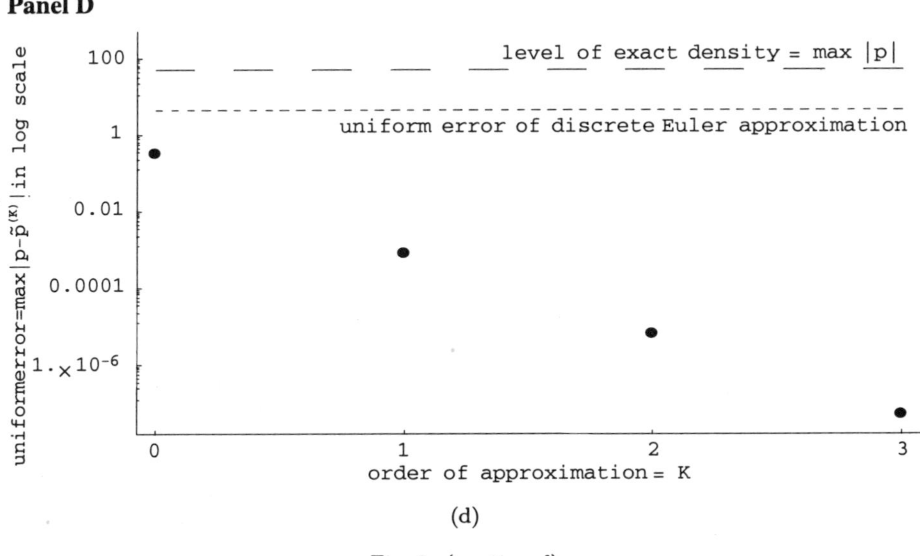

(d)

Fig. 3. (*continued*)

to models recently proposed in the literature to describe the time series properties of the short-term interest rate, and the final example illustrates the applicability of the method to a double-well model where the stationary density is bimodal.

Example 4 (Linear Drift, CEV Diffusion). Chan *et al.* (1992) have proposed the specification

$$dX_t = \kappa(\alpha - X_t)\,dt + \sigma X_t^\rho\,dW_t\,, \tag{24}$$

with $\theta \equiv (\alpha, \kappa, \sigma, \rho)$. X is distributed on $(0, +\infty)$ when $\alpha > 0$, $\kappa > 0$ and $\rho > 1/2$ (if $\rho = 1/2$; see Example 2 for an additional constraint). This model does not admit a closed-form density unless $\alpha = 0$ (see Cox, 1996), which then makes it unrealistic for interest rates. I will concentrate on the case where $\rho > 1$, which corresponds to

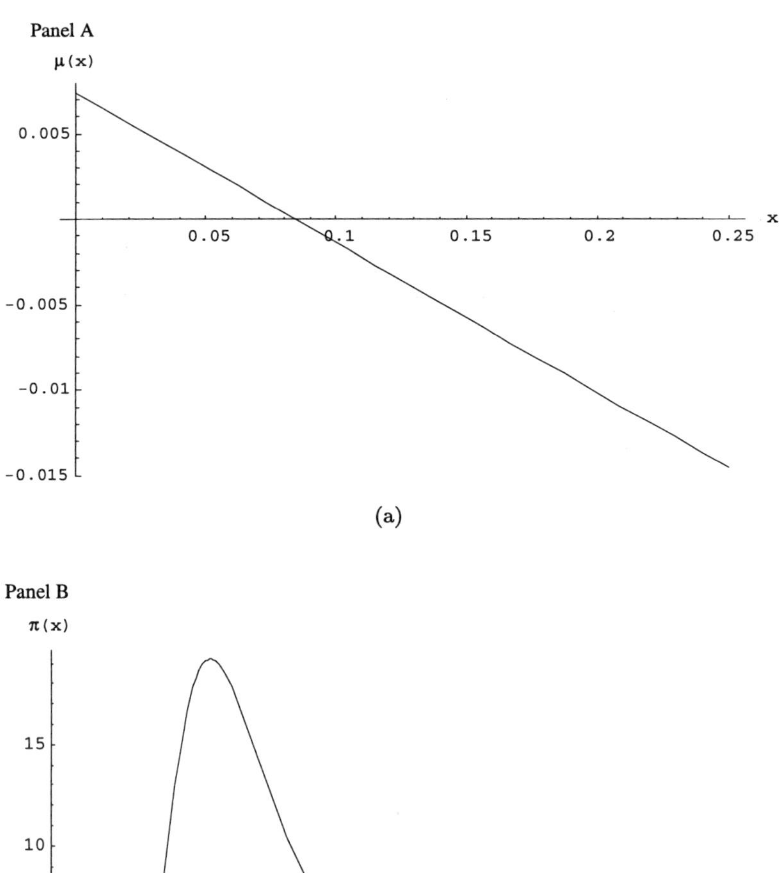

Fig. 4. Conditional density approximations for the linear drift, CEV diffusion model. These figures plot for the linear drift, CEV diffusion model of Chan *et al.* (1992) (see Example 4 and Table 3) the drift function, $\mu(X_t, \theta) = \kappa(\alpha - X_t)$ (Fig. 4(a)), the marginal density $\pi(X_t, \theta)$ (Fig. 4(b)), and the conditional density approximations, p_X^{Euler} and $\bar{p}_X^{(1)}$ as functions of the forward variable x, for two values of the conditional variable x_0 in Figs. 4(c) and 4(d) respectively. The sampling frequency is monthly ($\Delta = 1/12$) and the parameter vector θ is evaluated at the MLE reported in Table 6.

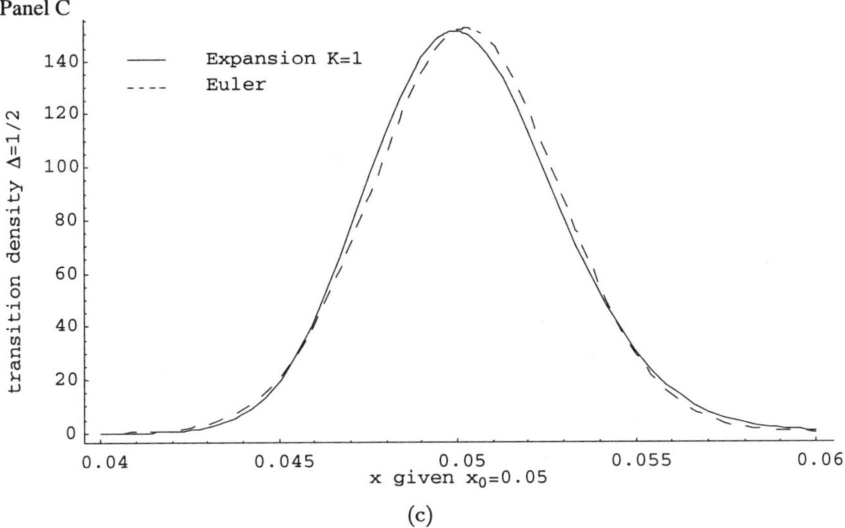

(c)

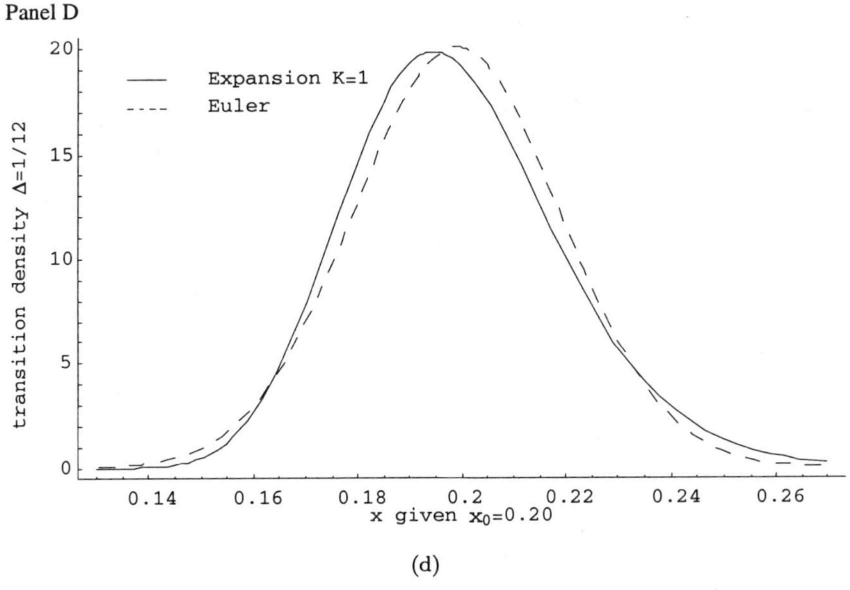

(d)

Fig. 4. (*continued*)

the empirically plausible estimate for U.S. interest rate data. The transformation from X to Y is given by $Y_t = \gamma(X_t; \theta) = X_t^{1-\rho}/\{\sigma(\rho - 1)\}$ and

$$\mu_Y(y; \theta) = \frac{\rho}{2(\rho - 1)y} - \kappa(\rho - 1)y + \alpha\kappa\sigma^{1/(\rho-1)}(\rho - 1)^{\rho/(\rho-1)}y^{\rho/(\rho-1)}. \qquad (25)$$

The first term in the expansion is given in Table 3. The corresponding formulas can be derived analogously for the transformation $Y_t = \gamma(X_t; \theta) = X_t^{1-\rho)}/\{\sigma(1 - \rho)\}$, which is appropriate if $1/2 < \rho < 1$. I plot in Fig. 4(a) the drift function

Table 3. Explicit sequence for the linear drift, CEV diffusion model. This table contains the coefficients of the density approximation for p_Y corresponding to the Chan *et al.* (1992) model in Example 4, $dX_t = \kappa(\alpha - X_t)dt + \sigma X_t^\rho dW_t$. The terms in the expansion are evaluated by applying the formulae in Eq. (12). From Eq. (11), the $K = 0$ term in this expansion is $\tilde{p}_X^{(0)}(\Delta, y|y_0; \theta)$, the $K = 1$ term is

$$\tilde{p}_Y^{(1)}(\Delta, y|y_0; \theta) = \tilde{p}_Y^{(0)}(\Delta, y|y_0; \theta)\{1 + c_1(y|y_0; \theta)\Delta\}.$$

Additional terms can be obtained by applying Eq. (12) further.

$$\tilde{p}_Y^{(0)}(\Delta, y|y_0, \theta) = \frac{1}{\sqrt{\Delta}\sqrt{2\pi}} \exp\left[-\frac{(y - y_0)^2}{2\Delta} + \kappa(\rho - 1) \right.$$

$$\times y^2(2\rho - 1) - 2y^{1+(\rho/(\rho-1))}\alpha(\rho - 1)^{\rho/(\rho-1)}\sigma^{1/(\rho-1)}$$

$$+ y_0(y_0 - 2\rho y_0 + 2\alpha(\rho - 1)^{\rho/(\rho-1)}\sigma^{1/(\rho-1)}y_0^{\rho/(\rho-1)}))/(4\rho - 2) \Bigg]$$

$$\times y^{\rho/(-2+2\rho)}y_0^{\rho/(2-2\rho)}.$$

$$c_1(y|y_0, \theta) \text{ for } y \neq y_0 = (-4y^4\kappa^2(\rho - 1)^4(2 - 9\rho + 9\rho^2)y_0 + 3\rho(4 + 20\rho + 27\rho^2 - 9\rho^3)$$

$$\times y_0 - 12y^2\kappa(\rho - 1)^2(13\rho - 27\rho^2 + 18\rho^3 - 2)$$

$$\times y_0 - 24y^{3+(\rho/(\rho-1))}\alpha\kappa^2(\rho - 1)^{4+(\rho/(\rho-1))}(3\rho - 1)\sigma^{1/(\rho-1)}$$

$$\times y_0 - 24y^{1+(\rho/(\rho-1))}\alpha\kappa(\rho - 1)^{3+(\rho/(\rho-1))}(2 - 9\rho + 9\rho^2)\sigma^{1/(\rho-1)}$$

$$\times y_0 - 12y^{2+2(\rho/(\rho-1))}\alpha^2\kappa^2(\rho - 1)^{5+(2/(\rho-1))}(3\rho - 2)\sigma^{2/(\rho-1)}$$

$$\times y_0 + y(3\rho(20\rho - 27\rho + 9\rho^3 - 4)$$

$$+ 12\kappa(\rho - 1)^2(13\rho - 27\rho^2 + 18\rho^3 - 2)y_0^2$$

$$+ 4\kappa^2(\rho - 1)^4(2 - 9\rho + 9\rho^2)y_0^4$$

$$- 24\alpha\kappa(\rho - 1)^{3+(1/(\rho-1))}(2 - 9\rho + 9\rho^2)\sigma^{1/(\rho-1)}y_0^{1+(\rho/(\rho-1))}$$

$$- 24\alpha\kappa^2(\rho - 1)^{4+(\rho/(\rho-1))}(3\rho - 1)\sigma^{1/(\rho-1)}y_0^{3+(\rho/(\rho-1))}$$

$$- 12\alpha^2\kappa^2(\rho - 1)^{5+(2/(\rho-1))}(3\rho - 2)\sigma^{2/(\rho-1)}$$

$$\times y_0^{2+(2\rho/(\rho-1))}))/(24y(\rho - 1)^2(3\rho - 2)(3\rho - 1)(y - y_0)y_0).$$

$$c_1(y|y_0, \theta) \text{ for } y = y_0 = \frac{1}{8(\rho - 1)^2 y_0^2}((\rho - 2)\rho - 4\kappa(\rho - 1)^2(2\rho - 1)y_0^2 - 4\kappa^2(\rho - 1)^4 y_0^4$$

$$+ 8\alpha\kappa(\rho - 1)^{2+(1/(\rho-1))}\rho\sigma^{1/(-1+\rho)}y_0^{1+(\rho/(\rho-1))}$$

$$+ 8\alpha\kappa^2(\rho - 1)^{3+(\rho/(\rho-1))}\sigma^{1/(\rho-1)}y_0^{3+(\rho(\rho-1))}$$

$$- 4\alpha^2\kappa^2(\rho - 1)^{4+(2/(\rho-1))}\sigma^{2/(\rho-1)}y_0^{2+(2\rho/(\rho-1))}).$$

---

corresponding to maximum-likelihood estimates (based on the expansion with $K = 1$, see Table 6 below), in Fig. 4(b) the unconditional density and in Figs. 4(c) and 4(d) the conditional density approximations for monthly sampling at $x_0 = 0.05$ and 0.20, respectively.

Example 5 (Nonlinear Mean Reversion). The following model was designed to produce very little mean reversion while interest rate values remain in the middle part of their domain, and strong nonlinear mean reversion at either end of the domain (see Aït-Sahalia, 1996b):

$$dX_t = (\alpha_{-1}X_t^{-1} + \alpha_0 + \alpha_1 X_t + \alpha_2 X_t^2)\,dt + \sigma X_t^\rho\,dW_t\,, \tag{26}$$

with $\theta \equiv (\alpha_{-1}, \alpha_0, \alpha_1, \alpha_2, \sigma, \rho)$. This model has been estimated empirically by Aït-Sahalia (1996b), Conley *et al.* (1997), and Gallant and Tauchen (1998) using a variety of empirical techniques. The new method in this paper makes it possible to

---

Table 4. Explicit sequence for the nonlinear drift model. This table contains the coefficients of the density approximation for p_Y corresponding to the model in Aït-Sahalia (1996b), Conley *et al.* (1997), and Tauchen (1997) given in Example 5, $dX_t = (\alpha_{-1} - X_t^{-1}) + \alpha_0 + \alpha_1 X_t + \alpha_2 X_t^2)dt + \sigma X_t^\rho dW_t$ with $\rho = 3/2$. The terms in the expansion are evaluated by applying the formulas in Eq. (12). From Eq. (11), the K=0 term in this expansion is $\tilde{p}_X^{(0)}(\Delta, y|y_0; \theta)$, the $K = 1$ term is

$$\tilde{p}_Y^{(1)}(\Delta, y|y_0; \theta) = \tilde{p}_Y^{(0)}(\Delta, y|y_0; \theta)\{1 + c_1(y|y_0; \theta)\Delta\}\,.$$

Additional terms can be obtained by applying Eq. (12) further.

$$\tilde{p}_X^{(0)}(\Delta, y|y_0, \theta) = \frac{1}{\sqrt{\Delta}\sqrt{2\pi}}\exp\left[-\frac{(y - y_0)^2}{2\Delta} + \frac{1}{192}(\sigma^4(-y^6 + y_0^6)\alpha_{-1}\right.$$

$$\left. - 6(y^2 - y_0^2)(\sigma^2(y^2 + y_0^2)\alpha_0 + 8\alpha_1))\right]$$

$$\times\, y^{(3/2)-(2\alpha_2/\sigma^2)}y_0^{-(3/2)+(2\alpha_2/\sigma^2)}\,.$$

$$c_1(y|y_0, \theta) = \frac{1}{7096320y\sigma^4 y_0}(315y\sigma^{12}y_0(y^{10} + y^9 y_0 + y^8 y_0^2 + y^7 y_0^3 + y^6 y_0^4 + y^5 y_0^5 + y^4 y_0^6$$

$$+ y^3 y_0^7 + y^2 y_0^8 + y y_0^9 + y_0^{10})\alpha_{-1}^2 + 88y\sigma^6 y_0\alpha_{-1}$$

$$\times\, (35\sigma^4(y^8 + y^7 y_0 + y^6 y_0^2 + y^5 y_0^3 + y^4 y_0^4 + y^3 y_0^5 + y^2 y_0^6 + y y_0^7 + y_0^8)$$

$$\times\, \alpha_0 + 36(-56y^4\sigma^2 - 56y^3\sigma^2 y_0 - 56y^2\sigma^2 y_0^2 - 56y\sigma^2 y_0^3$$

$$- 56\sigma^2 y_0^4 + 5y^6\sigma^2\alpha_1 + 5y^5\sigma^2 y_0\alpha_1 + 5y^4\sigma^2 y_0^2\alpha_1$$

$$+ 5y^3\sigma^2 y_0^3\alpha_1 + 5y^2\sigma^2 y_0^4\alpha_1 + 5y\sigma^2 y_0^5\alpha_1$$

$$+ 5\sigma^2 y_0^6\alpha_1 + 28y^4\alpha_2 + 28y^3 y_0\alpha_2 + 28y^2 y_0^2\alpha_2 + 28y y_0^3\alpha_2 + 28y_0^4\alpha_2))$$

$$+ 528(15y\sigma^8 y_0(y^6 + y^5 y_0 + y^4 y_0^2 + y^3 y_0^3 + y^2 y_0^4 + y y_0^5 + y_0^6)$$

$$\times\, \alpha_0^2 + 56y\sigma^4 y_0\alpha_0(-30y^2\sigma^2 - 30y\sigma^2 y_0 - 30\sigma^2 y_0^2 + 3y^4\sigma^2\alpha_1 + 3y^3\sigma^2 y_0\alpha_1$$

$$+ 3y^2\sigma^2 y_0^2\alpha_1 + 3y\sigma^2 y_0^3\alpha_1 + 3\sigma^2 y_0^4\alpha_1 + 20y^2\alpha_2 + 20y y_0\alpha_2 + 20y_0^2\alpha_2)$$

$$+ 560(9\sigma^4 - 24y\sigma^4 y_0\alpha_1 + y^3\sigma^4 y_0\alpha_1^2 + y^2\sigma^4 y_0^2\alpha_1^2$$

$$+ y\sigma^4 y_0^3\alpha_1^2 - 48\sigma^2\alpha_2 + 24y\sigma^2 y_0\alpha_1\alpha_2 + 48\alpha_2^2)))\,.$$

Panel A

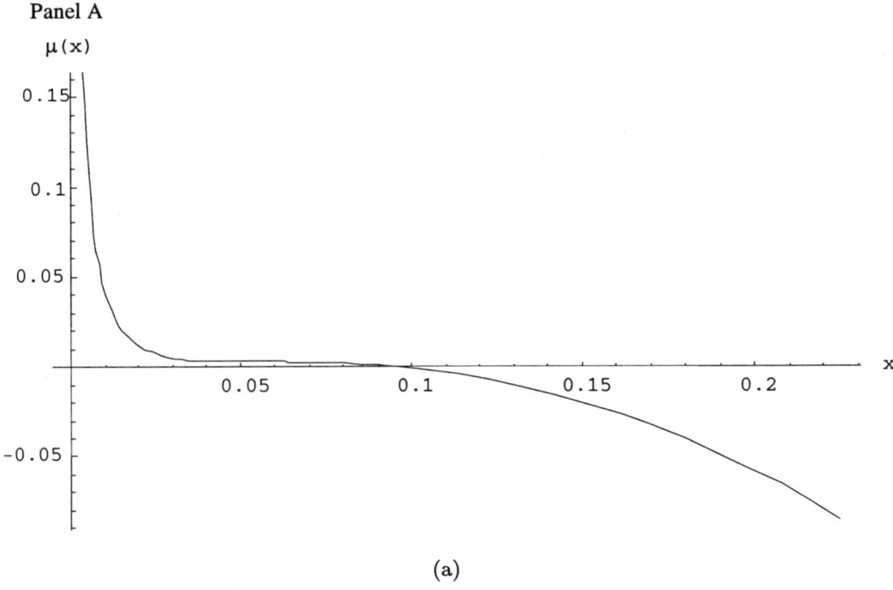

(a)

Panel B

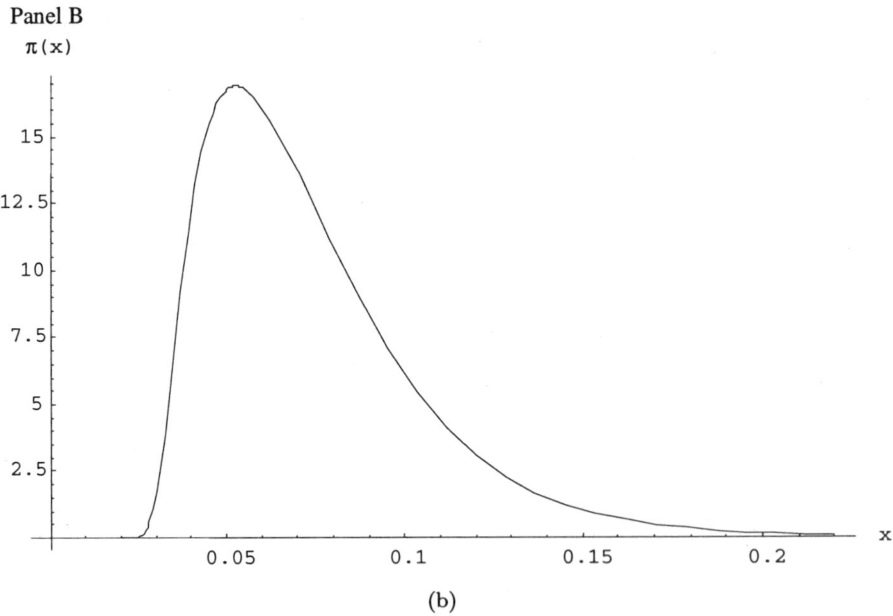

(b)

Fig. 5. Drift and density and approximations for the nonlinear drift model. These figures report for the nonlinear drift model of Aït Sahalia (1996b) (also estimated by Conley *et al.*, 1997 and Gallant and Tauchen, 1998) described in Example 5 and Table 4. Figure 5(a) plots the drift function, $\mu(X_t, \theta) = \alpha_{-1} X_t^{-1} + \alpha_0 \alpha_1 X_t + \alpha_2 X_t^2$ and Fig. 5(b) the marginal density $\pi(X_t, \theta)$. This model does not have a closed-form solution for p_X. Figures 5(c) and 5(d) plot the conditional density approximations, p_X^{Euler} and $\tilde{p}_X^{(1)}$ as functions of the forward variable x, for two different values of the conditional variable x_0. The sampling frequency is monthly ($\Delta = 1/12$) and the parameter vector θ is evaluated at the the MLE reported in Table 6.

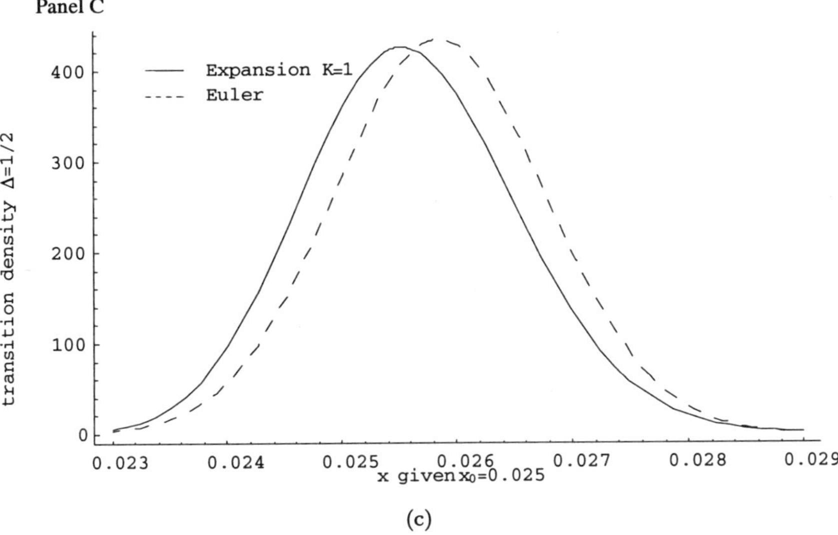

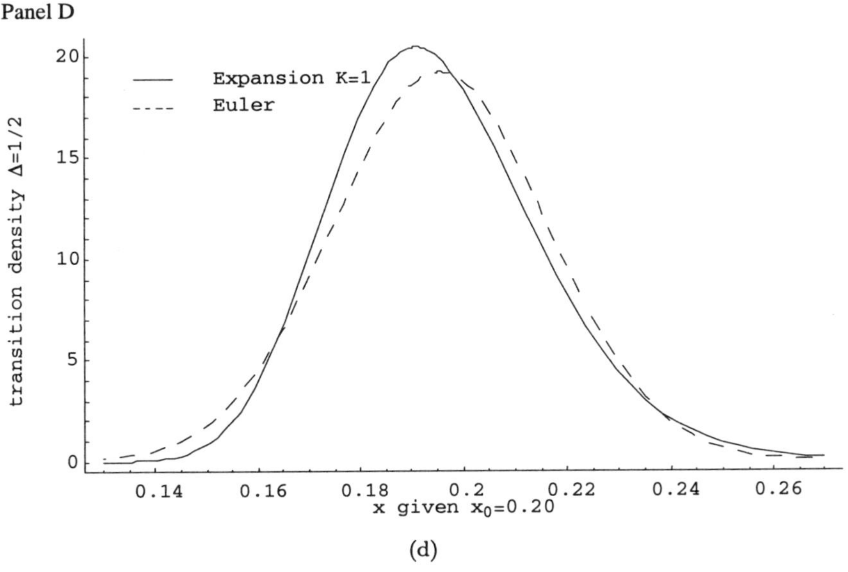

Fig. 5. (*continued*)

estimate it using maximum-likelihood. I will again concentrate on the case where $\rho > 1$, and to save space evaluate the formulas in Table 4 for $\rho = 3/2$. This process has $D_X = (0, +\infty)$, $Y_t = \gamma(X_t; \theta) = 2/\sigma\sqrt{X_t})$, and

$$\mu_Y(y; \theta) = \frac{3/2 - 2\alpha_2\sigma^2}{y} - \frac{\alpha_1 y}{2} - \frac{\alpha_0\sigma^2 y^3}{8} - \frac{\alpha_{-1}\sigma^4 y^5}{32}. \tag{27}$$

Figure 5(a) plots the drift evaluated at the maximum-likelihood parameter estimates (corresponding to $K = 1$). Figure 5(b) plots the unconditional or marginal density of the process: in the specification test in Aït-Sahalia (1996b), this density is matched against a nonparametric kernel estimator. Figures 5(c) and 5(d) contain the conditional density approximations for $K = 1$, compared with the Euler approximation, for the two values $x_0 = 0.025$ and 0.20, respectively. As before, sampling is at the monthly frequency.

Example 6 (Double-Well Potential). In this example, I generate a bimodal stationary density through the specification

$$dX_t = (X_t - X_t^3)\, dt + dW_t\,. \tag{28}$$

This model is distributed on $D_X = (-\infty, +\infty)$. Since the model is already set in unit diffusion, no transformation is needed $(Y = X)$.

Table 5 contains the first two terms of the expansion; Fig. 6(a) plots its drift, Fig. 6(b) its marginal density and Figs. 6(c) and 6(d) the transition density for

Table 5. Explicit sequence for the double-well model. This table contains the coefficients of the density approximation for p_Y corresponding to the model in Example 6, $dX_t = (X_t - X_t^3)dt + dWt$. The terms in the expansion are evaluated by applying the formulas in Eq. (12). From Eq. (11), the $K = 0$ term in this expansion is $\tilde{p}_Y^{(0)}(\Delta, y|y_0; \theta)$, the $K = 1$ term is

$$\tilde{p}_Y^{(1)}(\Delta, y|y_0; \theta) = \tilde{p}_Y^{(0)}(\Delta, y|y_0; \theta)\{1 + c_1(y|y_0; \theta)\Delta\}\,,$$

and the $K = 2$ term is

$$\tilde{p}_Y^{(2)}(\Delta, y|y_0; \theta) = \tilde{p}_Y^{(0)}(\Delta, y|y_0; \theta)\{1 + c_1(y|y_0; \theta)\Delta + c_2(y|y_0; \theta)\Delta^2/2\}\,.$$

Additional terms can be obtained in the same manner by applying Eq. (12) further.

$$\tilde{p}_Y^{(0)}(\Delta, y|y_0, \theta) = \frac{1}{\sqrt{\Delta}\sqrt{2\pi}} \exp\left[-\frac{(y - y_0)^2}{2\Delta} + \frac{y^2}{2} - \frac{y^4}{4} - \frac{y_0^2}{2} + \frac{y_0^4}{4} \right]\,.$$

$$c_1(y|y_0, \theta) = \frac{1}{210}\left(-105 + 70y^2 + 42y^4 - 15y^6 + (70y + 42y^3 - 15y^5)y_0\right.$$
$$\left. + (70 + 42y^2 - 15y^4)y_0^2 + (42y - 15y^3)y_0^3 + (42 - 15y^2)y_0^4 - 15yy_0^5 - 15y_0^6\right).$$

$$c_2(y|y_0, \theta) = \frac{1}{44100}\left(25725 + 11760y^2 - 19670y^4 + 9030y^6 - 336y^8 - 1260y^8 - 1260y^{10}\right.$$
$$+ 225y^{12} + 2y(10290 - 12110y^2 + 7455y^4 - 336y^6 - 1260y^8 + 225y^{10})y_0$$
$$+ 3(3920 - 7490y^2 + 6930y^4 - 336y^6 - 1260y^8 + 225y^{10})y_0^2$$
$$+ 2y(-12110 + 10395y^2 + 378y^4 - 2520y^6 + 450y^8)y_0^3$$
$$+ 5(-3934 + 4158y^2 + 504y^4 - 1260y^6 + 225y^8)y_0^4$$
$$+ 6y(2485 + 126y^2 - 1050y^4 + 225y^6)y_0^5$$
$$+ 21(430 - 48y^2 - 300y^4 + 75y^6)y_0^6 + 6y(-112 - 840y^2 + 225y^4)y_0^7$$
$$+ 3(-112 - 1260y^2 + 375y^4)y_0^8 + 180y(-14 + 5y^2)y_0^9$$
$$\left. + 45(-28 + 15y^2)y_0^{10} + 450yy_0^{11} + 225y_0^{12}\right).$$

Panel A

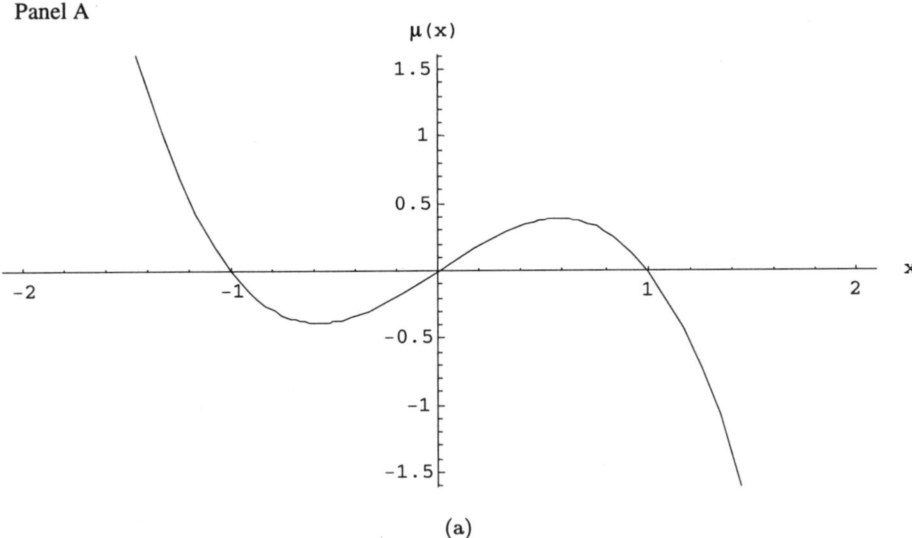

(a)

Panel B

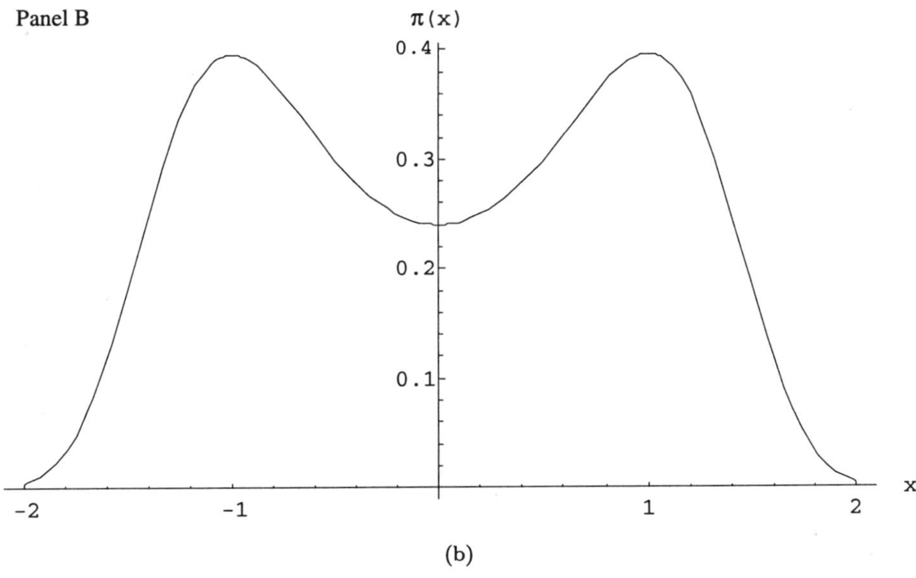

(b)

Fig. 6. Drift and density for the double-well model. Results for the double-well model of Example 6 and Table 5 are reported. The drift function, $\mu(x) = x - x^3$ (Fig. 6(a)) is such that the process avoids staying near 0 and is attracted to either -1 or $+1$, a fact reflected by the bimodality of the marginal density $\pi(x)$ in Fig. 6(b). This model does not have closed-form solutions for p_X. Figures 6(c) and 6(d) plot the conditional density approximations, p_X^{Euler} and $\bar{p}_X^{(2)}$ as functions of the forward variable x, for two different values of the conditional variable with $\Delta = 1/2$. As is clear from these figures, the Euler approximation cannot reflect the substantial nonnormality captured by the density approximation of this paper. Figure 6(e) plots the conditional density surface, $(x, x_0) \mapsto \bar{p}_X^{(2)}(\Delta, x|x_0, \theta)$ for $(\Delta = 1/2)$, and θ replaced by the MLE.

Panel C

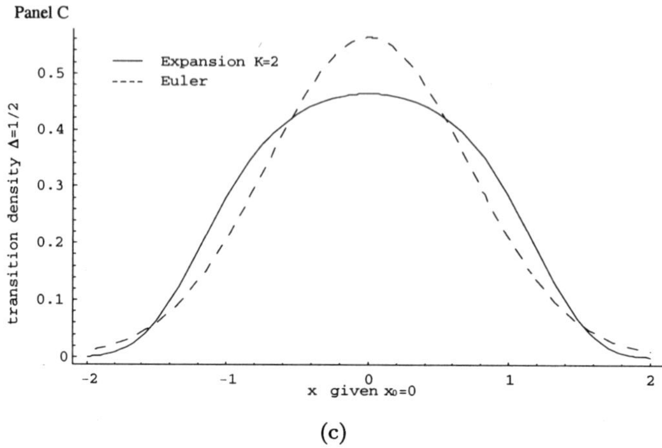

(c)

Panel D

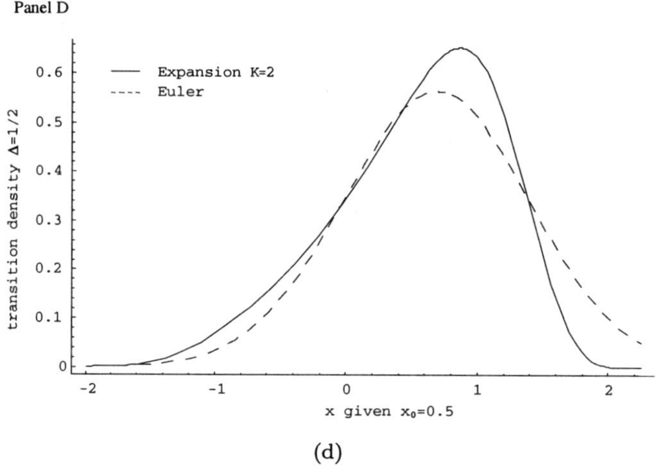

(d)

Panel E

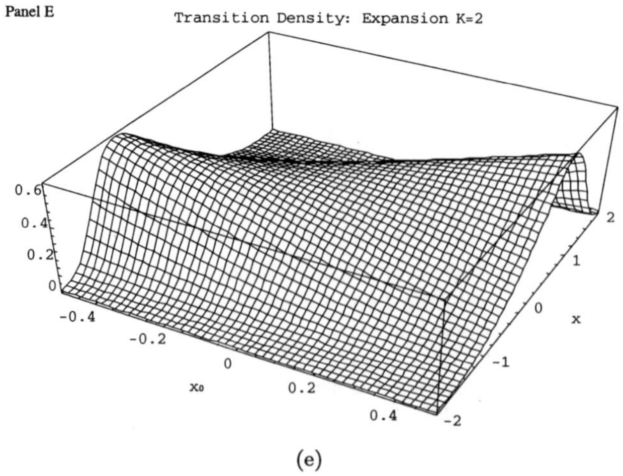

(e)

Fig. 6. (*continued*)

Table 6. Maximum-likelihood estimates for the monthly federal funds data, 1963–1998. This table reports the MLE for the parameters of five interest rate models estimated using the Federal funds data, monthly from January 1963 through December 1998. The estimates are calculated using the Euler approximation, the density approximation of this paper with $K = 1$ and when the transition density is available in closed-form (Examples 1, 2 and 3), the expansion with $K = 2$ and the true density. In the table, "ln L" refers to the maximized value of the log-likelihood. The formulas for the density expansion can be found in the respective tables indicated in the third column. The asymptotic standard errors in the last column are computed from Eq. (30), with Fisher's Information Matrix in Eq. (31) replaced by the sample averages evaluated at the second derivative of the log-likelihood expansion with $K = 1$, and confirmed with the average of the first derivative squared.

Model	Example Number	Density Expansion Table	Figure	Parameter Estimates: Euler	Parameter Estimates: Expansion $K = 1$	Parameter Estimates: Expansion $K = 2$	Parameter Estimates: True Density	Asymptoic Standard Error
$dX_t = \kappa(\alpha - X_t)dt + \sigma dW_t$	1	I	1	$\alpha = 0.0717$	$\alpha = 0.0719$	$\alpha = 0.0717$	$\alpha = 0.0717$	$\alpha : 0.014$
				$\kappa = 0.258$	$\kappa = 0.257$	$\kappa = 0.261$	$\kappa = 0.261$	$\kappa : 0.12$
				$\sigma = 0.02213$	$\sigma = 0.02237$	$\sigma = 0.02237$	$\sigma = 0.02237$	$\sigma : 0.00078$
				$\ln L = 3.634$	$\ln L = 3.634$	$\ln L = 3.634$	$\ln L = 3.634$	
$dX_t = \kappa(\alpha - X_t)dt + \sigma\sqrt{X_t}dW_t$	2	II	2	$\alpha = 0.0732$	$\alpha = 0.0742$	$\alpha = 0.0742$	$\alpha = 0.0721$	$\alpha : 0.016$
				$\kappa = 0.145$	$\alpha = 0.189$	$\kappa = 0.189$	$\kappa = 0.219$	$\kappa : 0.10$
				$\sigma = 0.06521$	$\sigma = 0.06658$	$\sigma = 0.06658$	$\sigma = 0.06665$	$\sigma : 0.0023$
				$\ln L = 3.917$	$\ln L = 3.918$	$\ln L = 3.918$	$\ln L = 3.918$	
$dX_t = X_t(\kappa - (\sigma^2 - \kappa\alpha)X_t)dt$ $+ \sigma X_t^{3/2}dW_t$	3	II	3	$\sigma = 15.019$	$\alpha = 15.157$	$\alpha = 15.150$	$\alpha = 15.141$	$\alpha : 2.9$
				$\kappa = 0.177$	$\kappa = 0.181$	$\kappa = 0.182$	$\kappa = 0.182$	$\kappa : 0.1$
				$\sigma = 0.8059$	$\sigma = 0.8211$	$\sigma = 0.8211$	$\sigma = 0.8211$	$\sigma : 0.03$
				$\ln L = 4.171$	$\ln L = 4.158$	$\ln L = 4.158$	$\ln L = 4.158$	
$dX_t = \kappa(\alpha - X_t)dt + \sigma X_t^{\rho}dW_t$	4	III	4	$\alpha = 0.0808$	$\alpha = 0.0844$			$\alpha : 0.05$
				$\kappa = 0.0972$	$\kappa = 0.0876$			$\kappa : 0.11$
				$\sigma = 0.7224$	$\sigma = 0.7791$			$\sigma : 0.16$
				$\rho = 1.46$	$\rho = 1.48$			$\rho : 0.08$
				$\ln L = 4.172$	$\ln L = 4.159$			

Table 6 (continued)

Model	Example Number	Density Expansion Table	Figure	Parameter Estimates: Euler	Parameter Estimates: Expansion $K=1$	Parameter Estimates: Expansion $K=2$	Parameter Estimates: True Density	Asymptoic Standard Error
$dX_t = (\alpha_{-1}X_t^{-1} + \alpha_0 + \alpha_1 X_t + \alpha_2 X_t^2)dt$ $+ \sigma X_t^{3/2} dW_t$	5	IV	5	$\alpha_{-1} = 0.00107$ $\alpha_0 = -0.0517$ $\alpha_1 = 0.877$ $\alpha_2 = -4.604$ $\sigma = 0.8047$ $\ln L = 4.173$	$\alpha_{-1} = 0.000693$ $\alpha_0 = -0.0347$ $\alpha = 0.676$ $\alpha_2 = -4.059$ $\sigma = 0.8214$ $\ln L = 4.160$			$\alpha_{-1} : 0.002$ $\alpha : 0.09$ $\alpha : 1.3$ $\alpha : 6.4$ $\sigma = 0.03$

$K = 2$, monthly sampling and $x_0 = 0.0$ and 0.5, respectively, with $\Delta = 1/2$. As is apparent from the figures, the densities in this model exhibit strong nonnormality, which obviously cannot be captured by the Euler approximation of Eq. (16).

3. The Estimation of Interest Rate Diffusions

3.1. *The data and maximum-likelihood estimates*

To calculate approximate maximum-likelihood estimates, I maximize the approximate log-likelihood function

$$\ell_n^{(K)}(\theta) \equiv n^{-1} \sum_{i=1}^{n} \ln\{\tilde{p}_X^{(K)}(\Delta, X_{i\Delta} | X_{(i-1)\Delta}; \theta)\} \tag{29}$$

(with the convention that $\ln(\alpha) = -\infty$ if $\alpha < 0$) over θ in Θ. This results in an estimator $\tilde{\theta}_n^{(K)}$, which, as shown in Aït-Sahalia (1998), is close to the exact (but uncomputable in practice) maximum-likelihood estimator $\tilde{\theta}_n$.

The data consist of monthly sampling of the Federal Funds rate between January 1963 and December 1998 (see Fig. 7). The source for the data is the H-15 Federal Reserve Statistical Release (Selected Interest Rate Series). Though the Fed Funds rate series exhibits strong microstructure effects at the daily frequency (due for instance to the second Wednesday settlement effect; see Hamilton, 1996), these effects are largely mitigated at the monthly frequency. On the other hand, this rate represents one of the closest possible proxies for what is meant by an "instantaneous"

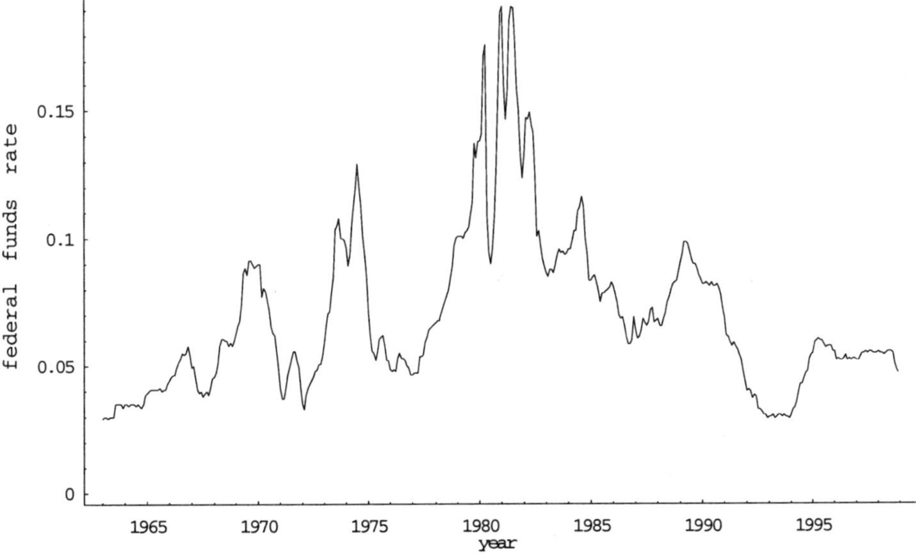

Fig. 7. Federal funds rate, monthly frequency, 1963–1998. This figure plots the time series of the Federal funds data used for the estimation of the parameters in Table 6.

short rate in theoretical models. Since the method in this paper does not rely on the sampling interval being small, the trade-off between a larger sampling interval and the virtual absence of microstructure effects seems worthwhile. Of course, the implicit (unrealistic) assumption is made that a single diffusion specification can represent the evolution of the short rate for the entire period. Naturally, nothing prevents the estimation to be conducted on a shorter time period at the expense of reducing the sample size. One advantage of the long time series used here is that it contains different episodes of U.S. interest rate history, such as the Volcker period, as well as the low interest rate environments that preceded it and followed it. It is therefore interesting to see how different models would accommodate these different regimes.

The results for the five models of Examples 1 to 5, compared to the Euler approximation and, when available (Examples 1 to 3), the true log-likelihood, are reported in Table 6. The last column of the table reports the asymptotic standard deviations for the estimated parameters, derived as explained below.

The results in Table 6 confirm those of Section 2: the expansion used with $K = 1$ or 2 produces estimates $\tilde{\theta}_n^{(K)}$ that are very close to $\tilde{\theta}_n$. It is interesting to note that because the model evaluated at the true parameter values often display very little drift (hence their near unit root behavior), and because interest rates are not particularly volatile, the fitted densities over a one-month interval are often fairly close to a Gaussian density. In other words, for these data, $\Delta =$ one month is a "small" time interval. Hence the Euler approximation performs relatively well in this specific context (except in the nonlinear drift model of Example 5, where the estimated parameters can be off by as much as 30% (although the standard deviation in this case is large), in the inverse Feller process of Example 3 where they are off by 5% to 10%, and in the Chan *et al.* (1992) specification of Example 4 where the drift parameters are off by 10%).

3.2. *Estimation of the asymptotic variance and how many terms to include*

I consider here only the situation where the process admits a stationary distribution. For the more general case, see Aït-Sahalia (1998). The asymptotic variance of the maximum-likelihood estimator is given by the inverse of Fisher's Information Matrix, which is the lowest possible achievable variance among the competing estimators discussed in the introduction.

Define $L(\theta) \equiv \ln(p_X(\Delta, X_\Delta | X_0; \theta))$, the $d \times 1$ vector $\dot{L}(\theta) \equiv \partial L(\theta)/\partial\theta$, and the $d \times d$ matrix $\ddot{L}(\theta) \equiv \partial^2 L(\theta)/\partial\theta\partial\theta^T$ where T denotes transposition. We have that

$$n^{1/2}(\hat{\theta}_n - \theta_0) \xrightarrow{d} N(0, i(\theta_0)^{-1}), \tag{30}$$

where Fisher's Information Matrix is

$$i(\theta) \equiv E[\dot{L}(\theta)\dot{L}(\theta)^T] \equiv -E[\ddot{L}(\theta)]. \tag{31}$$

Note that it is necessary that the transition function p_X not be uniformly flat in the direction of any one of the parameters θ_m, $m = 1, \ldots, d$, otherwise $\partial p_X(\Delta, x|x_0; \theta)/\partial \theta_m \equiv 0$ for all (x, x_0) and the model cannot be identified. In other words, no parameter entering the likelihood function can be redundant. The asymptotic standard deviations from Eq. (30) are reported in the last column of Table 6 for the interest rate models estimated above, with the expected values in Eq. (31) replaced by the sample averages evaluated at the parameter MLE.

Test statistics can be derived. Suppose that the model is given by Eq. (1) and that we wish to test $H_0{:}\theta = \theta_0$ against the two-sided alternative $H_a{:}\theta \neq \theta_0$. The likelihood ratio test statistic evaluated behaves under H_0 as:

$$2\{\ell_n(\hat{\theta}_n) - \ell_n(\theta_0)\} \xrightarrow{d} \chi^2_d. \tag{32}$$

Distributional results can be also be obtained for tests of a nested model which only allows for $\bar{d}$ free parameters from the d parameters in θ, and one can also consider Rao's efficient score statistic, which depends only on the restricted estimator $\bar{\theta}_n$, and Wald's test statistic, which depends only on the unrestricted estimator $\hat{\theta}_n$.

In all the results below, one can then replace $\hat{\theta}_n$ (respectively $\bar{\theta}_n$) by $\hat{\theta}_n^{(K)}$ (respectively $\bar{\theta}_n^{(K)}$). As the examples above have shown, it is not necessary to go much beyond $K = 2$ in the relevant financial examples to estimate the true density with a high degree of precision. More generally, to select an appropriate K at which to stop adding terms to the expansion, the following approach can be adopted: take K large enough so that the *approximation error* made in replacing p_X by $\tilde{p}_X^{(K)}$ is smaller than the *sampling error* due to the random character of the data, by a predetermined factor.

That is, in

$$\|\hat{\theta}_n^{(K)} - \theta_0\| \leq \|\hat{\theta}_n^{(K)} - \hat{\theta}_0\| + \|\hat{\theta}_n - \theta_0\| \tag{33}$$

we can estimate the asymptotic standard variance of $\hat{\theta}_n$ about θ_0 by Eq. (30). By Chebyshev's Inequality, one can then bound the second term on the right-hand-side of Eq. (33). We can then stop considering higher order approximations at an order K such that the distance between the two successive estimates $\hat{\theta}_n^{(K)}$ and $\hat{\theta}_n^{(K-1)}$ is an order of magnitude smaller than the distance between $\hat{\theta}_n$ and θ_0. In practice, this is unlikely to make much of a difference and in most cases one can safely restrict attention to the first two terms, $K = 1$ and $K = 2$.

4. Conclusion

This paper has demonstrated how to obtain very accurate closed-form approximations to the respective transition densities of a variety of models commonly used to represent the dynamics of the short term interest rate. Applications to derivative pricing, consisting in obtaining pricing formulas for any underlying price process, have been briefly outlined and will be developed in future work. Finally, an extension of these results to multivariate diffusions will be investigated.

Appendix A. Regularity Conditions

Assumption 1 (Smoothness of the Coefficients). *The functions $\mu(x;\theta)$ and $\sigma(x;\theta)$ are infinitely differentiable in x in D_X, and twice continuously differentiable in θ in the parameter space $\Theta \subset R^d$.*

Assumption 2 (Nondegeneracy of the Diffusion).

1. *If $D_X = (-\infty, +\infty)$, there exists a constant c such that $\sigma(x;\theta) > c > 0$ for all $x \in D_X$ and $\theta \in \Theta$.*

2. *If $D_X = (0, +\infty)$, I allow for the possible local degeneracy of σ at $x = 0$: If $\sigma(0;\theta) = 0$, then there exist constants $\xi_0, \omega \geq 0, \rho \geq 0$ such that $\sigma(x;\theta) \geq \omega x^\rho$ for all $0 < x < \xi_0$ and $\theta \in \Theta$. Away from 0, σ is nondegenerate; that is, for each $\xi > 0$, there exists a constant c_ξ such that $\sigma(x;\theta) \geq c_\xi > 0$ for all $x \in [\xi + \infty)$ and $\theta \in \Theta$.*

Assumption 3 below restricts the behavior of the function μ_Y and its derivatives near the boundaries of D_Y. It is formulated in terms of the function μ_Y for reasons of convenience, but the equivalent formulation directly in terms of the original functions μ and σ can be obtained from Eq. (8). Recall that $\lambda_Y(y;\theta) \equiv -(\mu_Y^2(y;\theta) + \partial\mu_Y(y;\theta)/\partial y)/2$.

Assumption 3 (Boundary Behavior). *For all $\theta \in \Theta$, $\mu_Y(y;\theta)$, $\partial\mu_Y(y;\theta)/\partial y$, and $\partial^2\mu_Y(y;\theta)/\partial y^2$ have at most exponential growth near the infinity boundaries and $\lim_{y\to\underline{y} \text{ or } \bar{y}} \lambda_Y(y;\theta) < +/\infty$.*

1. *Left Boundary:*

    i. *If $\underline{y} = 0^+$, there exist constants $\epsilon_0, \kappa, \alpha$ such that for all $0 < y \leq \epsilon_0$ and $\theta \in \Theta, \mu_Y(y;\theta) \geq \kappa y^{-\alpha}$ where either $\alpha > 1$ and $\kappa > 0$, or $\alpha = 1$ and $\kappa \geq 1$.*

    ii. *If $\underline{y} = -\infty$, there exist constants $E_0 > 0$ and $K > 0$ such that for all $y \leq -E_0$ and $\theta \in \Theta, \mu_Y(y;\theta) \geq Ky$.*

2. *Right Boundary: If $\bar{y} = +\infty$, there exist constants $E_0 > 0$ and $K > 0$ such that for all $y \geq E_0$ and $\theta \in \Theta, \mu_Y(y;\theta) \leq Ky$.*

The following remarks can help demonstrate the generality of these assumptions:

1. The upper bound $\lim_{y\to\underline{y} \text{ or } \bar{y}} \lambda_Y(y;\theta) < +\infty$ does not restrict λ_Y from going to $-\infty$ near the boundaries.

2. Similarly, Assumption 3 does not preclude μ_Y from going to $-\infty$ very fast near $\bar{y}$, and similarly, from going to $+\infty$ very fast near $\underline{y}$. Assumption 3 only restricts how large μ_Y can grow if it has the "wrong" sign; that is, if it is positive near $\bar{y}$ and negative near $\underline{y}$ then linear growth is at the maximum possible growth rate. If μ_Y has the "right" sign then the process is being pulled back away from the boundary and I do not restrict how fast mean reversion occurs (up to an exponential rate for technical reasons). The admissible behavior of the drift function μ_Y under these assumptions is summarized in Fig. A1.

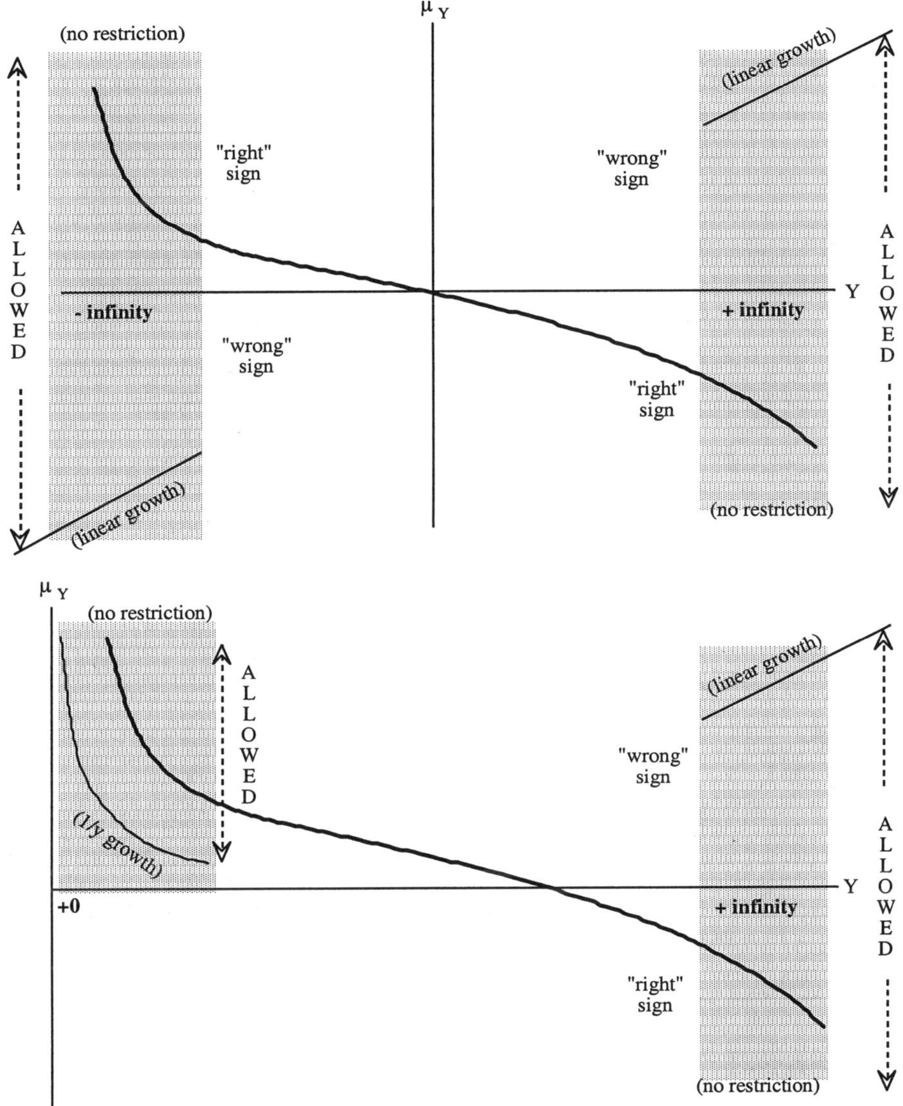

Fig. A1. Growth conditions for the drift $\mu_Y(Y;\theta)$. This figure translates graphically the assumptions made in the appendix regarding the shape of the function μ_Y. The admissible shape of the function is substantially less restricted than under the standard growth conditions. In particular, I only restrict the growth of μ_Y when it has the "wrong" sign (positive near $+\infty$, negative near $-\infty$).

3. The constraints on the behavior of the function μ_Y are essentially the best possible. For example, if μ_Y has the "wrong" sign near an infinity boundary, and grows faster than linearly, then Y explodes in finite time. Near a zero boundary at 0^+, if there exists $\kappa > 0$ and $\alpha < 1$ such that $\mu_Y(y;\theta) \leq ky^{-\alpha}$ in a neighborhood of 0^+ then 0 and negative values become attainable.

4. I can now fully characterize the boundary behavior of the diffusion Y implied by the assumptions made: if $+\infty$ is a boundary then it is natural if, near $+\infty$, $|\mu_Y(y;\theta)| \leq Ky$ and entrance if $\mu_Y(y;\theta) \leq -Ky^\beta$ for some $\beta > 1$. If $-\infty$ is a boundary then it is natural if, near $-\infty$, $|\mu_Y(y;\theta)| \leq K|y|$ and entrance if $\mu_Y(y;\theta) \geq K|y|^\beta$ for some $\beta > 1$. If 0^+ is a boundary, then it is entrance. Both entrance and natural boundaries are unattainable (see Feller, 1952 or Karlin and Taylor, 1981, Section 15.6 for the definition of boundaries). Natural boundaries can neither be reached in finite time, nor can the diffusion be started from there. Entrance boundaries, such as 0^+, cannot be reached starting from an interior point in $D_Y = (0, +\infty)$, but it is possible for Y to begin there. In that case, the process moves quickly away from zero and never returns there. Typically, economic intuition says little about how the process would behave if it were to start at the boundary, or whether that is even possible, and hence it is sensible to allow both types of boundary behavior.

5. Assumption 3 neither requires nor implies that the process is stationary. When *both* boundaries of the domain D_Y are entrance boundaries then the process is necessarily stationary with unconditional (marginal) density,

$$\pi(y;\theta) \equiv \exp\left\{2\int^y \mu_Y(u;\theta)\,du\right\} \Big/ \int_{\underline{y}}^{\bar{y}} \exp\left\{2\int^v \mu_Y(u;\theta)\,du\right\} dv\,,$$

$$\text{(A.1)}$$

provided that the initial random variable Y_0 is itself distributed with the same density π. When at least one of the boundaries is natural, stationarity is neither precluded nor implied. For instance, both an Ornstein–Uhlenbeck process, where $\mu_Y(y;\theta) = \kappa(\alpha - y)$, and a standard Brownian motion, where $\mu_Y(y;\theta) = 0$, satisfy the assumptions made, and both have natural boundaries at $-\infty$ and $+\infty$. Yet the former process is stationary, due to mean reversion, while the latter (null recurrent) is not.

Finally, the following assumption is needed for the purpose of maximizing the log-likelihood function only, not for the purpose of constructing the density expansion in Eq. (11).

Assumption 4 (Strengthening of Assumption 2 in the limiting case where $\alpha = 1$ and the diffusion is degenerate at 0): *Recall the constant ρ in Assumption 2(2), and the constants α and κ in Assumption 3(1.i). If $\alpha = 1$, then either $\rho \geq 1$ with no restriction on κ, or $\kappa \geq 2\rho/(1-\rho)$ if $0 < \rho < 1$. If $\alpha > 1$, no restriction is required.*

References

Ahn, Dong-Hyun and Bin Gao, "A parametric nonlinear model of term structure dynamics", Working Paper, University of North Carolina at Chapel Hill, 1998.

Aït-Sahalia, Yacine, "Non parametric pricing of interest rate derivative securities", *Econometrica* **64** (1996) 527–560.

Aït-Sahalia, Yacine, "Testing continuous-time models of the spot interest rate", *Review of Financial Studies* **9** (1996) 385–426.

Aït-Sahalia, Yacine, "Maximum-likelihood estimation of discretely sampled diffusions: A closed-form approach", Working Paper, Princeton University, 1998.

Bibby, Bo M. and Michael Sørensen, "Martingale estimation functions for discretely observed diffusion processes", *Bernoulli* **1** (1995) 17–39.

Black, Fisher and Myron Scholes, "The pricing of options and corporate liabilities", *Journal of Political Economy* **81** (1973) 637–654.

Chan, K. C., G. Andrew Karolyi, Francis A. Longstaff and Anthony B. Sanders, "An empirical comparison of alternative models of the short-term interest rate", *Journal of Finance* **47** (1992) 1209–1227.

Conley, Timothy G., Lars P. Hansen, Erzo G. J. Luttmer and José A. Scheinkman, "Short-term interest rates as subordinated diffusions", *Review of Financial Studies* **10** (1997) 525–578.

Cox, John C., "The constant elasticity of variance option pricing model", *The Journal of Portfolio Management*, Special Issue, 1996.

Cox, John C. and Stephen A. Ross, "The valuation of options for alternative stochastic processes", *Journal of Financial Economics* **3** (1995) 145–166.

Cox, John C., John E. Ingersoll and Stephen A. Ross, "A theory of the term structure of interest rates", *Econometrica* **53** (1985) 385–407.

Cox, John C. and Stephen A. Ross, "The valuation of options for alternative stochastic processes", *Journal pf Financial Economics* **3** (1976) 145–166.

Dacunha-Castelle, Didier and Danielle Florens-Zmirou, "Estimation of the coefficients of a diffusion from discrete observations", *Stochastics* **19** (1986) 263–284.

Duffie, Darrell and Peter Glynn, "Estimation of continuous-time Markov processes sampled at random time intervals", Working Paper, Stanford University, 1997.

Duffie, Darrell and Kenneth Singleton, "Simulated moments estimation of Markov models of asset prices", *Econometrica* **61** (1993) 929–952.

Elerian, Ola, Sidartha Chib and Neil Shephard, "Likelihood inference for discretely observed nonlinear diffusions", Working Paper, Oxford University, 1998.

Eraker, Bjorn, "MCMC analysis of diffusion models with application to finance", Working Paper, Norwegian School of Economics, Bergen, 1997.

Feller, William, "The parabolic differential equations and the associated semi-groups of transformations", *Annals of Mathematics* **55** (1952) 468–519.

Florens, Jean-Pierre, Eric Renault and Nizar Touzi, "Testing for embeddability by stationary scalar diffusions, *Econometric Theory* forthcoming.

Gallant, A. Ronald and George Tauchen, "Reprojecting partially observed systems with an application to interest rate diffusions", *Journal of the American Statistical Association* **93** (1998) 10–24.

Gihman, I. I. and A. V. Skorohod, *Stochastic Differential Equations*, Springer-Verlag, New York, 1972.

Gouriéroux, Christian, Alain Monfort and Eric Renault, "Indirect inference", *Journal of Applied Econometrics* **8** (1993) S85–S118.

Hamilton, James D, "The daily market for Federal funds", *Journal of Political Economy* **104** (1996) 26–56.

Hansen, Lars P. and José A. Scheinkman, "Back to the future: Generating moment implications for continuous time Markov processes", *Econometrica* **63** (1995) 767–804.

Hansen, Lars P., José A. Scheinkman and Nizar Touzi, "Identification of scalar diffusions using eigenvectors", *Journal of Econometrics* **86**(1) (1998) 1–32.

Honoré, Peter, "Maximum-likelihood estimation of nonlinear continuous-time term structure models", Working Paper, Aarhus University, 1997.

Huggins, Douglas J., "Estimation of a diffusion process for the U.S. short interest rate using a semigroup pseudo-likelihood", PhD Dissertation, University of Chicago, 1997.

Jarrow, Robert and Andrew Rudd, "Approximate option valuation for arbitrary stochastic processes", *Journal of Financial Economics* **10** (1982) 347–349.

Jones, Christopher S., "Bayesian analysis of the short-term interest rate", Working Paper, The Wharton School, University of Pennsylvania, 1997.

Karlin, Samuel and Howard M. Taylor, *A Second Course in Stochastic Processes*, Academic Press, New York, 1981.

Kloeden, Peter E. and Eckhardt Platen, *Numerical Solution of Stochastic Differential Equations*, Springer-Verlag, New York, 1982.

Lo, Andrew W., "Maximum likelihood estimation of generalized Itô processes with discretely sampled data", *Econometric Theory* **4** (1988) 231–247.

Melino, Angelo, "Estimation of continuous-time models in finance", in *Advances in Econometrics, Sixth World Congress*, Vol. II, Christopher S. Sims, ed., (Cambridge University Press, Cambridge, England, 1994).

Merton, Robert C., "On estimating the expected return on the market: An exploratory investigation", *Journal of Financial Economics* **8** (1980) 323–361.

Pedersen, Asger R., "A new approach to maximum-likelihood estimation for stochastic differential equations based on discrete observations", *Scandinavian Journal of Statistics* **22** (1995) 55–71.

Santa-Clara, Pedro, "Simulated likelihood estimation of diffusions with an application to the short term interest rate", Working Paper, UCLA, 1995.

Stanton, Richard, "A nonparametric model of term structure dynamics and the market price of interest rate risk", *Journal of Finance* **52** (1997) 1973–2002.

Vasicek, Oldrich, "An equilibrium characterization of the term structure", *Journal of Financial Economics* **5** (1997) 177–188.

Wong, Eugene, "The construction of a class of stationary Markov processes", in *Stochastic Processes in Mathematical Physics and Engineering*, R. Bellman, ed., Proceedings of Symposia in Applied Mathematics, 16 (American Mathematical Society, Providence, RI, 1964).

HIDDEN MARKOV EXPERTS

ANDREAS S. WEIGEND

ShockMarket Corporation, 151 Lytton Avenue, Palo Alto, CA 94301, USA
E-mail: andreas@weigend.com

SHANMING SHI

J. P. Morgan & Co. Inc., 60 Wall Street, New York, NY 10260, USA
E-mail: shi_shanming@jpmorgan.com

Most approaches in forecasting merely try to predict the next value of the time series. In contrast, this paper presents a framework to predict the full probability distribution. It is expressed as a mixture model: the dynamics of the individual states is modeled with so-called "experts" (potentially non-linear neural networks), and the dynamics between the states is modeled using a hidden Markov approach. The full density predictions are obtained by a weighted superposition of the individual densities of each expert. This model class is called "hidden Markov experts".

Results are presented for daily S&P500 data. While the predictive accuracy of the mean does not improve over simpler models, evaluating the prediction of the full density shows a clear out-of-sample improvement both over a simple GARCH(1,1) model (which assumes Gaussian distributed returns) and over a "gated experts" model (which expresses the weighting for each state non-recursively as a function of external inputs). Several interpretations are given: the blending of supervised and unsupervised learning, the discovery of hidden states, the combination of forecasts, the specialization of experts, the removal of outliers, and the persistence of volatility.

Keywords: Forecasting, density prediction, conditional distribution, mixture models, time series analysis, hidden Markov models, gated experts, hidden Markov experts, model comparison, density evaluation, computational finance, risk management.

Data: Daily S&P500 (January 1977 to December 1997).

Code: Please contact the first author for the MATLAB implementation.

1. Introduction

The introduction reviews several approaches to density forecasting in time series, informally introduces the model class of "hidden Markov experts" (HME), discusses methods for density evaluation, and relates HME to previous work. A brief overview of the sections of the paper are given at the end of the introduction.

1.1. *Tasks in time series prediction*

A time series is a sequence of observations $\mathcal{Y}^T = \{y^t | t = 1, \ldots, T\}$. t enumerates the elements of the sequence, and T is the total number of the observations. Our methodology is to split the entire set of available data into at least two sets. The first part is used to estimate the parameters of the model and is called the training set. The second part of the data is only used at the very end of the entire modeling process to compute performance measures and referred to as the test set.[a] The test set thus serves as the out-of-sample set, since waiting for genuinely new observations would just take too long for daily data.

Many forecasting methods (in particular almost all non-linear forecasting methods) focus on predicting the next value or *point* of the time series. Such point predictions are appropriate on problems where the signal is only distorted with a small amount of noise, as typically the case in non-linear dynamics.[b] However, in financial time series, the noise is often larger than the signal itself, requiring methods that predict not just a point but a density. This paper focuses on such density prediction, addresses the problems of a small signal to noise ratio, and includes non-Gaussian density forecasts.

We start by briefly discussing a path through various tasks for prediction.

(1) Model (1) uses the mean of the training set as point prediction. However, with sufficiently precise experimental resolution, the exact value of the prediction is almost always wrong: probability densities are needed.

(2) An interpretation of Model (1) is that of a single Gaussian (instead of a sharp point) whose constant variance is that of the training set. We call this density implicit in the point forecast Model (2) and use it as the baseline model in the empirical evaluations.

The two possible next steps are (a) to allow the mean to vary, or (b) to allow the variance to vary.

(3a) The predicted mean varies (i.e., it is a function of some inputs, x) but the variance remains constant. The input variables can be other time series (exogenous variables), or they can be lagged values of the series to be predicted (autoregression). The functional mapping from these variables to the output (expected mean) is, in the simplest case, linear. Our framework allows for general non-linear functions, typically be expressed as neural networks. The parameters of the model can be estimated by minimizing the squared error between the prediction and the observed value. In machine learning, the observed value is called the "target", and an input-output pair is called a "pattern".

(3b) Rather than varying the mean and keeping the variance constant, Model (3b) fixes the mean (to the mean of the training set) but allows for conditional variance. Estimating the parameters becomes more complicated

[a]Additional sets can be set aside from periods earlier than the test set if there are meta-parameters, such as the number of experts.

[b]This is not a coincidence. Many non-linear systems can generate so-called chaotic behavior where the time series continues in an "interesting" way forever. This is an important difference to linear systems that die out if they are not driven by noise.

than minimizing a squared error since, in contrast to Model (3a), we here do not have a desired value or target for each pattern. In order to estimate the parameters of the model, a more complicated statistical framework is needed. We use a maximum likelihood approach: the predicted conditional variance is written as a function of the inputs. The coefficients of this function are estimated such that the likelihood of the observed data given the model and the inputs is maximized.

Model (3b) — mean fixed, variance varying — has important applications in finance. While it is very difficult to predict the once-differenced time series of prices (i.e., returns) better than a constant Model (1), more accurate predictions of the variance than a constant are often possible. This reflects the well-known property of many financial time series called volatility clustering or volatility persistence: there are time periods with large (positive and negative) returns, which should be predicted with a larger variance, and there are time periods where the market is quiet and the predicted variance should be smaller.

(4) The fourth level of complexity predicts Gaussian densities with conditional means and conditional variances, combining the two degrees of freedom from Models (3a) and (3b).

So far, the form of the density in all the models has been Gaussian. Now we want to generate density predictions that are **non-Gaussian**. To achieve this goal, there are two philosophies: using expansions (e.g., Edgeworth expansion), and using mixture distributions. The *expansion approach* has the advantage of orthogonality. The computation of increasing orders of approximation is sequential; the term of order $(n + 1)$ is not effected by terms of order n and below. The corresponding weakness is that the term of order $(n + 1)$ can only patch up problems the lower orders have left for it, rather than all $(n + 1)$ terms joining together and trying to find a better overall solution. Another aspect of the relationship between mixture models and moments is discussed in Timmermann (1993).[c]

The *mixture approach* expresses the density $P(y^{t+1})$ as a sum of M distributions:

$$P(y^{t+1}|\text{ information set at time } t, \text{ model parameters})$$

$$= \sum_{j=1}^{M} \gamma_j^{t+1}(\bullet)P(y^{t+1}|x^{t+1}, \ldots, \text{model parameters}) .$$

In the context of non-linear sub-models for the mixture components, the sub-models are called "experts". This follows the notation introduced for mixture models to

---

[c]A parallel can be drawn between the two philosophies for modeling densities (expansions vs. mixtures), and the two philosophies for function approximation (polynomials vs. neural networks). Expansions and polynomials are computationally cheap, have incremental updates, and are often amenable for an analytical treatment of convergence properties. Mixtures and neural networks are computationally expensive, since the entire model needs to be re-estimated when the number of components changes.

the neural network community by Jacobs, Jordan, Nowlan and Hinton (1991) who applied it to a classification problem, see also Jordan and Jacobs (1994). Their choice of the term expert does not imply any connotations to human experts. In the economics literature, the experts are called "states".

Several choices need to be made:

- How many experts should the model have? Hansen (1992) has developed a pseudo-likelihood ratio criterion to determine this nuisance parameter, and Baxter (1996) has developed an alternative based on the Minimum Message Length principle (Wallace and Boultor, 1968). This paper takes a more pragmatic approach: since the ultimate goal is to predict densities of financial time series, we evaluate the quality of the model on out-of-sample predictions. We typically choose between three and ten mixture experts, estimate the model, convince ourselves of its performance, and finally analyze the resulting experts. Their interpretation is part of the creativity of the modeling process and is hard to do automatically.
- What is the functional form of the individual densities each expert generates? It can be any member of the exponential family, and this article keeps the theoretical derivations general. However, when a specific distribution needs to be chosen (in the computer implementation and the comparisons), we assume these individual distributions to be Gaussian.[d]
- γ_j^{t+1} is the weight given to Gaussian j for the prediction for time $(t+1)$, with $\sum_{j=1}^{M} \gamma_j = 1$. What should $\gamma(\bullet)$ depend on?

Three possible answers to the last question form the basis for the remaining three model classes.

- **(5)** In the simplest case of an unconditional density, the γ_j's do not depend on anything: a mixture of Gaussians is fitted to the training set and all parameters are constants. The parameters (the mixture weights γ_j, and the means and variances of the individual Gaussians) are estimated in a maximum likelihood framework using the EM algorithm (explained in Sec. 2.5). This unconditional mixture will be one model in empirical comparisons.
- **(6)** The mixture weights γ_j depend on a set of external variables. Based on the performance of all the experts on each pattern of the training set, a "gate" learns the mapping from its inputs, the exogenous variables, to the γ_j's. This model class is called **gated experts** (GE) (Weigend, Mangeas and Srivastava, 1995) and represents a regression model. When used in forecasting, the temporal structure of the time series enters only through

---

[d]The idea of a mixture model can be traced back to Pearson (1894) who "mined" a data set consisting of measurements of the forehead size of crabs with a mixture of two Gaussians, thus "discovering" two sub-populations.

the construction of the patterns (the input-output pairs). Note that once these patterns have been generated from the raw data, randomizing the order of the training data has no effect on the resulting model. In the real world, there are time series problems where a regression approach is appropriate. A successful application of this architecture is energy demand forecasting where the inputs into the gate represent cloud coverage, temperature, special tariff days, and other exogenous variables (Weigend *et al.*, 1995). However, there are other time series problems where the nature of the problem requires time to be taken into account in a more fundamental way. One such example is given by the model class of HME.

(7) This model class is called **hidden Markov experts** (HME). It is best described by its underlying assumptions:

- There are several discrete states. Their corresponding functional input-output mapping can be expressed as feedforward networks. These sub-models are called experts.

- At each time step, a single expert is responsible for generating the corresponding observation. We do not know which of the experts actually generated the observation — the probabilities of the experts for each time step need to be estimated from the data.

- Modeling the sequence of the hidden states, we assume that the dynamics of the hidden states can be described by a first-order Markov process, i.e., the next state depends only on the current state. This is expressed as a matrix of transition probabilities between the hidden states. We do not know these transition probabilities either; they also must be estimated from the data.

Fortunately, the statistically solid framework of hidden Markov models (Baum and Eagon, 1963; Hamilton, 1989) provides algorithms to estimate the unknown quantities. We combine this framework with connectionist techniques. We show how we can learn the potentially non-linear functions of each expert, the parameters of the transition matrix, and the probability vector across the states at each time step.

The distinction made above emphasizes that GE and HME model time in a fundamentally different way. We now focus on the common aspects and consequences thereof. Both model classes share the goal to generate non-Gaussian density forecasts, and both are based on mixture models. The implications that hold for both cases include:

- **Discovering hidden states.** Conventional data analysis, data mining, and knowledge discovery often do not have a clearly defined concept of what it means to "discover hidden knowledge". This paper clearly defines *hidden states* as the components of the mixture density. The solid statistical basis allows for a principled interpretation in terms of probabilities, enabling the discovery of interesting relations. In the case of predicting financial returns, the hidden states can be related to volatilities. Methodologically, it is important to clarify that the HME approach does not insert knowledge that volatilities are important for characterizing regimes, but it does make statistical assumptions that in turn yield this knowledge.

- **Blending supervised with unsupervised learning.** Approaches to learning from data and computational intelligence are traditionally dichotomized into supervised learning (regression and classification where the desired outcome is known for the training data) and unsupervised learning (clustering where no target is available and the goal is to discover the underlying structure). Both GE and HME combine the strengths of supervised learning with those of unsupervised learning: They build on the advantage of supervised learning that allows for performance evaluation, while providing the flexibility of unsupervised learning that has the advantage of discovering and interpreting hidden states.

- **Combining forecasts.** The idea of combining forecasts, going back to (Bates and Granger, 1969), has become increasingly important in areas ranging from applied forecasting (Clemen, Murphy and Winkler, 1995) to computational learning theory (Cesa-Bianchi, Freund, Helmbold, Haussler, Schapire and Warmuth, 1997). Both GE and HME softly combine the forecasts of the experts. Also, the relative weights for each expert vary at each time step. These weights are the estimates of the posterior probabilities. They reflect the training set performance for similar situations. For GE, the similarity is given through the gate, and for HME through the previous state and the transition matrix.

- **Becoming experts through competition.** In most approaches to forecast combination, the individual models give equal weight to all their training points. GE and HME use *competitive learning*. For each training pattern, all experts compete. If one expert's prediction is better than the predictions of the other experts, it receives a larger share of the data point to update its parameters than the others. It thus learns to improve its predictions in areas where it is already quite good, and learns to ignore areas where some of its competitors are better. For both GE and HME, the experts become true experts and the algorithm learns about their area of applicability. Since we use unconditional variances for each expert, one delineation of the experts is according to the local noise level. Weigend *et al.* (1995) show that the adaptation of each expert to its (overall) local noise level helps to avoid overfitting. The standard assumption of constant variance often leads to local underfitting in some regions, and to local overfitting in others. When predicting financial returns, the different noise levels correspond to different volatility regimes. Given volatility clustering, this pulls the solution in the same direction as the Markov assumption of staying in a regime rather than switching to another one. In general, the grouping depends both on similar noise levels and on similar functional forms of the experts.

- **Modeling outliers.** Many practical problems in data mining use some heuristic to remove outliers. Given the strong effect outliers have on the model, the specific heuristic can determine the resulting model. As an alternative to removing outliers, robust statistics uses an influence function that downweighs patterns where the observation and prediction are far apart. This practice can be dangerous in risk management, a new area of increasing importance for financial firms. Risk management focuses on rare events and on tails of distributions. Removing outliers or reducing their influence leads to an underestimation of risk that can be detrimental. In contrast, GE and HME model outliers naturally. In our experience, one expert has a relatively large variance compared to the others. Its role is thus to become the "garbage-collector", effectively removing the outliers and "explaining" them much better than all of the other experts whose likelihoods vanish at that point. In turn, the remaining experts have cleaner data which often allows the models to be interpreted more easily.

- **Saving inputs.** When learning from data, one can never be sure that one has the "best" set of inputs. In many cases there is no shortcut to the creative process of arguing for several sets of inputs, building the model, and then evaluating the out-of-sample performance to learn which inputs are important. This paper does not address the problem of input selection. However, once a set of inputs has been decided on, it is often possible to have different experts look at different subsets of the full set of inputs. When linear experts suffice, standard linear theory helps determine the significance of the inputs, which often leads to further reduction. Individual experts can end up with only a fraction of the union of the inputs. This simpler structure also lends clearer interpretations of the individual models. Note that the formalism matches the noise level of each expert to the noise level of its corresponding data. This has been shown to be an important aspect against overfitting of GE.

This part of the introduction showed several angles on the proposed architecture that complement the rigorous evaluation of out-of-sample performance that any data driven modeling has to follow.

1.2. *Evaluating predictions*

In the model comparison, GE Model (6) and HME Model (7) are chosen to have identical assumptions whenever possible. They have the same number of experts, the same inputs, the same functional form for the experts (e.g., linear or neural network), and, on the output side, the same noise model and degrees of freedom (i.e., expert-specific variances, and expert and input-dependent means). The only difference is the gate. Since many financial time series exhibit volatility clustering, the gate inputs should include some volatility proxy such as exponentially smoothed square returns.

In addition to the comparison between the mixture architectures HME and GE, we also compare them with several simpler architectures: unconditional Gaussian Model (1), unconditional mixture of Gaussians Model (5), and a simple GARCH(1,1) model (constant mean but varying variance, Model (3b)). The main two questions the empirical evaluation tries to answer are:

- HME vs. GE: are there hidden states in the market that cannot be observed directly? The answer is positive if the assumption of an underlying hidden Markov process improves predictive accuracy compared to conditioning on exogenous variables. Of course, any model contains assumptions! Of particular importance here is the specific choice for the input variables. It is always possible that there is yet another model with more suited inputs that gives better out-of-sample performance. In our empirical study, however, we try to be as fair as possible in comparing the two cases on the S&P500 returns.
- HME vs. GARCH: do HME predicting non-Gaussian densities generate better forecasts than a GARCH model predicting Gaussian densities?

To answer these questions, we compare the out-of-sample performance on a test set, i.e., data from a time period after the end of the training period. No single measure suffices: we use several measures that capture different aspects of the density prediction.

- The first measure focuses on the predicted probability density function (pdf) and computes the average log-likelihood of the test data given the model. This measure, evaluated on test data, allows us to compare the performance of different architectures.
- The second measure focuses on the predicted cumulative density function (cdf). This integral transform method was suggested by Diebold, Gunther and Tay (1998).
- In addition, we also provide the normalized mean squared error. Note that it only evaluates the quality of the point forecast, but does not measure the quality of the density forecast, thus missing the central goal of this paper.

While point forecasting is predominant in the forecasting literature, some studies discuss interval forecasts (Chatfield, 1993; Christoffersen, 1997) and probability forecasts (Murphy and Winkler, 1992; Clemen *et al.*, 1995). As Diebold *et al.* (1998) point out, the reasons for the relative neglect of density forecasting and evaluation include: uncertainty about the specific distribution, difficulty in evaluation, and the lack of demands from practice. This has changed in the recent past: risk management has become central for financial firms, and trading and pricing models increasingly depend on good density estimates.

1.3. *Related work*

A hidden Markov model is a parametric stochastic probability model with which a time series can be generated or analyzed. A hidden Markov model has two interrelated processes: a finite-state Markov chain that cannot be observed, and an emission model associated with each state. The Markov chain is characterized by the matrix of transition probabilities between states. The output probability densities given by the emission model can be characterized along two axes:

1. The output probability densities can be represented non-parametrically or parametrically.
2. The output probability densities may depend on an input (conditional) or they may be a constant for each expert (unconditional).

The mathematical representation that describes the observation probabilities is called the **emission model.** Viewed from the perspective of time series *generation*, the Markov chain generates a sequence of discrete states that we call a path. Based on this path, the emission model generates the probability density for each time step. The specific realization (the "observation") is then generated from this probability density for each time step.

Viewed from the perspective of time series *analysis*, the output probabilities impose a "veil" between the states and the observer of the time series (Ferguson, 1980). The task is to lift that veil. The term *hidden* is used because these states cannot be seen directly from the observed data. It is called *Markov* since it assumes that the probability of the next state depends only on the current state and the transition probabilities between the states. Both the states and the observed process can be either discrete or continuous. In state space models, the states and the observations are both continuous (Harvey, 1989; Timmer and Weigend, 1997). HME use discrete states (corresponding to the experts) and continuous variables (corresponding to the observed time series).

Next, we need to address the question of how to estimate the parameters of the model from the observed sequence. Baum and Eagon (1963) solved this problem for hidden Markov models with discrete observation densities. Baum, Petrie, Soules and Weiss (1970) extend the algorithm to many of the classical distributions. Hidden Markov models have been widely used in speech recognition. In the neural network community, Bengio and Frasconi (1996) proposed the "Input-Output Markov model" which allows for non-constant transition probabilities in addition to non-linear emission models. The concept of the transition among states can also be used to model the time dependency of regime switching. Poritz (1982) first combined hidden Markov models with linear prediction. Hamilton (1989) introduced switching models to economics and econometrics, spawning a large body of research (Engel and Hamilton, 1990; Hamilton, 1990; Hansen, 1992; Hamilton, 1994; Lahiri and Wang, 1994).

Most of these applications focus on point predictions but not on densities. Fraser and Dimitriadis (1994), predicting one of the data sets of the Santa Fe Competition (Weigend and Gershenfeld, 1994), used a hidden Markov model and generated non-Gaussian through a Monte Carlo approach (generating many continuations and then essentially presenting a histogram for each time step.) Hamilton and Susmel (1994) proposed an approach to model the conditional variances within Markov switching framework, where they combined the regime switching process with an autoregressive conditional heteroskedasticity (ARCH) model by allowing the parameters of the ARCH process to come from different regimes. Gray (1996) proposed a more comprehensive method to nest the generalized ARCH (GARCH) model into regime switching model. However, these two models are limited to the first and second conditional moment of the distribution.

None of these approaches focused on the prediction and evaluation of more general densities. We emphasize the fact that the Markov switching models by their nature of being mixture models generate densities, and that these densities should be evaluated with appropriate measures. Furthermore, we allow for non-linear experts.

This paper is organized as follows: Sec. 2 explains the notation, describes the likelihood function, and illustrates the Expectation Maximization (EM) algorithm used in HME. Section 3 explains how to generate density predictions using HME and describes methods to evaluate the density. Section 4 shows what can be learned from computer generated data for the HME approach to density forecasting. Section 5 presents the empirical results on comparing HME with GE, GARCH, an unconditional mixture, and an unconditional Gaussian for the daily density forecasts of S&P500 returns. Some conclusions are drawn in Sec. 6.

2. The Assumptions and the Algorithm

2.1. *Notation*

1. **Observations.** $\mathcal{Y}^T = \{y^t | t = 1, \ldots, T\}$ refers to the observed time series data. T is the number of the observations and t is the time index. Similarly, $\mathcal{X}^T = \{x^t | t = 1, \ldots, T\}$ represents the *input* to the emission model. x^t itself can be a vector or a scalar. In the example of auto-regression, x^t is given by the previous d values, $x^t = \{y^{t-1}, y^{t-2}, \ldots, y^{t-d}\}$, where d is the dimension of the input. x^t can also consist of exogenous variables.

2. **States.** $\mathcal{S} = \{1, 2, \ldots, j, \ldots, M\}$ denotes the state. M is the number of states in the model and j refers to a specific state. The analysis of the model usually provides interpretations for the states in terms of physical significance or economic meaning such as relations to market sentiment, growth, recession, interest rates or volatility.

3. **Transition probabilities.** a_{ij} is the transition probability of switching from state i to j,

$$\mathbf{A} = \{a_{ij}, \quad i, j \leq M, \quad a_{ij} = P(s^{t+1} = j | s^t = i)\}$$

where $a_{ij} \geq 0$, $\sum_j a_{ij} = 1$, and s^t describes the state at time t.[e]

---

[e]This paper assumes that the transition probabilities are constant over time. This assumption can be relaxed, allowing the transition probabilities to vary over time (Durland and McCurdy, 1994; Filardo, 1994; Shi and Weigend, 1997).

4. **Emission probabilities.** b_j^t is the probability of observing y^t given the state and the model. In GE and HME this probability depends on the inputs x^t into the experts at time t through the conditional mean

$$\mathbf{B} = \{b_j^t, \quad j \leq M, \quad t \leq T, \quad b_j^t = P(y^t|s^t = j, x^t)\}.$$

5. **Initial probabilities of each state.** $\Pi = \{\pi_i, i = 1, \ldots, M\}$, where the probabilities have to sum to unity, $\sum_{i=1}^{M} \pi_i = 1$.

For convenience, $\theta = \{\mathbf{A}, \mathbf{B}, \Pi\}$ denotes the entire set of parameters of the model. The emission probability can thus be written as $P(y^t|s^t, x^t, \theta)$.

2.2. *The likelihood function*

To define the likelihood function, we impose the constraint that the probability of the current state depends only on the previous state:

$$P(s^t|s^{t-1}, s^{t-2}, \ldots, s^1, \mathcal{X}^{t-1}, \mathcal{Y}^{t-1}) = P(s^t|s^{t-1}). \tag{1}$$

With q^T denoting a specific sequence or path of states from $t = 1$ to T, this first-order Markov assumption enables us to write the probability of path $q^T = (s^1, s^2, \ldots, s^T)$ as

$$P(q^T) = P(s^T, s^{T-1}, \ldots, s^t, \ldots, s^1)$$

$$= P(s^1) \prod_{t=2}^{T} P(s^t|s^{t-1}). \tag{2}$$

Given the current input x^t and the previous state s^{t-1}, earlier values of s and y are irrelevant,

$$P(y^t, s^t|q^{t-1}, \mathcal{X}^{t-1}, \mathcal{Y}^{t-1}) = P(y^t, st|s^{t-1}, x^t). \tag{3}$$

With Eq. (1) this expression can be transformed in the following way:

$$P(y^t, s^t|s^{t-1}, x^t) = P(y^t|s^t, x^t)P(s^t|s^{t-1}). \tag{4}$$

The central problem of hidden Markov models is to find the entire set of parameters of the model. Using Eq. (3) and Eq. (4), the likelihood $P(\mathcal{Y}^T|\theta)$ is then given as

$$P(\mathcal{Y}^T|\theta) = \sum_{q^T} P(\mathcal{Y}^T, q^T|\theta)$$

$$= \sum_{q^T} P(y^T, s^T|q^{T-1}, \mathcal{Y}^{T-1}, \theta)P(\mathcal{Y}^{T-1}, q^{T-1}|\theta) \qquad \text{conditional probability}$$

$$= \sum_{q^T} P(y^T, s^T | s^{T-1}, x^T, \theta) P(\mathcal{Y}^{T-1}, q^{T-1} | \theta) \qquad \text{using Eq. (3)}$$

$$= \sum_{q^T} P(y^T | s^T, x^T, \theta) P(s^T | s^{T-1}) P(\mathcal{Y}^{T-1}, q^{T-1} | \theta) \ \text{using Eq. (4)}$$

$$= \sum_{q^T} \underbrace{P(y^1 | s^1, x^1, \theta)}_{b^1} \underbrace{P(s^1)}_{\text{initial}} \prod_{t=2}^{T} \underbrace{P(y^t | s^t, x^T, \theta)}_{=:b_j^t} \underbrace{P(s^t | s^{t-1})}_{=:a_{ij}}. \tag{5}$$

To obtain the probability $P(\mathcal{Y}^T | \theta)$, two probabilities need to be estimated. First, the emission probability given the current state, $P(y^t | s^t, x^T, \theta)$; it varies at each time step. Second, the transition probability $P(s^t | s^{t-1})$; it is a parameter of the model.

The product $b_j^t a_{ij}$ is at the heart of the hidden Markov framework. If there was no Markov assumption, the second term a_{ij} was absent, and the observation at time t would be attributed to state j with probability $b_j^t / \sum_i^M b_i^t$. Model based clustering is (without Markov assumption, no a_{ij}) the unconditional case (no input x). The presence of the second term, a_{ij}, however introduces the trade-off with the first term towards the entire likelihood. In most applications, the main diagonal elements a_{ii}, describing the self-transitions (i.e., the probability of staying in a state) typically have values above 0.9, corresponding to an average time of staying in the state of above ten steps. Only if the next observation in the sequence can be explained much better by a state different from the current state does the model switch to the next state.

2.3. *Modeling the conditional emission probabilities: the experts*

- **Independence.** Given the input of the emission model, the likelihood of observing y^t given the current state and given the current input is $b_j^t = P(y^t | s^t = j, x^t, \theta)$. They are independent for each t. We call each of the specified emission models an **expert**, and each individual expert corresponds to a state.
- **Density Function.** We can assume different forms for the distribution of the "noise". In the specific example of a Gaussian, the emission probability of expert j becomes

$$b_j^t = P(y^t | s^t = j, x^t, \theta)$$

$$= \frac{1}{\sqrt{2\pi\sigma_j^2}} \exp\left(-\frac{(y^t - \hat{y}_j^t(x^t))^2}{2\sigma_j^2} \right)$$

  $\hat{y}_j^t(x^t)$ is the conditional mean, and σ_j^2 is the variance of the predicted Gaussian density.
- **Architecture.** The functional dependence of the conditional mean $\hat{y}_j^t(x^t)$ on its input (x^t) can potentially be non-linear and is expressed through a

feedforward neural network with non-linear hidden units and a linear output transfer function. Removing the hidden units reduces the functional form of each expert to a linear relation. In the autoregressive case with a single lag, $\hat{y}^t$ is given by $\hat{y}^t = k_0 + k_1 x^t$. We can use linear autoregressive models as well as non-linear neural networks as experts. The emission probability, **B**, is determined by the set of parameters, θ_j, of expert j, according to the architecture of the emission model.

Different experts can have different sets of inputs. Typically, the number of inputs to each expert is a subset of the full set of inputs that would be useful. When different dynamics modeled by the different experts "live" on subsets of the full set of inputs, this approach can help reduce the "curse of dimensionality".

2.4. *Computing the likelihood: the forward-backward procedure*

Rather than computing $P(\mathcal{Y}|\theta)$ directly using Eq. (5), Baum (1972) proposed an elegant algorithm called the **forward-backward** procedure. Dempster, Laird and Rubin, (1977) subsequently introduced the so-called "Expectation Maximization" or EM algorithm to maximize this probability.

Let α_i^t be the *entire sequence* probability of having observed y from time 1 to time t and of being in state i at time t,

$$\alpha_i^t = P(y^1, y^2, \ldots, y^t, s^t = i|\theta)$$

where $1 \leq t \leq T$ and θ denotes the model parameters. The probability of the *entire sequence* of observations is given by the sum over the states at the end of the sequence (at time T):

$$P(\mathcal{Y}|\theta) = \sum_{i=1}^{M} \alpha_i^T . \tag{6}$$

The breakthrough of this idea is the computational complexity. Rather than being exponential in T (as one might expect, given the consideration of paths), it is only linear in time: α_i^T can be computed recursively

$$\alpha_j^{t+1} = \left[\sum_{i=1}^{M} \alpha_i^t a_{ij} \right] b_j^{t+1} . \tag{7}$$

At the beginning of the sequence, the α's are initialized with probability $\alpha_i^1 = \pi_i b_i^1$. This recursion is called the *forward procedure*. Given initial estimates of π_i and b_i^1, Eq. (7) prescribes the computation of the probability $P(\mathcal{Y}|\theta)$, and, for $t = T$, the entire likelihood. Similarly, the backward variable β_i^t is defined as the *conditional* probability of observing y from $t+1$ to T given state i at time t (and, as always, the parameters):

$$\beta_i^t = P(y^{t+1}, y^{t+2}, \ldots, y^T | s^t = i, \theta) .$$

The recursive induction for β starts at the end of the sequence ($t = T$) and can be written as:

$$\beta_i^t = \sum_{j=1}^{M} a_{ij} b_j^{t+1} \beta_j^{t+1} \quad (\forall i) . \tag{8}$$

With $t = T - 1, T - 2, \ldots, 2, 1$, we obtain the β's for all t. Combining α and β, we now obtain the important posterior probability of being in state i at time t given the entire set of observations and parameters

$$\begin{aligned}
\gamma_i^t &= P(s^t = i | \mathcal{Y}, \theta) \\[2mm]
&= \frac{P(\mathcal{Y}, s^t = i | \theta)}{P(\mathcal{Y} | \theta)} \\[2mm]
&= \frac{\alpha_i^t \beta_i^t}{\sum_{k=1}^{M} P(\mathcal{Y}, s^t = k | \theta)} \\[2mm]
&= \frac{\alpha_i^t \beta_i^t}{\sum_{k=1}^{M} \alpha_k^t \beta_k^t}
\end{aligned} \tag{9}$$

γ_i^t is a key quantity that will serve as the estimate for $P(s^t = i | \theta)$.

Finally, the joint probability of the conjunction, $\xi_{ij}^{t,t+1} = P(s^t = i, s^{t+1} = j | \mathcal{Y}, \theta)$, is also computed from α and β:

$$\begin{aligned}
\xi_{ij}^{t,t+1} &= \frac{P(s^t = i, s^{t+1} = j, \mathcal{Y} | \theta)}{P(\mathcal{Y} | \theta)} \\[2mm]
&= \frac{\alpha_i^t a_{ij} b_j^{t+1} \beta_j^{t+1}}{\sum_{i=1}^{M} \sum_{j=1}^{M} \alpha_i^t a_{ij} b_j^{t+1} \beta_j^{t+1}} .
\end{aligned} \tag{10}$$

We have defined the important *variable* γ_i^t, the probability of being in state I at time t, and showed how it can be computed from α_i^t and β_i^t capturing the likelihoods of the beginning of the sequence through t, and from t to the end of the sequence, respectively. The variable ξ will serve as an auxiliary quantity in the computation of the transition probabilities, discussed in the next section that discusses how the *parameters* of the model are estimated.

2.5. *The Baum–Welch algorithm: EM algorithm for hidden Markov models*

The likelihood as given by Eq. (5) cannot be maximized directly since the hidden states are not known. The solution of this problem goes back to Baum *et al.* (1970), see Liporace (1982) and Juang (1984). Excellent tutorial expositions are Poritz (1988), Rabiner (1989) and the corresponding chapter in the book Rabiner and Juang (1993).

Their main result is a re-estimation algorithm, called the **Baum–Welch algorithm**. Its key idea is to go back and forth between two steps, the E-step and the M-step.

- The **E-step** ("Expectation Step") assumes that the parameters of the model are known, and computes for each time step t the likelihoods α_i^t and β_i^t, and in turn, the posterior probabilities γ_i^t and $\xi_{ij}^{t,t+1}$.
- The **M-step** ("Maximization Step") takes the variables computed in the E-step and updates the parameters of the model under the constraints $\sum_{i=1}^{M} \pi_i = 1$ and $\sum_{j=1}^{M} a_{ij} = 1$.

The new *transition probabilities* are given by:

$$a_{ij} = \frac{\text{expected number of transitions from state } i \text{ to state } j}{\text{expected number of transitions from state } i \text{ (to anywhere)}} = \frac{\sum_t \xi_{ij}^{t,t+1}}{\sum_t \gamma_i^t} \, .$$

The new *initial probabilities* of state i are $\pi_i = \gamma_i^1$.

While the original work by Baum *et al.* (1970) estimated only the *unconditional* density for each state, this paper allows for *conditional* densities. The formulae for the re-estimation of the *emission parameters* depend both on the specific noise model and the specific functional form for the parameters of the noise model (e.g., linear dependence, neural network) of the experts. For each expert, the following function G is maximized (cf., Fraser and Dimitriadis, 1994):

$$G = \sum_{t=1}^{T} \sum_{j=1}^{M} \gamma_j^t \log P(y^t | x^t, s^t = j, \theta_j) \, . \tag{11}$$

θ_j represents the parameters of the emission model of state j. Equation (11) can be interpreted as the negative of a cost function for the emission model. The estimation of the parameter θ_j depends on the specific form of the emission model. To be able to write down specific formulae for updating the parameters, the errors are assumed to be Gaussian. We first discuss the update for the variance of expert j, σ_j^2. Assuming that σ_j^2 depends only depends only on the expert and not on any inputs, the likelihood is maximized when

$$\frac{\partial G}{\partial \sigma_j^2} = \sum_{t=1}^{T} \gamma_j^t \frac{1}{P(y^t | x^t, s^t = j, \theta_j)} \frac{\partial P(y^t | x^t, s^t = j, \theta_j)}{\partial \sigma_j^2}$$

takes the value of zero, yielding

$$\sigma_j^2 = \frac{\sum_{t=1}^{T} \gamma_j^t \left(y^t - \hat{y}_j^t \right)^2}{\sum_{t=1}^{T} \gamma_j^t} \, . \tag{12}$$

This is the γ_j^t-weighted squared error between observation y^t and prediction $\hat{y}_j^t$. It describes the "local" noise level for expert j.[f]

The mean of expert j, $\hat{y}_j^t(x_j^t)$ is a function of the inputs into the expert, x_j^t. This (linear or non-linear) dependence is parameterized with θ_j. To maximize Eq. (11), its partial derivative with respect to the parameters θ_j has to vanish:

$$\frac{\partial G}{\partial \theta_j} = \sum_{t=1}^{T} \gamma_j^t \frac{1}{P(y^t|x^t, s^t = j, \theta_j)} \frac{\partial P(y^t|x^t, s^t = j, \theta_j)}{\partial \theta_j}$$

$$= \sum_{t=1}^{T} \gamma_j^t \frac{y^t - \hat{y}_j^t}{\sigma_j^2} \frac{\partial \hat{y}_j^t}{\partial \theta_j} \tag{13}$$

where the mean of the Gaussian of the jth expert is given by $\hat{y}_j^t = \hat{y}_j^t(x_j^t, \theta_j)$. In the special case where this functional form is linear, the parameters for expert j can be estimated directly by regressing $\sqrt{\gamma_j^t} y_j^t$ onto $\sqrt{\gamma_j^t} x_j^t$. In the general non-linear case, each pattern still has the importance γ_j^t, but the parameters are to be estimated iteratively, as an additional inner loop within each M-step.[g] Interpreting it as a cost function for a neural network, each expert minimizes the weighted squared error

$$\sum_{t=1}^{T} \gamma_j^t (y^t - \hat{y}_j^t(x_j^t, \theta_j))^2 \, .$$

The parameters θ_j can be estimated through weighted error backpropagation (local linearization around θ_j and taking a small step towards a better solution). This justifies viewing γ_j^t as an *effective learning rate*.

3. Predicting and Evaluating Densities

The previous section emphasized the underlying assumptions and algorithms for estimating the model. This section discusses how density predictions are generated and evaluated.

3.1. *Generating the density forecasts*

To generate a predictive density from a given HME model, one might be tempted to use the state as estimated by Eq. (9). This, however, is cheating: γ_j^t is estimated

---

[f]The corresponding formulae for the case of vector-valued predictions are the γ_j^t-weighted covariances for dimension m and n:

$$\frac{1}{\sum_{t=1}^{T} \gamma_j^t} \sum_{t=1}^{T} \gamma_j^t (y^t(m) - \hat{y}_j^t(m))(y^t(n) - \hat{y}_j^t(n)) \, .$$

In many applications it is reasonable to consider only a diagonal covariance matrix. This implies that the noise is drawn independently for the different outputs, and can often be interpreted more easily than the general case that allows for a rotation.

[g]The non-linear case is sometimes called the *generalized* EM (or GEM) algorithm.

using the entire sequence of observations, including future information. However, given the sequence of observations through time t, we can estimate the predictive probability of a state in terms of the transition probabilities a_{ij} and the joint α_i^t probability of state $s = j$ at time $t + 1$ and the observations through time t, as

$$P(s^{t+1} = j | \mathcal{Y}^t, \theta) = \frac{P(\mathcal{Y}^t, s^{t+1} = j | \theta)}{P(\mathcal{Y}^t | \theta)}$$

$$= \frac{\sum_{i=1}^{M} \alpha_i^t a_{ij}}{\sum_{j=1}^{M} (\sum_{i=1}^{M} \alpha_i^t a_{ij})} =: g_j^{t+1} . \tag{14}$$

Using the same notation for HME as for GE (Weigend *et al.*, 1995), we use g_j^{t+1} as an abbreviation for $P(s^{t+1} = j | \mathcal{Y}^t, \theta)$. Note that g is a *causal* version of the γ — it is based only on past information (through α) but does not use any future information (that enters γ through β).

The density for y^{t+1} is the linear g_j-weighted superposition of the densities of the individual experts:

$$P(y^{t+1} | \mathcal{X}^t, \mathcal{Y}^t, \theta) = \sum_{j=1}^{M} P(y^{t+1} | \mathcal{X}^t, s^{t+1} = j, \theta_j) P(s^{t+1} | \mathcal{Y}^t, \theta)$$

$$= \sum_{j=1}^{M} g_j^{t+1} P(y^{t+1} | \mathcal{X}^t, s^{t+1} = j, \theta_j) . \tag{15}$$

$\mathcal{X}^t$ summarizes the set of exogenous variables that are available at time t. For the specific case of Gaussian distributions for the individual noise models, the individual densities are described by their conditional means $\hat{y}_j^{t+1}$ and the variances σ_j^2. This completes the discussion of the ingredients needed to generate the full distribution for y^{t+1}.

Should one be interested in the overall mean of the predicted density at time $t+1$, due to its linearity, it is the g_j-weighted superposition of each individual mean:

$$\hat{y}^{t+1} = \sum_{j=1}^{M} g_j^{t+1} \hat{y}_j^{t+1} . \tag{16}$$

However, recall that the key goal is to generate a forecast of the density, and not just its mean. The emphasis on densities requires special care in evaluating the forecasts. The next subsection presents different evaluation methods.

3.2. *Evaluating the density forecasts*

We use two different methods to evaluate density forecasts that complement each other well. While the first method is based on the probability density function itself (pdf), the second method is based on the integral of the pdf, i.e., the cumulative distribution function (cdf).

- For each time step in the test set, the **pdf-based evaluation** records $\log P(y^{t+1}|\mathcal{X}^t, \mathcal{Y}^t, \theta)$, the value of the logarithm of the predicted density at the corresponding observation y^{t+1}. The average of these $\log P$ over the test set is used as measure for evaluation.[h] This average (or "per-pattern") likelihood of out-of-sample data given the density predictions of the model allows direct comparison between different model classes. Since the value on a specific test set is only an estimate for the true value, it is important to use identical training and test sets in the comparisons.[i]

- The **cdf-based evaluation** computes for each time step the cumulative probability distribution from the predicted density and records the value of the cdf at the observed data point for each day. This probability integral transform was recently used by Diebold *et al.* (1998) and goes back at least to Rosenblatt (1952). Z^{t+1} denotes the value that the predicted cdf takes at the observation y^{t+1}:

$$Z^{t+1} = \int_{-\infty}^{y^{t+1}} P(\eta^{t+1}|\mathcal{X}^t, \mathcal{Y}^t, \theta)d\eta$$

  η is the integration variable. The key idea is that these values of Z should be uniformly distributed. Diebold *et al.* (1998) point out that standard procedures (e.g., Kolmogorov-Smirnov) test jointly for uniformity and independence. If the test is rejected, it is not clear what conclusions should be drawn. We follow their suggestion and first evaluate unconditional uniformity using a simple histogram. Second, to evaluate whether Z is iid, we show the correlogram of the centered $(Z - \bar{Z})$, where $\bar{Z}$ is the mean of Z. To explore dependencies beyond linearity, we also show the correlogram of the powers of $(Z - \bar{Z})$.

For completeness, we also give the normalized mean squared error defined as

$$E_{\mathrm{NMS}} = \frac{\sum_t (\mathrm{observation}^t - \mathrm{prediction}^t)^2}{\sum_t (\mathrm{observation}^t - \mathrm{mean_{train}})^2} = \frac{\sum_t (y^t - \hat{y}^t)^2}{\sum_t (y^t - \mathrm{mean_{train}})^2} \tag{17}$$

where t usually enumerates the points in the withheld test set. E_{NMS} compares the model's point predictions to simply predicting the mean of the training set.

[h]To avoid possible confusion, it might be worth pointing out that there are two very different likelihood functions in estimation and in evaluation. The likelihood function maximized in the model estimation or search, Eq. (6), considers the likelihood of the *sequence* — this includes the trade-off between staying in a regime and allowing for somewhat worse predictions vs. changing regimes and obtaining better predictions. Note that this likelihood includes the transition matrix Eq. (7). In contrast, the likelihood function used for evaluation does not take transitions into account but only measures for each time step how well the observation was predicted by the pdf. It is important that this likelihood does not contain aspects of the sequence or the transition probabilities, but only the predicted densities. This allows for clean comparisons between approaches to density prediction.

[i]We thank Art Owen for pointing out that average log-likelihood can be very sensitive to a few extreme values. We computed trimmed means, but it turns out that outliers in log-likelihood are not a problem in the experiments reported here.

Note, however, that the normalized mean squared error only evaluates the point prediction and thus requires that we collapse the density prediction for each time step onto its mean. When predicting financial returns, many people do not expect a significant improvement over predicting the mean of past data. When the mean of the time series shifts, E_{NMS} can actually be larger than unity. This is also the case when the daily S&P500 forecasts are reduced to the mean and evaluated with E_{NMS}. However, the first two methods, using the pdf and the cdf, both evaluating the density, exhibit strong differences between the model classes.

4. Example 1: Computer Generated Data

For complicated model classes, it is important to understand the behavior of the model and to build up some intuitions about what happens when the model assumptions deviate from those of the generating process. Since mixture models contain an unsupervised part in learning, this section investigates whether the states found by the model actually correspond to the true hidden states.[j] We generate data from a hidden Markov model with two states, and analyze these data with HME, GE and, as a naive sanity check, an unconditional Gaussian.

4.1. *Generation and recognition models*

The **data generation** consists of two distinct and different processes: the Markov chain of the hidden states, and the dynamics of the individual experts.

- **Dynamics of the Markov model.** The transition probabilities are given by the matrix

$$A = \begin{pmatrix} 0.98 & 0.02 \\ 0.03 & 0.97 \end{pmatrix}.$$

  This allows us to generate a realization for the time series of the hidden states.
- **Dynamics of the individual experts.** With financial processes in mind, we pick the first process as trending, and the second process as mean reverting:

$$y^{t+1} = \begin{cases} 0.5y^t + 0.8\varepsilon^{t+1} & \text{if in state 1} \\ -0.3y^t + 0.5\zeta^{t+1} & \text{if in state 2} \end{cases}$$

ε and ζ are $N(0,1)$ iid.

We first generated a sequence of length 15,000 of the (eventually hidden) states. This sequence determined which of the two processes was used for each time step to

[j]In the real world, such as when modeling S&P500 densities (Sec. 5), we do not know the true model. This problem is particularly serious in finance for two reasons. First, while in the sciences experiments are usually carried out under carefully controlled conditions, finance does not allow for carefully controlled experiments. Second, the high amount of noise tends to mask subtle differences between competing models. This is again quite different to, say, physics, where some predictions are made with incredible accuracy and the data can distinguish between two models that make almost the same predictions.

generate an "observation". From the generated data, we use the first 10,000 points as the training set, and the remaining 5,000 points as the test set.

The **recognition models** are HME, GE and the unconditional Gaussian, defined by the mean and variance of the training set. In the case of HME, it is possible to choose the recognition process to perfectly match the data generating process by using two experts that are linear autoregressive models with one lag, AR(1).

The GE model used for the comparison also has two linear AR(1) experts, chosen to be as similar to the HME model as possible. The difference is that the probability of being in state j at time t in the GE model is learned as a feed-forward function of some variables, as opposed to recursively from the series itself using the hidden Markov assumption. One of the two gate inputs is the input that is also used in the experts, i.e., the current value of the time series, y^t. The other gate-input is an exponential moving average of the squared values of the observations $(y^t)^2$

$$\xi^t = \lambda \xi^{t-1} + (1 - \lambda)(y^t)^2 \tag{18}$$

with a decay constant $\lambda = 0.95$. The gate is implemented as a non-linear neural network with three hidden units (tanh transfer function) and two outputs. The outputs are constrained to be positive and to sum to unity, using the "softmax" architecture as discussed in Weigend *et al.* (1995).

4.2. *Results and interpretation*

We present selected results on the computer generated data for several purposes:

- Illustrate to what degree HME recover the hidden regimes on these fairly noisy time series. Note that true regimes are known in the computer generated example, but not in real world examples.
- Show how the partitioning of the gate-input space performed by the GE differs from the segmentation of the HME. For real applications, it is important to recognize signatures that indicate the wrong model class.
- Build up some intuitions for interpreting results of the analysis based on the probability integral evaluation proposed by Diebold *et al.* (1997).

Figures 1 and 2 show in the time domain the same 1,000 points of the test set for HME and for GE, respectively. In both figures, the top panel shows the true data. The bottom panel shows the probability that the model predicts for one of the two experts. The probability of the other expert is not shown but corresponds to the difference between unity and the probability shown. The dash-dotted line indicates the true regimes used in the generation of the test data. Despite the high noise level in both training and test data, HME discover the regimes adequately.

The corresponding results for GE are shown in Fig. 2. The training and test data are identical to those used for the HME. The GE architecture is chosen to be as similar to the HME architecture as possible, as discussed in Sec. 4.1. The main difference is in the segmentation. Comparing the bottom panels of Figs. 1 and 2,

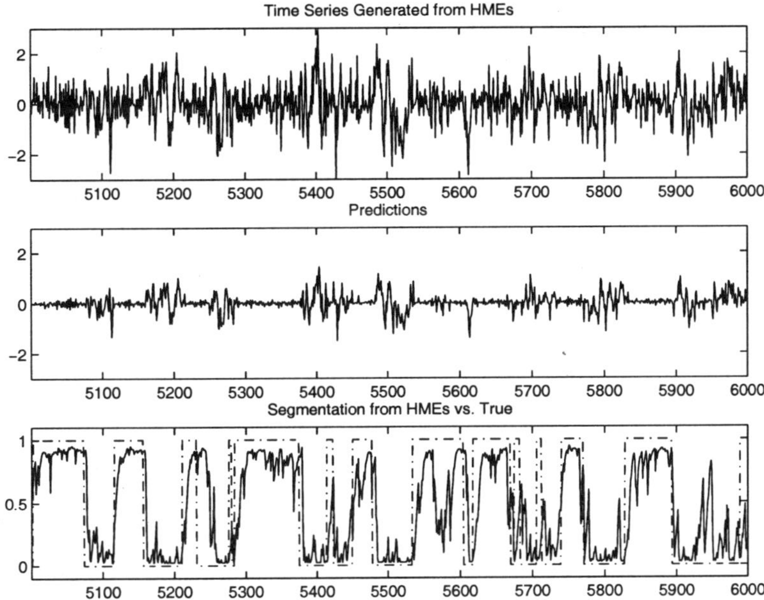

Fig. 1. Time series of the computer generated data modeled with HME. From top to bottom, the panels display 1,000 true values of the out-of-sample data, the point forecasts, and the probability of one expert, g_1^t. It sums with the other expert (not shown) to unity for each time step. The true regime used in the generation of the test data is indicated in the bottom panel as dash-dotted line.

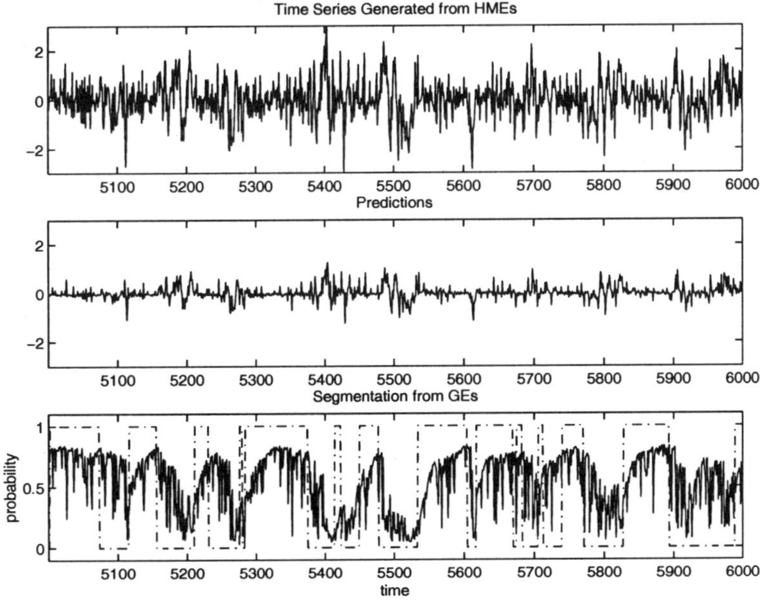

Fig. 2. Time series of the computer generated data modeled with GE. The top panel shows 1,000 points of the test set, the middle panel the one-step-ahead point predictions of the GE for the same period, and the bottom panel the probability g_1^t. These probabilities are not as clean as those found by HME since GE ignore the difference between adjacent and distant patterns in time.

note that the regime assignments are cleaner for HME than for GE. This can be explained by the different likelihood functions: while GE represent a feedforward architecture that necessarily produces solutions that are invariant under re-shuffling of the input-output patterns, HME "know" about the sequence of the patterns through the assumption of the hidden Markov structure. HME can be said to trade-off the switching with the likelihood of the observation.

Although this paper focuses on density prediction, we included the mean of each one-step-ahead prediction as the middle panels. For these point predictions, the normalized mean squared errors (defined in Eq. (17)) on a 5,000 point test set are for the two model classes $E_{\mathrm{NMS}}(\mathrm{HME}) = 0.826$ and $E_{\mathrm{NMS}}(\mathrm{GE}) = 0.886$ with the ratio (squared error(HME))/(squared error(GE)) = 0.93.

We now turn to the evaluation of the densities, first using the predicted probability densities directly. On the same test set as above, the log-likelihood ratios are:

$$\frac{\text{log-likelihood(HME)}}{\text{log-likelihood(GE)}} = 0.96 \quad \text{and} \quad \frac{\text{log-likelihood(HME)}}{\text{log-likelihood(Gaussian)}} = 0.57 \,.$$

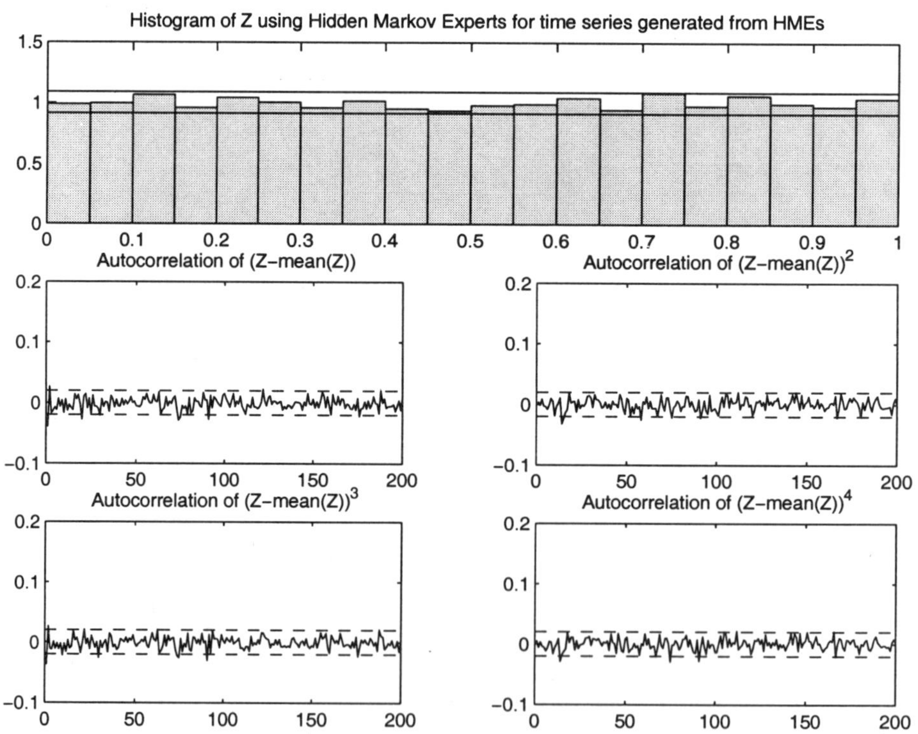

Fig. 3. Evaluation using the integral transform, Z, of the probability density predictions generated by HME. The histogram of Z indicates that the distribution of Z is uniformly distributed between 0 and 1, indicating good density predictions. The absence of autocorrelations indicates that there is no residual time structure in the mean corrected Z and its powers. The horizontal lines indicate two standard deviations.

While there is a clear improvement of the conditional mixtures over the unconditional Gaussian, the difference between the mixtures is not significant.

This second approach uses the cdf-based integral transform (Diebold *et al.*, 1998). This analysis focuses on Z^t, the area of the pdf to the left of the observation, i.e., the probability that a value below the observation was predicted. The qualitative aspects of the density forecasts are exposed in Figs. 3, 4, and 5 for HME, GE and the naive unconditional Gaussian, respectively. In these figures, the top panels give the histogram of Z on the test set. As discussed in Sec. 3.2, Z should be uniformly distributed between 0 and 1. The remaining four panels in each figure show the correlograms of powers of the mean-subtracted Z-series, i.e., the empirical autocorrelations of $(Z - \bar{Z})$, $(Z - \bar{Z})^2$, $(Z - \bar{Z})^3$, and $(Z - \bar{Z})^4$.

Analyzing the density predictions obtained with HME, Fig. 3 indicates that the histogram of the Z series is consistent with a uniform distribution. Furthermore, there are no significant autocorrelations in the powers of the Z-series after substracting the mean. These good density forecasts are reassuring — but not surprising since the structure of the HME recognition model was chosen to be identical to that of the generating process.

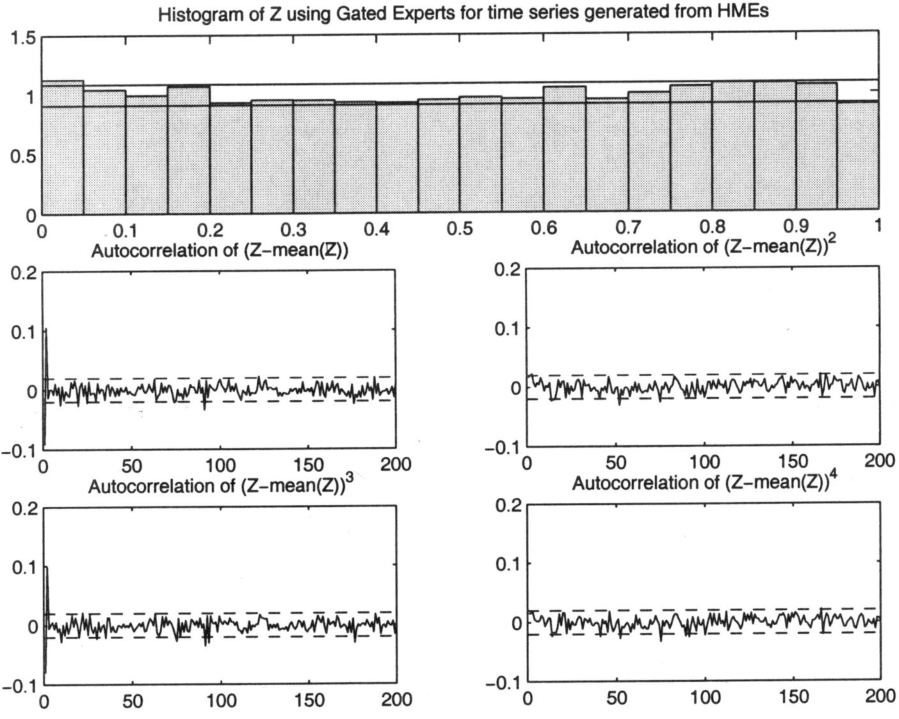

Fig. 4. Evaluation of density predictions using GE on the computer generated data. Note the appearance of significant autocorrelations for the odd powers of $(Z - \bar{Z})$ compared to the correctly specified HME.

Figure 4 shows the effect of a misspecified model. While the structure of the emission models (the experts) is still identical to the data generating process, GE cannot model correctly the underlying Markov structure of the sequence. In comparison to HME (Fig. 3), the histogram for GE is less uniform, and there are some short but significant autocorrelations in $(Z - \bar{Z})$ and $(Z - \bar{Z})^3$.

To put the qualitative aspects of the HME and GE predictions into perspective, Fig. 5 presents the histogram and the correlograms of Z when the model is a single unconditional Gaussian. In this model, more observations occur than were predicted in the central region of the histogram of about one standard deviation, and fewer observations in the areas around the 10 and 90 percentiles.[k] Furthermore, there are long autocorrelation dependencies in the Z-series. The non-uniformity of the histogram and the Z-autocorrelations are consistent with the poor performance on the quantitative measures of squared errors and of the log-likelihood.

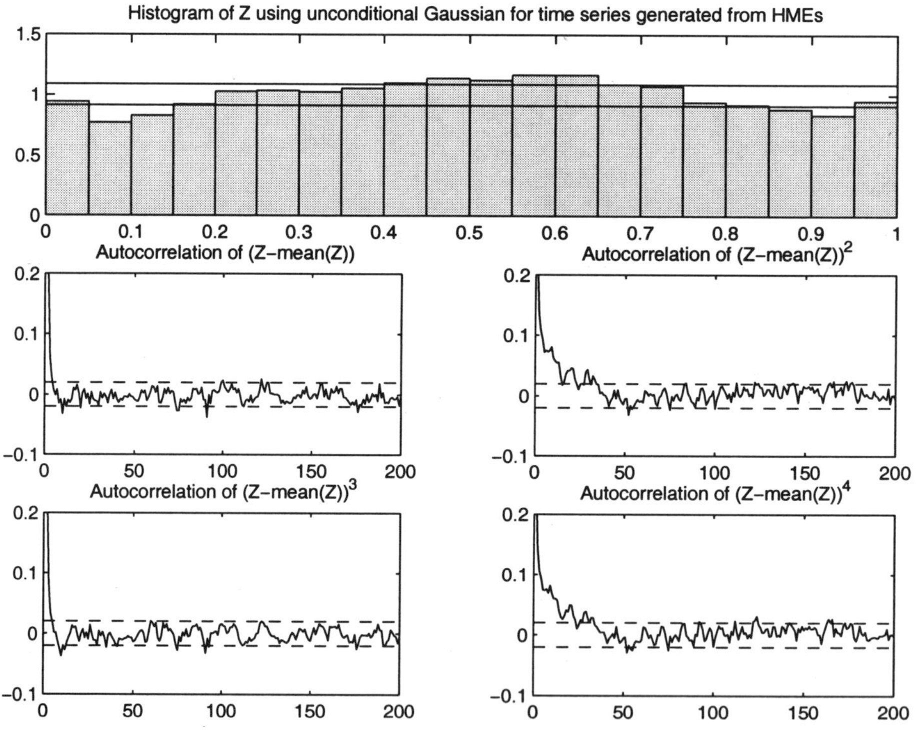

Fig. 5. Evaluation of density predictions using an unconditional Gaussian on the computer generated data. The model mismatch is indicated by both the non-uniformity of the histogram and the significant autocorrelations in the correlograms.

[k]The histogram focuses on the central part of the distribution since each bin has roughly the same number of points. The histogram can be viewed as expanding the center and compressing the tails. To focus on the tails of the distribution, a quantile-quantile plot (qq-plot) is more appropriate. It shows that there are too many observations in the extreme tails.

Table 1. Estimated parameters for HME and GE along with the true values used in the data generating process. a_{ii} denotes the self-transition probabilities of staying in regime i; the off-diagonal terms are the complements to unity. k_i denotes the autoregressive coefficients of the individual experts, and σ_i the noise levels of the individual experts.

	a_{11}	a_{22}	k_1	k_2	σ_1	σ_2
true value	0.980	0.970	0.500	-0.300	0.800	0.500
HME	0.976	0.969	0.507	-0.269	0.808	0.492
GE	N/A	N/A	0.466	0.003	0.867	0.528

Knowing the true model in this first example of computer generated data allows us to compare the estimated parameters with the true parameters: the diagonal elements of transition probability matrix $\mathbf{A}$, the autoregressive coefficients k_i, and the noise level σ_i. Table 1 gives the true values of the parameter and the estimates of the models. The correctly specified HME found the correct parameters. In contrast, the estimation of the corresponding GE experts is significantly worse than that of HME. In this specific run, the second expert does not even learn the mean-reverting dynamics but predicts an essentially unconditional Gaussian.

5. Example 2: S&P500 returns

This section applies HME to the real-world problem of forecasting the density of daily S&P500 returns. To provide a perspective and a deeper understanding of HME, comparisons are carried out to several other model classes: an unconditional Gaussian, an unconditional mixture of Gaussians, a generalized autoregressive conditional heteroskedastic GARCH(1,1) model, and the GE model that is as similar as possible to the HME model.

We first describe the data and models and analyze the estimated HME model. We then present the segmentation obtained by HME and by GE and explain the difference. Among the performance comparisons, the most important metric is the direct evaluation of the out-of-sample likelihood of the test data given each of the models. We also include the graphs of the probability integral transform evaluation of the density forecasts.

5.1. *Data and model classes*

For the data, we start with 21 years of daily S&P500 prices, p^t, and compute the series we try to predict, y^t, by taking the difference between the logarithms of the prices at adjacent days

$$y^t = \log p^t - \log p^{t-1} = \log \frac{p^t}{p^{t-1}} \approx \frac{p^t - p^{t-1}}{p^{t-1}} \, .$$

The Taylor expansion used in the last step, $\log(1+\epsilon) \approx \epsilon$, gives the interpretation of y^t as the relative price change, i.e., as the difference between today's and yesterday's price with respect to yesterday's price. This series corresponds to continuously compounded returns.

We use the first ten years (from 3 January 1977 to 31 December 1986) of the data as the training set, and the last ten years (from 2 March 1988 to 31 December 1997) as the test set. To avoid possible artifacts of the Oct 1987 crash, we do not use the data from 3 January 1987 to 1 March 1988 in this study.

Both HME and GE have four experts. We chose that number based on the predictive performance of the resulting model and the interpretability. The experts are simple linear autoregressive models that predict the mean based on the values of the previous seven lagged returns. (In other studies, not reported here, different sets of inputs, and different neural networks architectures are chosen.)

For GE, we need to also specify the structure of the gate: we use a non-linear neural network with five tanh hidden units and four "softmax" outputs. The inputs into the gate include the seven lagged values of the returns given to the experts, in addition to seven lagged values of the exponential moving average of the squared returns Eq. (18).

5.2. *Results*

After discussing the data and the models, we now turn to the results. We first inspect the segmentation obtained with HME and GE, then discuss the estimated parameters and their meaning, and finally turn to the evaluation of the densities, using the pdf and the cdf methods.

5.2.1. *Segmentation of the S&P500 series*

Figures 6 and 7 respectively show HME and GE models for daily S&P500 returns. Training and test periods are indicated by the arrows in the center of the figure. The lower half characterizes the importance of the individual experts for each day. The top panel shows the time series y^t, the daily S&P500 returns for the period from 1977 through 1997. The bottom four panels give the probability g_i^t for each expert $i(i = 1, \ldots, 4)$. The experts are ordered in terms of decreasing σ_i. The expert with the lowest noise level corresponds to the lowest panel.

Figure 7 indicates that GE cannot generate clear regimes. Note, for example, that the probability of the expert with the second smallest variance (the second plot from the bottom) hardly ever leaves the range between 0.1 and 0.4. One reason for this poor segmentation is that the smoothed squared returns Eq. (18) as gate-inputs do not characterize volatility as well as the recursively computed HME variances, σ_j^2. Another interpretation for the very noisy nature of the regimes is the absence of regime information from the neighboring pattern for GE, in contrast to HME.

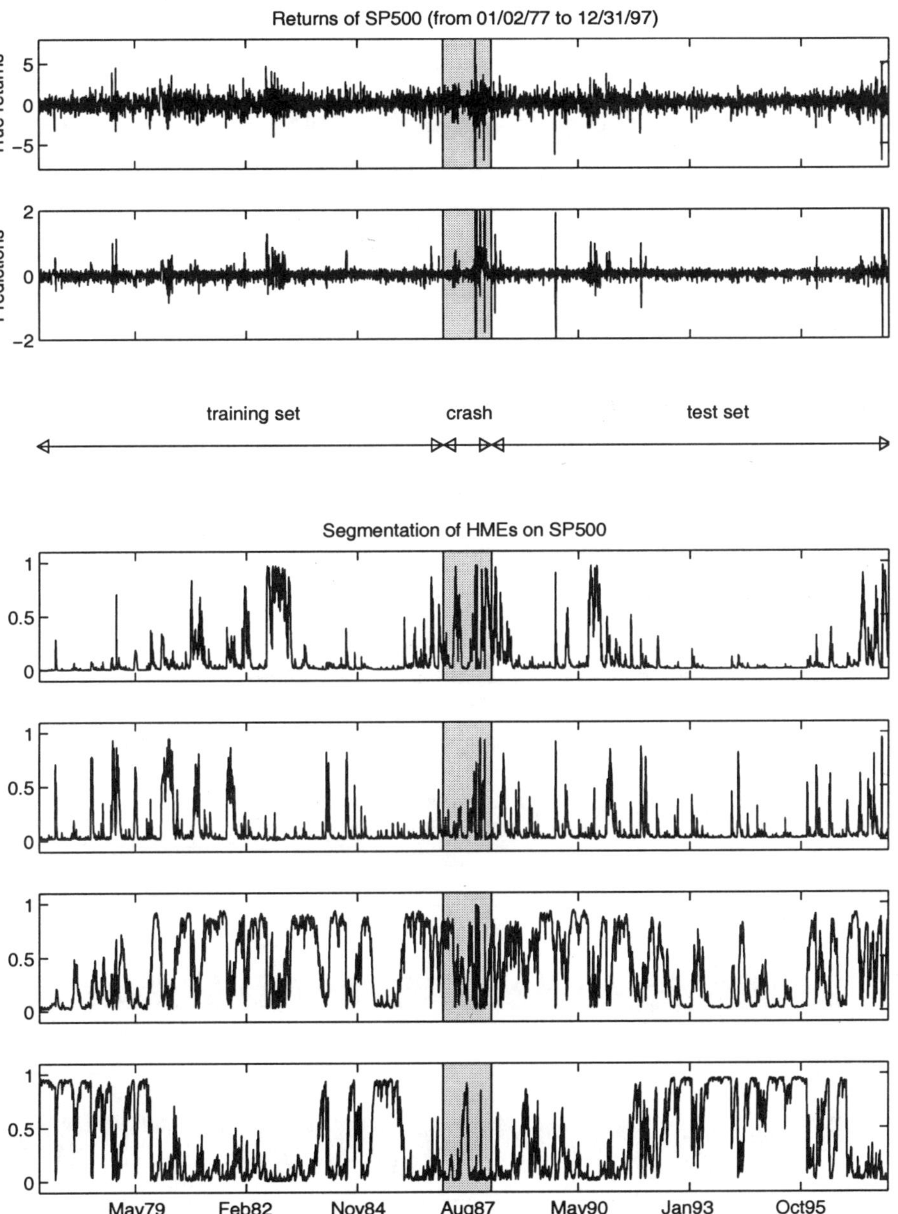

Fig. 6. Time series of S&P500 returns modeled with HME. The returns (top panel) have been normalized to zero mean and unit variance. The four plots at the bottom show the probabilities of the experts for each time step. The experts are arranged by decreasing noise level: the expert with the lowest noise level is at the bottom of the figure. (For completeness, the mean of each day's density is shown in the remaining panel, labeled "Predictions".)

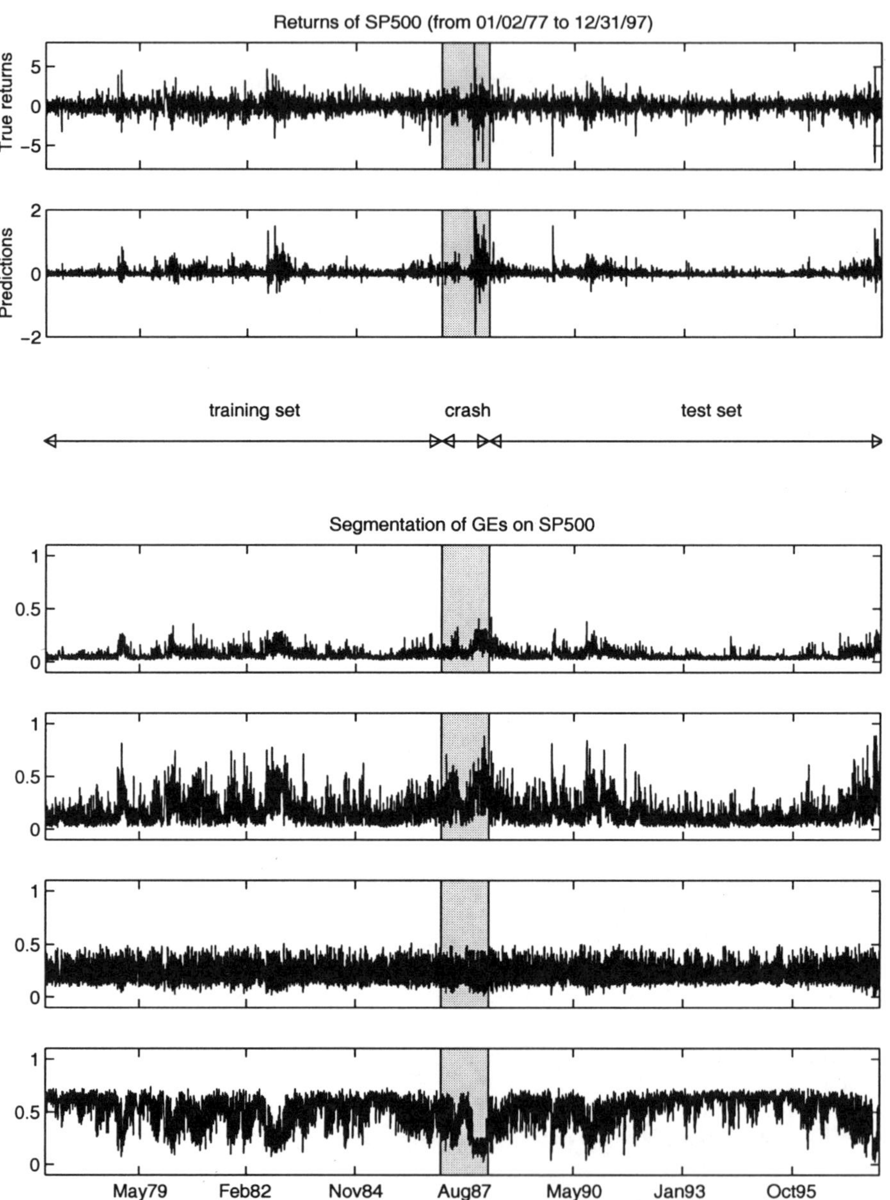

Fig. 7. Time series of S&P500 returns modeled with GE. The description of the panels is the same as in the preceding figure. Note the poorer and more noisy segmentation in comparison with the previous figure.

5.2.2. *Estimated parameters and interpretation*

The dynamics of the hidden Markov process is characterized by the matrix of transition probabilities between the states. For the four states assumed in our model of the S&P returns, we obtain

$$
\mathbf{A} = \begin{pmatrix}
0.904 & 0.031 & 0.023 & 0.042 \\
0.014 & 0.950 & 0.029 & 0.007 \\
0.011 & 0.014 & 0.969 & 0.007 \\
0.011 & 0.002 & 0.004 & 0.983
\end{pmatrix}.
$$

The elements of this matrix are averages over 200 runs with different initializations. To make the averaging meaningful, we sort in each run the states by decreasing σ_i, i.e., the first state becomes the one with the largest noise level, etc., and the fourth state the one with the smallest noise level. For example, $a_{11} = 0.904$ is the average of the self-transitions of the expert with the largest noise level. Note that a_{11} is the smallest of all the self-transitions (the elements on the diagonal): on average, the system stays uses expert 1 for only ten days. Looking back to Fig. 6, we can see that this expert takes responsibility for some of the large returns in the training set, as well as for the region of high volatility in late 1982.

Table 2 lists the noise levels of the experts for both HME and GE. For each run, the experts were ordered in terms of decreasing noise levels, and means and standard deviations of the square roots of the variances of the Gaussians are shown.

Table 2. The average noise levels σ_i of the individual experts for HME and GE for the S&P500 density predictions. In each run, i.e., for each set of initial conditions, the expert with the largest variance is assigned the label "Expert 1", etc. The table gives the means of the square roots of the variances of the Gaussians. The standard deviations are indicated in parentheses. High-noise experts have more relative variation in the noise levels than the low-noise experts than in those of high-noise experts. Furthermore, GEs are more sensitive to initial conditions than HMEs.

σ_i	Expert 1	Expert 2	Expert 3	Expert 4
HME	1.37	0.92	0.74	0.61
	(0.05)	(0.12)	(0.04)	(0.01)
GE	2.18	0.98	0.52	0.33
	(1.10)	(0.44)	(0.16)	(0.07)

5.2.3. *Evaluation of the S&P500 density predictions*

The function optimized in training is significantly different for HME and GE — we have emphasized that HMEs include the transitions between states, whereas GEs do not. For prediction, we are ultimately interested in how well the next day's density is predicted. This function can indeed be different from the one optimized in the estimation of the model. We compare all architectures with the same measure: the likelihood of the observations of the test set given the predicted densities. In our comparison, we do not take any uncertainty of the observed values into account, i.e., we assume a delta distribution whose integral is unity on infinitesimally small support. For each pattern in the test set, we compute the log-likelihood, and show in Fig. 8 the average over the test set.

On the bottom of Fig. 8 three benchmark log-likelihoods are shown: (i) of a single Gaussian, (ii) of an unconditional mixture of four Gaussians, and (iii) of a GARCH(1,1) model. All models are estimated on the same training set as HME and GE, and averaged over the same test set as HME and GE. The figure shows that the single Gaussian, representing the hypothesis of a random walk with fixed variance, is worst. The unconditional mixture is better, and the GARCH(1,1) model is again slightly better.

For HME and GE, any specific solution depends on the initialization. Rather than only showing the "typical" performance of one run, Fig. 8 gives the individual

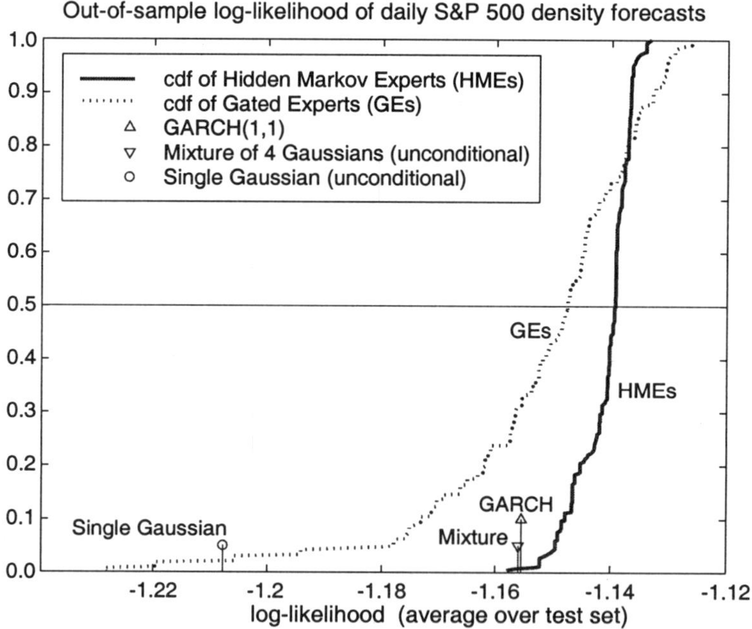

Fig. 8. S&P500 density forecasts evaluated for several models based on the predicted pdf. The horizontal axis gives the log-likelihood averaged over the test set. For HME and GE the empirical cumulative distribution of 200 runs each is plotted. For comparison, we also indicate the log-likelihood averaged over the same test set for a single Gaussian, an unconditional mixture of four Gaussians, and a GARCH(1,1) model. Note that the GEs do not give an acceptable solution for this hard learning problem since the resulting models show a large variance in performance. In contrast, the distribution of quality of the HME is relatively sharp, indicated by a relatively steep curve. Considering only the uncertainty stemming from the initialization, about 98% percent of the HME have a better out-of-sample likelihood than the GARCH model and than the unconditional mixture model. This indicates that all of HME's aspects (conditional model *and* mixture model *and* hidden Markov model) are needed for the improvement.

results of 200 runs for HME and GE each. For both architectures, it displays the cumulative probability distribution of the 200 runs. The medians correspond to the locations where the dashed line (GEs) and solid line (HMEs) intersect with the horizontal line. Note that GEs have an unacceptable large range of performance variation, ranging from worse than a single Gaussian to better than the best model. In contrast, the log-likelihood of HME is better behaved. Only about 2% of the runs are worse than the unconditional mixture or the GARCH(1,1) model, and the mean and median performance is clearly above the benchmark models. This indicates that the *combination* of the conditional variance *and* the mixture aspect is needed for the improvement of the quality of the density predictions.

To rule out the possibility that the results are due to a few outliers, we analyzed the trimmed means of the log-likelihood. The ranking of the different methods remains the same when the means are trimmed; we removed up to 2% on each side. This establishes that HME give better predictions than the alternatives we considered when comparing the out-of-sample likelihood of new data.

We now turn from the analysis based on the predicted probability distributions to an analysis based on the predicted cumulative distributions. Figures 9, 10 and 11 show the results for HME, GE, and unconditional Gaussian, respectively. In

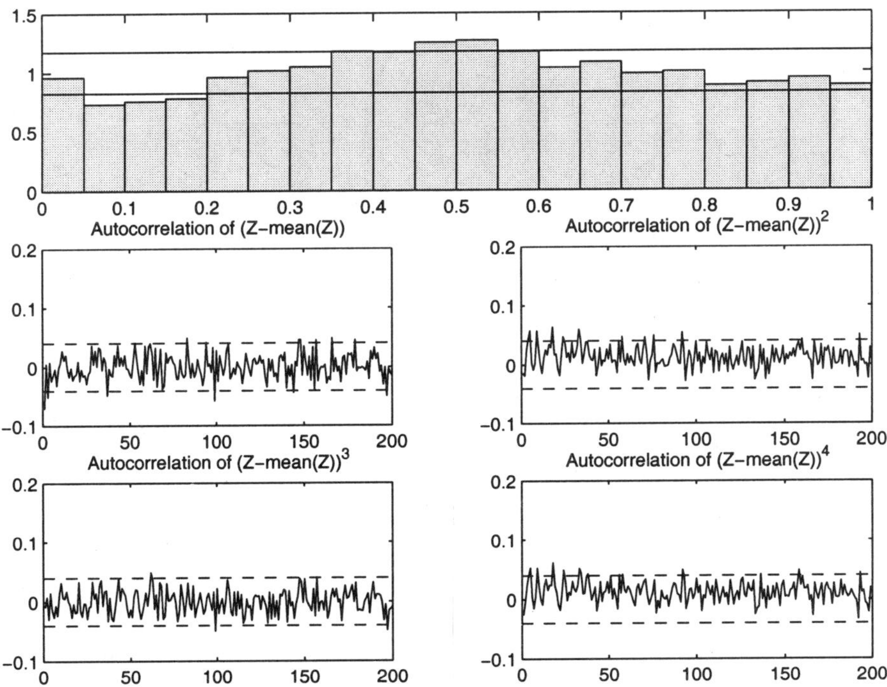

Fig. 9. Evaluating the probability density predictions of HME for S&P500 returns. The top panel plots the histogram of the probability integral transform on S&P500: the Z series is reasonably close to uniform. The four bottom panels show the correlograms: there are not many significant auto-correlations in the Z series and its powers. The dashed lines correspond to two standard deviations.

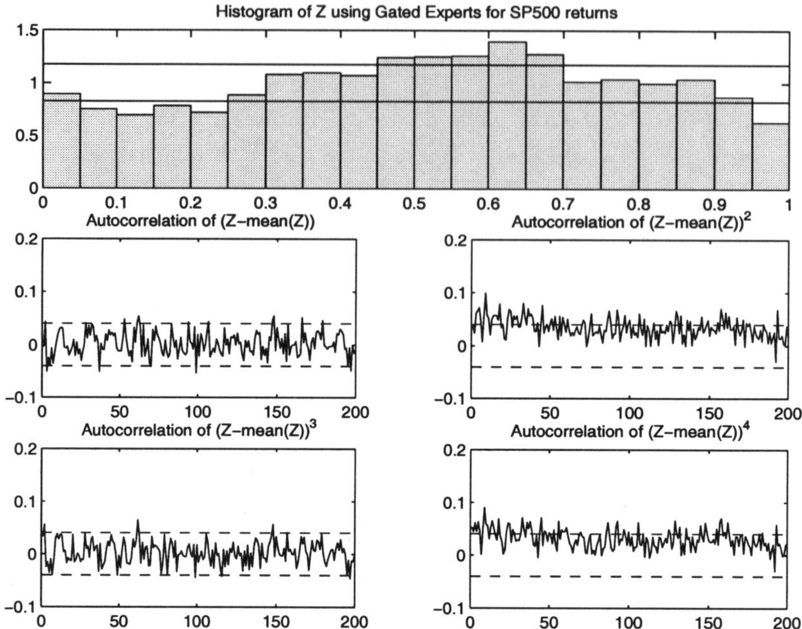

Fig. 10. Evaluating the probability density predictions of GE for S&P500 returns. The Z series is less uniformly distributed as in the previous figure (HME), and auto-correlations remain in the Z series.

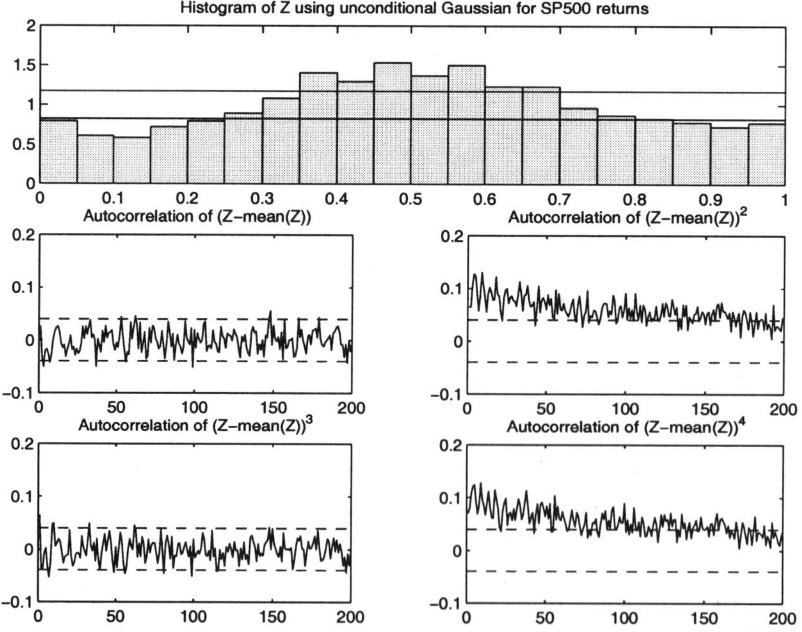

Fig. 11. Evaluating the probability density predictions of an unconditional Gaussian for S&P500 returns. The Z series is far from uniformly distributed, and the auto-correlations are large.

all cases, the top panel shows the histogram of the probability integral transform Z, and the four bottom panels the correlograms of the Z series and its powers. The results are acceptable for HME, slightly worse for GE and, as expected, a lot worse for the unconditional Gaussian.

For completeness, we close by reporting the normalized mean squared errors that can be computed by collapsing the daily density predictions to their means: $E_{\text{NMS}}(\text{HME}) = 1.014$ (standard deviation for 200 runs with different initial emission probabilities is 0.002); $E_{\text{NMS}}(\text{GE}) = 1.043$ (standard deviation for 200 runs with different initial weights is 0.083). Values larger than unity indicate a drift in the mean.

6. Conclusions

This paper started out by discussing different tasks for prediction, and proceeded by presenting hidden Markov experts (HME) in detail. The main focus is the prediction of the full conditional density distribution. This is in contrast to the literature on Markov switching models that focuses on point predictions and segmentation, and on the literature on stochastic volatility and GARCH models that focuses on conditional variances. The density predictions we obtained as mixture models were evaluated in comparison to these standard approaches using several methods, including Diebold *et al.* (1998).

The approach was illustrated with two time series. Section 4 showed the results of a computer generated example where the true regimes are known. This helped us obtain intuitions for model misspecification, e.g., by revealing the signature of misapplying GE to data generated by HME. When the right model class is used (HME), the parameters are estimated correctly and the density is predicted well.

Section 5 applied the approach to the density of daily S&P500. On the test set, about 98% of the HMEs estimated (they differed by their initial conditions) outperformed a GARCH(1,1) model. While HME found a solution rather reliably, GE showed a large dispersion for two reasons: (i) in any task with very high noise levels it is very difficult for the gate to learn a mapping from some exogenous variables to the expected probabilities of the experts, and (ii) in the specific case of financial returns, volatility is often estimated better recursively (as in GARCH and stochastic volatility models) than with a feedforward architecture without memory, such as GE, see Timmer Weigend (1997).

This paper focused on introducing hidden Markov experts. The examples were chosen to communicate some intuitions and illustrate several methods to evaluate the performance of density predictions. An identical set of inputs, consisting of lags of the time series, was used to facilitate the comparisons between the methods. When using this architecture in trading, we find that carefully selected exogenous inputs lead to better predictions than autoregressive models. In addition to trading applications, we have also used HME in risk measurement to capture non-Gaussian tails and compute Value-at-Risk.

Acknowledgments

We thank Jens Timmer for sharing his insights into hidden Markov models during his visits to CU Boulder and NYU. Andreas Weigend would like to thank Frank Diebold, Jerry Friedman, Clive Granger, Helmut Lütkepohl, Mark Mueller, Art Owen, Morio Yoda, and the participants of the 1997 NBER Summer Institute for their valuable feedback, in particular Bruce Hanson, Blake LeBaron and Jim Stock. Shanming Shi would like to thank Fei Chen, Elion Chin, Mike Henderson, Christian Pirkner, Satinder Singh and Ashok Srivastava.

References

J. M. Bates and C. W. J. Granger, "The combination of forecasts", *Operations Research Quarterly* **20** (1969) 451–468.

L. E. Baum, "An inequality and associated maximization technique in statistical estimation for probabilistic of Markov process", *Inequalities* **3** (1972) 1–8.

L. E. Baum and J. A. Eagon, "An inequality with applications to statistical prediction for functions of Markov processes and to a model for ecology", *Bull. Amer. Math. Soc.* **73** (1963) 360–363.

L. E. Baum, T. Petrie, G. Soules and N. Weiss, "A maximization technique ocurring in the statistical analysis of probabilistic functions of Markov chains", *Annals of Mathematical Statistics* **41** (1970) 164–171.

R. Baxter, "Minimum message length inference: theory and applications", PhD thesis, Department of Computer Science, Monash University, Clayton, Australia, 1996. `www.ultimode.com/rohan/thesis.html`.

Y. Bengio and P. Frasconi, "Input-Output HMM's for sequence processing", *IEEE Transactions Neural Networks* **7** (1996) 1231–1249.

N. Cesa-Bianchi, Y. Freund, D. P. Helmbold, D. Haussler, R. E. Schapire and M. K. Warmuth, "How to use expert advice," *Journal of the ACM* **44**(3) (1997) 427–485.

C. Chatfield, "Calculating interval forecasts", *Journal of Business and Economics Statistics* **11** (1993) 121–135.

P. F. Christoffersen, "Evaluating interval forecasts", *International Economic Review*, forthcoming, 1997.

R. T. Clemen, A. H. Murphy and R. L. Winkler, "Screening probability forecasts: Contrasts between choosing and combining," *International Journal of Forecasting* **11** (1995) 133–146.

A. P. Dempster, N. M. Laird and D. B. Rubin, "Maximum likelihood from incomplete data via the EM algorithm", *J. Roy. Stat. Soc. B* **39** (1997) 1–38.

F. X. Diebold, T. A. Gunther and A. S. Tay, "Evaluating density forecasts, with applications to risk management", *International Economic Review*, forthcoming, 1998.

J. M. Durland and T. H. McCurdy, "Duration-dependent transitions in a Markov model of U.S. GNP growth", *Journal of Business and Economic Statistics* **12** (1994) 279–288.

C. Engel and J. D. Hamilton, "Long swings in the dollar: Are they in the data and do markets know it?" *The American Economic Review* **80** (1990) 689–713.

J. D. Ferguson, "Hidden Markov analysis: An introduction", in *The Symposium on the Applications of Hidden Markov Models to Text and Speech*, J. D. Ferguson, ed., (Princeton, NJ, 1980), pp. 143–179.

A. J. Filardo, "Business cycle phases and their transitional dynamics," *Journal of Business and Economic Statistics* **12**(3) (1994) 299–308.

A. M. Fraser and A. Dimitriadis, "Forecasting probability densities by using hidden Markov models", in *Time Series Prediction: Forecasting the Future and Understanding the Past*, A. S. Weigend and N. A. Gershenfeld, ed., (Addison-Wesley, Reading, MA, 1994), pp. 265–282.

S. F. Gray, "Modeling the conditional distribution of interest rates as a regime-switching process", *Journal of Financial Economics* **42** (1996) 27–62.

J. D. Hamilton, "A new approach to the economic analysis of non-stationary time series and the business cycle", *Econometrica* **57** (1989) 357–384.

J. D. Hamilton, "Analysis of time series subject to changes in regime", *Journal of Econometrics* **45** (1990) 39–70.

J. D. Hamilton, *Time Series Analysis*, Princeton University Press, Princeton, 1994.

J. D. Hamilton and R. Susmel, "Autoregressive conditional heteroskedasticity and changes in regime", *Journal of Econometrics* **64** (1994) 307–333.

B. E. Hansen, "The likelihood ratio test nonstandard conditions: Testing the Markov switching model of GNP", *Journal of Applied Economics* **7** (1992) s61–s82.

A. C. Harvey, *Forecasting Structural Time Series Models and the Kalman Filter*, Cambridge University Press, Cambridge, University Press, Cambridge, 1989.

R. A. Jacobs, M. I. Jordan, S. J. Nowlan and G. E. Hinton, "Adaptive mixtures of local experts", *Neural Computation* **3** (1991) 79–87.

M. I. Jordan and R. A. Jacobs, "Hierarchical mixtures of experts and the EM algorithm", *Neural Computation* **6** (1994) 181–214.

B. H. Juang, "On hidden Markov model and dynamic time warping for speech recognition — a unified view", *AT&T BLTJ* **63** (1984) 1213–1243.

K. Lahiri and J. G. Wang, "Predicting cyclical turning points with leading index in a Markov switching model", *Journal of Forecasting* **13** (1994) 245–263.

L. A. Liporace, "Maximum likelihood estimation for multivariate observations of Markov sources", *IEEE Transactions on Information Theory* **IT-28** (1982) 729–734.

A. H. Murphy and R. L. Winkler, "Diagnostic verification of probability forecasts", *International Journal of Forecasting* **7** (1992) 435–455.

K. Pearson, "Contributions to the mathematical theory of evolution", *Phil. Trans. Royal Soc.* **185** (1894) 71–110. See also V. **185A**, p. 195.

A. B. Poritz, "Linear predictive hidden Markov models and the speech signal", *Proc. ICASSP'82*, Paris, France, 1982, pp. 1291–1294.

A. B. Poritz, "Hidden Markov models: A guided tour", *IEEE International Conference on Acoustics, Speech and Signal Processing*, 1988, pp. 7–13.

L. Rabiner and B.-H Juang, *Fundamentals of Speech Recognition*, Prentice Hall, Englewood Cliffs, NJ., 1993.

L. R. Rabiner, "A tutorial on hidden Markov models and selected applications in speech recognition", *Proceedings of the IEEE* **77** (1989) 257–286.

M. Rosenblatt, "Remarks on multivariate transformations", *Annals of Mathematical Statistics* **23** (1952) 470–472.

S. Shi and A. S. Weigend, "Markov gated experts for time series analysis: Beyond regression", *International Conference on Networks (ICNN'97, Houston, TX)*, Vol. IV, 1997, pp. 2039–2044.

J. Timmer and A. S. Weigend, "Modeling volatility using state space models", *International Journal of Neural Systems* **8** (1997) 385–398.

A. Timmermann, "Moments of Markov switching models", *Journal of Econometrics*, forthcoming, 1999.

C. S. Wallace and D. M. Boulton, "An information for classification", *Comp. J.* **11** (1968) 185–195.

A. S. Weigend and N. A. Gershenfeld, eds., *Time Series Prediction: Forecasting the Future and Understanding the Past*, Addison-Wesley, Reading, MA, 1994.

A. S. Weigend, M. Mangeas and A. N. Srivastava, "Non-linear gated experts for time series: Discovering regimes and avoiding overfitting", *International Journal of Neural Systems* **6**(1995) 373–399.

Reprinted from the J. of Financial Economics
55(2) (2000) 173–204

WHEN IS TIME CONTINUOUS?[*]

DIMITRIS BERTSIMAS, LEONID KOGAN and ANDREW W. LO[†]

Received 22 September 1998

Contents

[*]This research was partially supported by the MIT Laboratory for Financial Engineering, the National Science Foundation (Grant No. SBR–9709976) and a Presidential Young Investigator Award (Grant No. DDM–9158118) with matching funds from Draper Laboratory. We thank Michael Brandt, George Martin, Matthew Richardson, Bill Schwert, Jiang Wang, an anonymous referee, and seminar participants at Columbia, the Courant Institute, the Federal Reserve Board, the London School of Economics, the MIT finance lunch group, the MIT LIDS Colloquium, the NBER Asset Pricing Program, NYU, Renaissance Technologies, the Wharton School, the University of Chicago, the University of Massachusetts at Amherst, the University of Southern California, and the 1997 INFORMS Applied Probability Conference for valuable discussions and comments.
[†]MIT Sloan School of Management. Corresponding author: Andrew Lo, 50 Memorial Drive, E52-432, Cambridge, MA 02142–1347, (617) 253–0920 (voice), (617) 258–5727 (fax), `alo@mit.edu` (e-mail).

Continuous-time stochastic processes have become central to many disciplines, yet the fact that they are approximations to physically realizable phenomena is often overlooked. We quantify one aspect of the approximation errors of continuous-time models by investigating the replication errors that arise from delta hedging derivative securities in discrete time. We characterize the asymptotic distribution of these replication errors and their joint distribution with other assets as the number of discrete time periods increases. We introduce the notion of *temporal granularity* for continuous-time stochastic processes, which allows us to quantify the extent to which discrete-time implementations of continuous-time models can track the payoff of a derivative security. We show that granularity is a function of the contract specifications of the derivative security, and of the degree of market completeness. We derive closed form expressions for the granularity of geometric Brownian motion and of an Ornstein–Uhlenbeck process for call and put options, and perform Monte Carlo simulations that illustrate the practical relevance of granularity.

Keywords: Derivatives, delta hedging, continuous-time models.

JEL classification codes: G13.

1. Introduction

Since Wiener's (1923) pioneering construction of Brownian motion and Itô's (1951) theory of stochastic integrals, continuous-time stochastic processes have become indispensible to many disciplines ranging from chemistry and physics to engineering to biology to financial economics. In fact, the application of Brownian motion to financial markets pre-dates Wiener's contribution by almost a quarter century (see Bachelier [1900]), and Merton's (1973) seminal derivation of the Black and Scholes (1973) option-pricing formula in continuous time and, more importantly, his notion of delta hedging and dynamic replication is often cited as the foundation of today's multi-trillion dollar derivatives industry.

Indeed, the mathematics and statistics of Brownian motion have become so intertwined with so many scientific theories that we often forget the fact that continuous-time processes are only approximations to physically realizable phenomena. In fact, for the more theoretically inclined, Brownian motion may seem more "real" than discrete-time discrete-valued processes. Of course, whether time is continuous or discrete is a theological question best left for philosophers. But a more practical question remains: under what conditions are continuous-time models good approximations to specific physical phenomena, i.e., when does time seem "continuous" and when does it seem "discrete"?

In this paper, we provide a concrete answer to this question in the context of continuous-time derivative-pricing models, e.g., Merton (1973), by characterizing the replication errors that arise from delta hedging derivatives in discrete time.

Delta-hedging strategies play a central role in the theory of derivatives and in our understanding of dynamic notions of spanning and market completeness. In particular, delta-hedging strategies are recipes for replicating the payoff of a complex security by sophisticated dynamic trading of simpler securities. When markets are dynamically complete (see, for example, Harrison and Kreps [1979] and Duffie and Huang [1985]) and continuous trading is feasible, it is possible to replicate certain derivative securities perfectly. However, when markets are not complete or when continuous trading is not feasible, e.g., trading frictions or periodic market closings, perfect replication is not possible and the usual delta-hedging strategies exhibit *tracking* errors. These tracking errors comprise the focus of our attention.

Specifically, we characterize the asymptotic distribution of the tracking errors of delta-hedging strategies using continuous-record asymptotics, i.e., we implement these strategies in discrete time and let the number of time periods increase while holding the time span fixed. Since the delta-hedging strategies we consider are those implied by continuous-time models like Merton (1973), it is not surprising that tracking errors arise when such strategies are implemented in discrete time, nor is it surprising that these errors disappear in the limit of continuous time. However, by focusing on the continuous-record asymptotics of the tracking error, we can quantify the discrepancy between the discrete-time hedging strategy and its continuous-time limit, answering the question "When is time continuous?" in the context of replicating derivative securities.

We show that the normalized tracking error converges weakly to a particular stochastic integral and that the root-mean-squared tracking error is of order $N^{-1/2}$ where N is the number of discrete time periods over which the delta hedging is performed. This provides a natural definition for *temporal granularity*: it is the coefficient that corresponds to the $\mathbf{O}(N^{-1/2})$ term. We derive a closed-form expression for the temporal granularity of a diffusion process paired with a derivative security, and propose this as a measure of the "continuity" of time. The fact that granularity is defined with respect to a derivative-security/price-process pair underscores the obvious: a need for specificity in quantifying the approximation errors of continuous-time processes. It is impossible to tell how good an approximation a continuous-time process is to a physical process without specifying the nature of the physical process.

In addition to the general usefulness of a measure of temporal granularity for continuous-time stochastic processes, our results have other, more immediate applications. For example, for a broad class of derivative securities and price processes, our measure of granularity provides a simple method for determining the approximate number of hedging intervals N^* needed to achieve a target root-mean-squared-error δ: $N^* = g^2/\delta^2$ where g is the granularity coefficient of the derivative-security/price-process pair. This expression shows that to halve the root-mean-squared-error of a typical delta-hedging strategy, the number of hedging intervals must be increased approximately fourfold.

Moreover, for some special cases, e.g., the Black–Scholes case, the granularity coefficient can be obtained in closed form, and these cases shed considerable light on several aspects of derivatives replication. For example, in the Black–Scholes case, does an increase in volatility make it easier or more difficult to replicate a simple call option? Common intuition suggests that the tracking error increases with volatility, but the closed-form expression for granularity (3.2) shows that the granularity achieves a maximum as a function of σ and that beyond this point, it becomes a decreasing function of σ. The correct intuition is that at lower levels of volatility, tracking error is an increasing function of volatility because an increase in volatility implies more price movements and a greater likelihood of hedging errors in each hedging interval. But at higher levels of volatility, price movements are so extreme that an increase in volatility in this case implies that prices are less likely to fluctuate near the strike price where delta-hedging errors are the largest, hence granularity is a decreasing function of σ. In other words, at sufficiently high levels of volatility, the nonlinear payoff function of a call option "looks" approximately linear and is therefore easier to hedge. Similar insights can be gleaned from other closed-form expressions of granularity (see, for example, Sec. 3.2).

In Sec. 2, we provide a complete characterization of the asymptotic behavior of the tracking error for delta hedging an arbitrary derivative security, and formally introduce the notion of granularity. To illustrate the practical relevance of granularity, in Sec. 3 we obtain closed-form expressions for granularity in two specific cases: call options under geometric Brownian motion, and under a mean-reverting process. In Sec. 4 we check the accuracy of our continuous-record asymptotic approximations by presenting Monte Carlo simulation experiments for the two examples of Sec. 3 and comparing them to the corresponding analytical expressions. We present other extensions and generalizations in Sec. 5 such as a characterization of the sample-path properties of tracking errors, the joint distributions of tracking errors and prices, a PDE characterization of the tracking error, and more general loss functions than root-mean-squared tracking error. We conclude in Sec. 6.

2. Defining Temporal Granularity

The relationship between continuous-time and discrete-time models in economics and finance has been explored in a number of studies. One of the earliest examples is Merton (1969), in which the continuous-time limit of the budget equation of a dynamic portfolio choice problem is carefully derived from discrete-time considerations (see also Merton [1975, 1982b]). Foley's (1975) analyis of "beginning-of-period" versus "end-of-period" models in macroeconomics is similar in spirit, though quite different in substance.

More recent interest in this issue stems primarily from two sources. On one hand, it is widely recognized that continuous-time models are useful and tractable approximations to more realistic discrete-time models. Therefore, it is important to establish that key economic characteristics of discrete-time models converge properly to the characteristics of their continuous-time counterparts. A review of recent research along these lines can be found in Duffie and Protter (1992).

On the other hand, while discrete-time and discrete-state models such as those based on binomial and multinomial trees, e.g., Cox, Ross, and Rubinstein (1979), He (1990, 1991), and Rubinstein (1994), may not be realistic models of actual markets, nevertheless they are convenient computational devices for analyzing continuous-time models. Willinger and Taqqu (1991) formalize this notion and provide a review of this literature.

For derivative-pricing applications, the distinction between discrete-time and continuous-time models is a more serious one. For all practical purposes, trading takes place at discrete intervals, and a discrete-time implementation of Merton's (1973) continuous-time delta-hedging strategy cannot perfectly replicate an option's payoff. The tracking error that arises from implementing a continuous-time hedging strategy in discrete time has been studied by several authors.

One of the first studies was conducted by Boyle and Emanuel (1980), who consider the statistical properties of "local" tracking errors. At the beginning of a sufficiently small time interval, they form a hedging portfolio comprised of options and stock according to the continuous-time Black–Scholes/Merton delta-hedging formula. The composition of this hedging portfolio is held fixed during this time interval, which gives rise to a tracking error (in continuous time, the composition of this portfolio would be adjusted continuously to keep its dollar value equal to zero). The dollar-value of this portfolio at the end of the interval is then used to quantify the tracking error.

More recently, Toft (1996) shows that a closed-form expression for the variance of the cash flow from a discrete-time delta-hedging strategy can be obtained for a call or put option in the special case of geometric Brownian motion. However, he observes that this expression is likely to span several pages and is therefore quite difficult to analyze.

But perhaps the most relevant literature for our purposes is Leland's (1985) investigation of discrete-time delta-hedging strategies motivated by the presence of transactions costs, an obvious but important motivation (why else would one trade discretely?) that spurred a series of studies on option pricing with transactions costs, e.g., Figlewski (1989), Hodges and Neuberger (1989), Bensaid *et al.* (1992), Boyle and Vorst (1992), Edirisinghe, Naik, and Uppal (1993), Henrotte (1993), Avellaneda and Paras (1994), Neuberger (1994), and Grannan and Swindle (1996). This strand of the literature provides compelling economic motivation for discrete delta-hedging: trading continuously would generate infinite transactions costs.

However, the focus of these studies is primarily the *tradeoff* between the magnitude of tracking errors and the cost of replication. Since we focus on only one of these two issues — the approximation errors that arise from applying continuous-time models discretely — we are able to characterize the statistical behavior of tracking errors much more generally, i.e., for large classes of price processes, payoff functions, and state variables.

Specifically, we investigate the discrete-time implementation of continuous-time delta-hedging strategies and derive the asymptotic distribution of the tracking error in considerable generality by appealing to continuous-record asymptotics. We introduce the notion of temporal granularity which is central to the issue of when time may be considered continuous, i.e., when continuous-time models are good approximations to discrete-time phenomena. In Sec. 2.1, we describe the framework in which our delta-hedging strategy will be implemented and define the tracking error and related quantities. In Sec. 2.2, we characterize the continuous-record asymptotic behavior of the tracking error and define the notion of temporal granularity. We provide an interpretation of granularity in Sec. 2.3 and discuss its implications.

2.1. *Delta-hedging in complete markets*

We begin by specifying the market environment. For simplicity, we assume that there are only two traded securities: a riskless asset (bond) and a risky asset (stock). Time t is normalized to the unit interval so that trding takes place from $t = 0$ to $t = 1$. In addition, we assume

(A1) *Markets are frictionless, i.e., there are no taxes, transactions costs, short-sales restrictions, or borrowing restrictions.*

(A2) *The riskless borrowing and lending rate is 0.*[a]

(A3) *The price P_t of the risky asset follows a diffusion process*

$$\frac{dP_t}{P_t} = \mu(t, P_t)\, dt + \sigma(t, P_t)\, dW_t, \quad \sigma(t, P_t) \geq \sigma_0 > 0 \qquad (2.1)$$

where the coefficients μ and σ satisfy standard regularity conditions that guarantee existence and uniqueness of the strong solution of (2.1) and market completeness (see Duffie [1996]).

We now introduce a European derivative security on the stock that pays $F(P_1)$ dollars at time $t = 1$. We will call $F(\cdot)$ the payoff function of the derivative. The equilibrium price of the derivative, $H(t, P_t)$, satisfies the following partial differential equation (PDE) (see, for example, Cox, Ingersoll, and Ross [1985]):

$$\frac{\partial H(t, x)}{\partial t} + \frac{1}{2}\sigma^2(t, x)x^2 \frac{\partial^2 H(t, x)}{\partial x^2} = 0 \qquad (2.2)$$

---

[a]This entails little loss of generality since we can always renormalize all prices by the price of a zero-coupon bond with maturity at time 1 (see, for example, Harrison and Kreps [1979]). However, this assumption does rule out the case of a stochastic interest rate.

with the boundary condition

$$H(1, x) = F(x). \tag{2.3}$$

This is a generalization of the standard Black–Scholes model which can be obtained as a special case when the coefficients of the diffusion process (2.1) are constant, i.e., $\mu(t, P_t) = \mu$, $\sigma(t, P_t) = \sigma$, and the payoff function $F(P_1)$ is given by $\max[P_1 - K, 0]$ or $\min[P_1, K]$.

The delta-hedging strategy was introduced by Black and Scholes (1973) and Merton (1973) and when implemented continuously on $t \in [0, 1]$, the payoff of the derivative at expiration can be replicated perfectly by a portfolio of stocks and riskless bonds. This strategy consists of forming a portfolio at time $t = 0$ containing only stocks and bonds with an initial investment of $H(0, P_0)$, and rebalancing it continuously in a *self-financing* manner — all long positions are financed by short positions and no money is withdrawn or added to the portfolio — so that at all times $t \in [0, 1]$ the portfolio contains $\partial H(t, P_t)/\partial P_t$ shares of the stock. The value of such a portfolio at time $t = 1$ is exactly equal to the payoff, $F(P_1)$, of the derivative. Therefore, the price, $H(t, P_t)$, of the derivative can also be considered the production cost of replicating the derivative's payoff $F(P_1)$ starting at time t.

Such an interpretation becomes important when continuous-time trading is not feasible. In this case, $H(t, P_t)$ can no longer be viewed as the equilibrium price of the derivative. However, the function $H(t, P_t)$, defined formally as a solution of (2.2)–(2.3), can still be viewed as the production cost $H(0, P_0)$ of an approximate replication of the derivative's payoff, and may be used to define the production process itself (we formally define a discrete-time delta-hedging strategy below).[b] Therefore, when we refer to $H(t, P_t)$ as the derivative's "price" below, we shall have in mind this more robust interpretation of production cost and approximate replication strategy.[c]

More formally, we assume:

(A4) *Trading takes place only at N regularly spaced times t_i, $i = 1, \ldots, N$, where*

$$t_i \in \left\{ 0, \frac{1}{N}, \frac{2}{N}, \ldots, \frac{N-1}{N} \right\}.$$

Under (A4), the difference between the payoff of the derivative and the end-of-period dollar-value of the replicating portfolio — the *tracking error* — will be non-zero.

---

[b]The term "approximate replication" indicates the fact that when continuous trading is not feasible, the difference between the payoff of the derivative and the end-of-period dollar-value of the replicating portfolio will be non-zero. See Bertsimas, Kogan, and Lo (1997) for a discussion of derivative replication in discrete time and the distinction between production cost and equilibrium price.

[c]Alternatively, we can conduct the following equivalent thought experiment: while some market participants can trade costlessly and continuously in time and thus ensure that the price of the derivative is given by the solution of (2.2)–(2.3), we will focus our attention on other market participants who can trade only a finite number of times.

Following Hutchinson, Lo, and Poggio (1994), let $V_{t_i}^{(N)}$ be the value of the replicating portfolio at time t_i. Since the replicating portfolio consists of shares of the stock and the bond, we can express $V_{t_i}^{(N)}$ as

$$V_{t_i}^{(N)} = V_{S,t_i}^{(N)} + V_{B,t_i}^{(N)} \tag{2.4}$$

where $V_{S,t_i}^{(N)}$ and $V_{B,t_i}^{(N)}$ denote the dollar amount invested in the stock and the bond, respectively, in the replicating portfolio at time t_i. At time $t = 0$ the total value of the replicating portfolio is equal to the price (production cost) of the derivative

$$V_0^{(N)} = H(0, P_0) \tag{2.5}$$

and its composition is given by

$$V_{S,0}^{(N)} = \left. \frac{\partial H(t, P_t)}{\partial P_t} \right|_{t=0} P_0 \,, \quad V_{B,0}^{(N)} = V_0^{(N)} - V_{S,0}^{(N)} \,, \tag{2.6}$$

hence the portfolio contains $\partial H(t, P_t)/\partial P_t|_{t=0}$ shares of stock. The replicating portfolio is rebalanced at time periods t_i so that

$$V_{S,t_i}^{(N)} = \left. \frac{\partial H(t, P_t)}{\partial P_t} \right|_{t=t_i} P_{t_i} \,, \quad V_{B,t_i}^{(N)} = V_{t_i}^{(N)} - V_{S,t_i}^{(N)} \,. \tag{2.7}$$

Between time periods t_i and t_{i+1}, the portfolio composition remains unchanged. This gives rise to non-zero tracking errors $\epsilon_{t_i}^{(N)}$:

$$\epsilon_{t_i}^{(N)} \equiv H(t_i, P_{t_i}) - V_{t_i}^{(N)} \,. \tag{2.8}$$

The value of the replicating portfolio at time $t = 1$ is denoted by $V_1^{(N)}$ and the end-of-period tracking error is denoted by $\epsilon_1^{(N)}$.

The sequence of tracking errors contains a great deal of information about the approximation errors of implementing a continuous-time hedging strategy in discrete time, and in Secs. 2.2 and 5 we provide a complete characterization of the continuous-time limiting distribution of $\epsilon_1^{(N)}$ and $\{\epsilon_{t_i}\}$. However, because tracking errors also contain noise, we also investigate the properties of the root-mean-squared-error (RMSE) of the end-of-period tracking error ϵ_1 (see Hutchinson, Lo, and Poggio [1994] for other alternatives):

$$\text{RMSE}^{(N)} = \sqrt{\mathbf{E}_0\left[(\epsilon_1^{(N)})^2\right]} \tag{2.9}$$

where $\mathbf{E}_0[\cdot]$ denotes the conditional expectation, conditional on information available at time $t = 0$. Whenever exact replication of the derivative's payoff is impossible, $\text{RMSE}^{(N)}$ is positive.

Of course, root-mean-squared-error is only one of many possible summary statistics of the tracking error. A more general specification is the expected loss of the tracking error

$$\mathbf{E}_0\left[U(\epsilon_1^{(N)})\right]$$

where $U(\cdot)$ is a general *loss function*, and we consider this case explicitly in Sec. 5.4.

2.2. *Asymptotic behavior of the tracking error and RMSE*

We characterize analytically the asymptotic behavior of the tracking error and RMSE by appealing to continuous-record asymptotics, i.e., by letting the number of trading periods N increase without bound while holding the time span fixed. This characterization provides several important insights into the behavior of the tracking error of general European derivative securities that previous studies have only hinted at indirectly (and only for simple put and call options).[d] A by-product of this characterization is a useful definition for the temporal granularity of a continuous-time stochastic process (relative to a specific derivative security).

We begin with the case of smooth payoff functions $F(P_1)$:

Theorem 1. *Let the derivative's payoff function $F(x)$ in (2.3) be six times continuously differentiable and all of its derivatives be bounded, and suppose there exists a positive constant K such that functions $\mu(\tau, x)$ and $\sigma(\tau, x)$ in (2.1) satisfy*

$$\left| \frac{\partial^{\beta+\gamma}}{\partial \tau^\beta \partial x^\gamma} \mu(\tau, x) \right| + \left| \frac{\partial^{\beta+\gamma}}{\partial \tau^\beta \partial x^\gamma} \sigma(\tau, x) \right| + \left| \frac{\partial^\alpha}{\partial x^\alpha} (x\sigma(\tau, x)) \right| \leq K \qquad (2.10)$$

where $(\tau, x) \in [0,1] \times [0,\infty)$, $1 \leq \alpha \leq 6$, $0 \leq \beta \leq 1$, $0 \leq \gamma \leq 3$, and all partial derivatives are continuous. Then under Assumptions (A1)–(A4):

(a) *The RMSE of the discrete-time delta-hedging strategy (2.7) satisfies*

$$\mathrm{RMSE}^{(N)} = \mathbf{O}\left(\frac{1}{\sqrt{N}}\right) . \qquad (2.11)$$

(b) *The normalized tracking error satisfies:*

$$\sqrt{N}\, \epsilon_1^{(N)} \Rightarrow G$$

where

$$G \equiv \frac{1}{\sqrt{2}} \int_0^1 \sigma^2(t, P_t) P_t^2 \frac{\partial^2 H(t, P_t)}{\partial P_t^2} \, dW_t' , \qquad (2.12)$$

W_t' is a Wiener process independent of W_t, and "$\Rightarrow$" denotes convergence in distribution.

(c) *The RMSE of the discrete-time delta-hedging strategy (2.7) satisfies*

$$\mathrm{RMSE}^{(N)} = \frac{g}{\sqrt{N}} + \mathbf{O}\left(\frac{1}{N}\right) \qquad (2.13)$$

where

$$g = \sqrt{\mathbf{E}_0\left[\mathcal{R}\right]}, \qquad (2.14)$$

$$\mathcal{R} = \frac{1}{2} \int_0^1 \left(\sigma^2(t, P_t) P_t^2 \frac{\partial^2 H(t, P_t)}{\partial P_t^2} \right)^2 dt . \qquad (2.15)$$

Proof. See the Appendix.

[d]See, for example, Boyle and Emanuel (1980) Hutchinson, Lo, and Poggio (1994), Leland (1985), and Toft (1996).

Theorem 1 shows that the tracking error is asymptotically equal in distribution to $G/\sqrt{N}$ (up to $\mathbf{O}(N^{-1})$ terms), where G is a random variable given by (2.12). The expected value of G is zero by the martingale property of stochastic integrals. Moreover, the independence of the Wiener processes W_t' and W_t implies that the asymptotic distribution of the normalized tracking error is symmetric, i.e., in the limit of frequent trading, positive values of the normalized tracking error are just as likely as negative values of the same magnitude.

This result might seem somewhat counterintuitive at first, especially in light of Boyle and Emanuel's (1980) finding that in the Black–Scholes framework the distribution of the local tracking error over a short trading interval is significantly skewed. However, Theorem 1(b) describes the asymptotic distribution of the tracking error over the *entire life* of the derivative, not over short intervals. Such an aggregation of local errors leads to a symmetric asymptotic distribution, just as a normalized sum of random variables will have a Gaussian distribution asymptotically under certain conditions, e.g., the conditions for a functional central limit theorem to hold.

Note that Theorem 1 applies to a wide class of diffusion processes (2.1) and to a variety of derivative payoff functions $F(P_1)$. In particular, it holds when the stock price follows a diffusion process with constant coefficients, as in Black and Scholes (1973).[e] However, the requirement that the payoff function $F(P_1)$ is smooth — six times differentiable with bounded derivatives — is violated by the most common derivatives of all: simple puts and calls. In the next theorem, we extend our results to cover this most basic set of payoff functions.

Theorem 2. *Let the payoff function $F(P_1)$ be continuous and piecewise linear, and suppose (2.10) holds. In addition, let*

$$\left| x^2 \frac{\partial^\alpha \sigma(\tau, x)}{\partial x^\alpha} \right| \le K_2 \tag{2.16}$$

for $(\tau, x) \in [0,1] \times [0, \infty)$, $2 \le \alpha \le 6$, and some positive constant K_2. Then under Assumptions (A1)–(A4):

(a) *The RMSE of the discrete-time delta-hedging strategy (2.7) satisfies*

$$\mathrm{RMSE}^{(N)} = \frac{g}{\sqrt{N}} + \mathbf{o}\left(\frac{1}{\sqrt{N}}\right)$$

  where g is given by (2.14)–(2.15).

(b) *The normalized tracking error satisfies*

$$\sqrt{N}\,\epsilon_1^{(N)} \Rightarrow \frac{1}{\sqrt{2}} \int_0^1 \sigma^2(t, P_t) P_t^2 \frac{\partial^2 H(t, P_t)}{\partial P_t^2} \, dW_t' \tag{2.17}$$

---

[e]For the Black–Scholes case, the formula for the RMSE (2.14)–(2.15) was first derived by Grannan and Swindle (1996). Our results provide a more complete characterization of the tracking error in their framework — we derive the asymptotic distribution — and our analysis applies to more general trading strategies than theirs, e.g., they consider strategies obtained by deterministic time deformations; our framework can accommodate deterministic and stochastic time deformations.

where W_t' is a Wiener process independent of W_t.[f]

Proof. See Bertsimas, Kogan, and Lo (1998). By imposing an additional smoothness condition (2.16) on the diffusion coefficient $\sigma(\tau, x)$, Theorem 2 assures us that the conclusions of Theorem 1 also hold for the most common types of derivatives, those with piecewise linear payoff functions.

Theorems 1 and 2 allows us to define the coefficient of *temporal granularity g* for any combination of continuous-time process $\{P_t\}$ and derivative payoff function $F(P_1)$ — it is the constant associated with the leading term of the RMSE's continuous-record asymptotic expansion:

$$g \equiv \sqrt{\tfrac{1}{2}\mathbf{E}_0 \left[\int_0^1 \left(\sigma^2(t, P_t) P_t^2 \frac{\partial^2 H(t, P_t)}{\partial P_t^2} \right)^2 dt \right]} \tag{2.18}$$

where $H(t, P_t)$ satisfies (2.2) and (2.3).

2.3. *Interpretation of granularity*

The interpretation for temporal granularity is clear: it is a measure of the approximation errors that arise from implementing a continuous-time delta-hedging strategy in discrete time. A derivative-pricing model — recall that this is comprised of a payoff function $F(P_1)$ and a continuous-time stochastic process for P_t — with high granularity requires a larger number of trading periods to achieve the same level of tracking error as a derivative-pricing model with low granularity. In the former case, time is "grainier", calling for more frequent hedging activity than the latter case. More formally, according to Theorems 1 and 2, to a first-order approximation the RMSE of an N-trade delta-hedging strategy is $g/\sqrt{N}$. Therefore, if we desire the RMSE to be within some small value δ, we require

$$N \approx \frac{g^2}{\delta^2}$$

trades in the unit interval. For a fixed error δ, the number of trades needed to reduce the RMSE to within δ grows quadratically with granularity. If one derivative-pricing model has twice the granularity of another, it would require four times as many delta-hedging transactions to achieve the same RMSE tracking error.

From (2.18) it is clear that granularity depends on the derivative-pricing formula $H(t, P_t)$ and the price dynamics P_t in natural ways. Equation (2.18) formalizes the intuition that derivatives with higher volatility and higher "gamma" risk (large second derivative with respect to stock price) are more difficult to hedge, since these cases imply larger values for the integrand in (2.18). Accordingly, derivatives on less volatile stocks are easier to hedge. Consider a stock price process which

---

[f]It is easy to show, using Hölder's inequality, that $g < \infty$ and this implies that the stochastic integral in (2.17) is well defined. See Bertsimas, Kogan, and Lo (1998) for further details.

is almost deterministic, i.e., $\sigma(t, P_t)$ is very small. This implies a very small value for g, hence derivatives on such a stock can be replicated almost perfectly, even if continuous trading is not feasible. Alternatively, such derivatives require relatively few rebalancing periods N to maintain small tracking errors.

Also, a derivative with a particularly simple payoff function should be easier to hedge than derivatives on the same stock with more complicated payoffs. For example, consider a derivative with the payoff function $F(P_1) = P_1$. This derivative is identical to the underlying stock, and can always be replicated perfectly by buying a unit of the underlying stock at time $t = 0$ and holding it until expiration. The tracking error for this derivative is always equal to zero, no matter how volatile the underlying stock is. This intuition is made precise by Theorem 1, which describes exactly how the error depends on the properties of the stock price process and the payoff function of the derivative: it is determined by the behavior of the integral $\mathcal{R}$, which tends to be large when stock prices "spend more time" in regions of the domain that imply high volatility and high convexity or gamma of the derivative.

We will investigate the sensitivity of g to the specification of the stock price process in Secs. 3 and 4.

3. Applications

To develop further intuition for our measure of temporal granularity, in this section we derive closed-form expressions for g in two important special cases: the Black–Scholes option pricing model with geometric Brownian motion, and the Black–Scholes model with a mean-reverting (Ornstein–Uhlenbeck) process.

3.1. *Granularity of geometric Brownian motion*

Suppose that stock price dynamics are given by:

$$\frac{dP_t}{P_t} = \mu\,dt + \sigma\,dW_t. \tag{3.1}$$

where μ and σ are constants. Under this assumption we obtain the following explicit characterization of the granularity g.

Theorem 3. *Under Assumptions* (A1)–(A4), *stock price dynamics* (3.1), *and the payoff function of simple call and put options, the granularity g in* (2.13) *is given by*

$$g = K\sigma \left(\int_0^1 \frac{\exp\left[-\dfrac{\left[\mu t + \ln\left(\frac{P_0}{K}\right) - \sigma^2/2\right]^2}{\sigma^2(1+t)} \right]}{4\pi\sqrt{1-t^2}}\, dt \right)^{\frac{1}{2}} \tag{3.2}$$

where K is the option's strike price.

Proof. See Bertsimas, Kogan, and Lo (1998).

It is easy to see that $g = 0$ if $\sigma = 0$ and g increases with σ in the neighborhood of zero. When σ increases without bound, the granularity g decays to zero, which means that it has at least one local maximum as a function of σ. The granularity g also decays to zero when P_0/K approaches zero or infinity. In the important special case of $\mu = 0$, we conclude by direct computation that g is a unimodal function of P_0/K, that achieves its maximum at $P_0/K = \exp(\sigma^2/2)$.

The fact that granularity is not monotone increasing in σ may seem counterintuitive at first — after all, how can delta-hedging errors become smaller for larger values of σ? The intuition follows from the fact that at small levels of σ, an increase in σ leads to larger granularity because there is a greater chance that the stock price will fluctuate around regions of high gamma (where $\partial^2 H(t, P_t)/\partial P_t^2$ is large, i.e., near the money), leading to greater tracking errors. However, at very high levels of σ, prices fluctuate so wildly that an increase in σ will decrease the probability that the stock price stays in regions of high gamma for very long — in these extreme cases, the payoff function "looks" approximately linear hence granularity becomes a decreasing function of σ.

Also, we show below that g is not very sensitive to changes in μ when σ is sufficiently large. This implies that, for an empirically relevant range of parameter values, g, as a function of the initial stock price, achieves its maximum close to the strike price, i.e., at $P_0/K \approx 1$. These observations are consistent with the behavior of the tracking error for finite values of N that we see in the Monte Carlo simulations of Sec. 4.

When stock prices follow a geometric Brownian motion, expressions similar to (3.2) can be obtained for derivatives other than simple puts and calls. For example, for a straddle, consisting of one put and one call option with the same strike price K, the constant g is twice as large as for the put or call option alone.

3.2. *Granularity of a mean-reverting process*

Let $p_t \equiv \ln(P_t)$ and suppose

$$dp_t = \left(-\gamma(p_t - (\alpha + \beta t)) + \beta \right) dt + \sigma \, dW_t \tag{3.3}$$

where $\beta = \mu - \sigma^2/2$ and α is a constant. This is an Ornstein–Uhlenbeck process with a linear time trend, and the solution of (3.3) is given by

$$p_t = (p_0 - \alpha)e^{-\gamma t} + (\alpha + \beta t) + \sigma \int_0^t e^{-\gamma(t-s)} \, dW_s \,. \tag{3.4}$$

Under these price dynamics, we have:

Theorem 4. *Under Assumptions* (A1)–(A4), *stock price dynamics* (3.3), *and the payoff function of simple call and put options, the granularity g in* (2.13) *is given by*

$$g = K\sigma \left(\int_0^1 \frac{\sqrt{\gamma}\exp\left[-\frac{\gamma[\alpha+\mu t+\left(\ln\left(\frac{P_0}{K}\right)-\alpha\right)\exp(-\gamma t)-\sigma^2/2]^2}{\sigma^2[\gamma(1-t)+1-\exp(-2\gamma t)]} \right]}{4\pi\sqrt{1-t}\sqrt{\gamma(1-t)+1-\exp(-2\gamma t)}}\, dt \right)^{\frac{1}{2}} \tag{3.5}$$

where K is the option's strike price.

Proof. See Bertsimas, Kogan, and Lo (1998).

Expression (3.5) is a direct generalization of (3.2): when the mean-reversion parameter γ is set to zero, the process (3.3) becomes a geometric Brownian motion and (3.5) reduces to (3.2). Theorem 4 has some interesting qualitative implications for the behavior of the tracking error in presence of mean-reversion. We will discuss them in detail in the next section.

4. Monte Carlo Analysis

Since our analysis of granularity is based entirely on continuous-record asymptotics, we must check the quality of these approximations by performing Monte Carlo simulation experiments for various values of N. The results of these Monte Carlo simulations are reported in Sec. 4.1. We also use Monte Carlo simulations to explore the qualitative behavior of the RMSE for various parameter values of the stock price process, and these simulations are reported in Sec. 4.2.

4.1. *Accuracy of the asymptotics*

We begin by investigating the distribution of the tracking error $\epsilon^{(N)}$ for various values of N. We do this by simulating the hedging strategy of Sec. 2.2 for call and put options,[g] assuming that price dynamics are given by a geometric Brownian motion (3.1).[h] We set the parameters of the stock price process to $\mu = 0.1$, $\sigma = 0.3$, $P_0 = 1.0$ and let the strike price be $K = 1$. We consider $N = 10$, 20, 50, 100, and simulate the hedging process 250,000 times for each value of N.

Figure 1(a) shows the empirical probability density function (PDF) of $\epsilon_1^{(N)}$ for each N. As expected, the distribution of the tracking error becomes tighter as the trading frequency increases. It is also apparent that the tracking error can be significant even for $N = 100$. Figure 1(b) contains the empirical PDFs of the normalized tracking error, $\sqrt{N}\epsilon_1^{(N)}$, for the same values of N. These PDFs are compared to the PDF of the asymptotic distribution (2.17), which is estimated by approximating the integral in (2.17) using a first-order Euler scheme. The functions

[g]According to Theorem 1, the asymptotic expressions for the tracking error and the RMSE are the same for put and call options since these options have the same second partial derivative of the option price with respect to the current stock price. Moreover, it is easy to verify, using the put-call parity relation, that these options give rise to identical tracking errors.

[h]When the stock price process P_t follows a geometric Brownian motion, the stock price at time t_{i+1} is distributed (conditional on the stock price at time t_i) as $P_{t_i}\exp((\mu - \sigma^2/2)\Delta t + \sigma\sqrt{\Delta t}\eta)$, where $\eta \sim \mathcal{N}(0, 1)$. We use this relation to simulate the delta-hedging strategy.

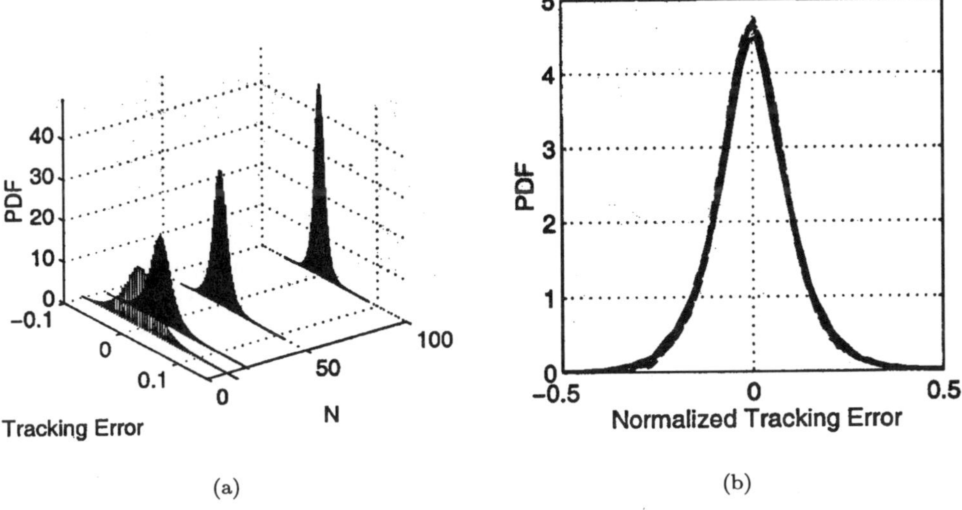

(a)   (b)

Fig. 1. Empirical probability density functions of (a) the tracking error and (b) the normalized tracking error (dashed line), are plotted for different values of the trading frequency N. Figure 1(b) also shows the empirical probability density function of the asymptotic distribution (2.17) (solid line). The stock price process is given by (3.1) with parameters $\mu = 0.1$, $\sigma = 0.3$, $P_0 = 1.0$. The option is a European call (put) option with strike price $K = 1$.

in Figure 1(b) are practically identical and indistinguishable, which suggests that the asymptotic expression for the distribution of $\sqrt{N}\epsilon_1^{(N)}$ in Theorem 1(b) is an excellent approximation to the finite-sample PDF for values of N as small as 10.

To evaluate the accuracy of the asymptotic expression $g/\sqrt{N}$ for finite values of N, we compare $g/\sqrt{N}$ to the actual RMSE from Monte Carlo simulations of the delta-hedging strategy of Sec. 2.2. Specifically, we simulate the delta-hedging strategy for a set of European put and call options with strike price $K = 1$ under geometric Brownian motion (3.1) with different sets of parameter values for (σ, μ, P_0). The tracking error is tabulated as a function of these parameters and the results are summarized in Tables 1, 2, and 3.

Tables 1–3 show that $g/\sqrt{N}$ is an excellent approximation to the RMSE across a wide range of parameter values for (μ, σ, P_0), even for as few as $N = 10$ delta-hedging periods.

4.2. *Qualitative behavior of the RMSE*

The Monte Carlo simulations of Sec. 4.1 show that RMSE increases with the diffusion coefficient σ in an empirically relevant range of parameter values (see Table 2), and that the RMSE is not very sensitive to the drift rate μ of the stock price process when σ is sufficiently large (see Table 3). These properties are illustrated in Figs. 2(a) and 3. In Fig. 2(a), the logarithm of RMSE is plotted against the logarithm of trading periods N for $\sigma = 0.1$, 0.2, 0.3 — as σ increases, the locus of points shifts upward. Figure 3 shows that granularity g is not a monotone function of σ and goes to zero as σ increases without bound.

Table 1. The sensitivity of the RMSE as a function of the initial price P_0. The RMSE is estimated using Monte Carlo simulation. Options are European calls and puts with strike price $K = 1$. 250,000 simulations are performed for every set of parameter values. The stock price follows a geometric Brownian motion (3.1). The drift and diffusion coefficients of the stock price process are $\mu = 0.1$ and $\sigma = 0.3$. $\text{RMSE}^{(N)}$ is compared to the asymptotic approximation $gN^{-1/2}$ in (2.13)–(3.2). The relative error (RE) of the asymptotic approximation is defined as $|gN^{-1/2} - \text{RMSE}^{(N)}|/\text{RMSE}^{(N)} \times 100\%$.

Parameters					Call Option		Put Option	
N	P_0	$gN^{-1/2}$	$\text{RMSE}^{(N)}$	RE	$H(0, P_0)$	$\frac{\text{RMSE}^{(N)}}{H}$	$H(0, P_0)$	$\frac{\text{RMSE}^{(N)}}{H}$
10	0.50	0.0078	0.0071	8.9%	7E-4	9.64	0.501	0.014
20	0.50	0.0055	0.0052	7.9%	7E-4	6.88	0.501	0.010
50	0.50	0.0035	0.0033	5.9%	7E-4	4.43	0.501	0.007
100	0.50	0.0025	0.0024	3.1%	7E-4	3.22	0.501	0.005
10	0.75	0.0259	0.0248	3.8%	0.023	1.08	0.273	0.091
20	0.75	0.0183	0.0177	2.9%	0.023	0.760	0.273	0.065
50	0.75	0.0116	0.0113	2.6%	0.023	0.490	0.273	0.041
100	0.75	0.0082	0.0082	2.3%	0.023	0.345	0.273	0.029
10	1.00	0.0334	0.0327	4.1%	0.119	0.269	0.119	0.269
20	1.00	0.0236	0.0227	3.4%	0.119	0.192	0.119	0.192
50	1.00	0.0149	0.0145	2.3%	0.119	0.122	0.119	0.122
100	1.00	0.0106	0.0104	1.9%	0.119	0.087	0.119	0.087
10	1.25	0.0275	0.0263	5.8%	0.294	0.088	0.044	0.588
20	1.25	0.0194	0.0187	3.9%	0.294	0.064	0.044	0.423
50	1.25	0.0123	0.0120	2.7%	0.294	0.041	0.044	0.271
100	1.25	0.0087	0.0087	1.8%	0.294	0.029	0.044	0.195
10	1.50	0.0181	0.0169	7.7%	0.515	0.033	0.015	1.130
20	1.50	0.0128	0.0122	5.3%	0.515	0.024	0.015	0.816
50	1.50	0.0081	0.0076	2.9%	0.515	0.015	0.015	0.528
100	1.50	0.0057	0.0056	3.0%	0.515	0.011	0.015	0.373

Table 2. The sensitivity of the RMSE as a function of volatility σ. The RMSE is estimated using Monte Carlo simulation. Options are European calls and puts with strike price $K = 1$. $250,000$ simulations are performed for every set of parameter values. The stock price follows a geometric Brownian motion (3.1). The drift coefficient of the stock price process is $\mu = 0.1$, and the initial stock price is $P_0 = 1.0$. $\text{RMSE}^{(N)}$ is compared to the asymptotic approximation $gN^{-1/2}$ in (2.13)–(3.2). The relative error (RE) of the asymptotic approximation is defined as $|gN^{-1/2} - \text{RMSE}^{(N)}|/\text{RMSE}^{(N)} \times 100\%$.

Parameters			Call and Put Options			
N	σ	$gN^{-1/2}$	$\text{RMSE}^{(N)}$	RE	$H(0, P_0)$	$\frac{\text{RMSE}^{(N)}}{H}$
10	0.3	0.0334	0.0327	4.1%	0.119	0.269
20	0.3	0.0236	0.0227	3.4%	0.119	0.192
50	0.3	0.0149	0.0145	2.3%	0.119	0.122
100	0.3	0.0106	0.0104	1.9%	0.119	0.087
10	0.2	0.0219	0.0212	3.4%	0.080	0.266
20	0.2	0.0155	0.0151	3.0%	0.080	0.189
50	0.2	0.0098	0.0096	2.1%	0.080	0.121
100	0.2	0.0069	0.0068	1.7%	0.080	0.086
10	0.1	0.0100	0.0102	1.6%	0.040	0.255
20	0.1	0.0071	0.0071	0.04%	0.040	0.177
50	0.1	0.0045	0.0044	1.1%	0.040	0.111
100	0.1	0.0032	0.0031	0.9%	0.040	0.078

Table 3. The sensitivity of the RMSE as a function of the drift μ. The RMSE is estimated using Monte Carlo simulation. Options are European calls and puts with strike price $K = 1$. $250,000$ simulations are performed for every set of parameter values. The stock price follows a geometric Brownian motion (3.1). The diffusion coefficient of the stock price process is $\sigma = 0.3$, and the initial stock price is $P_0 = 1.0$, the number of trading periods is $N = 20$. $\text{RMSE}^{(N)}$ is compared to the asymptotic approximation $gN^{-1/2}$ in (2.13)–(3.2). The relative error (RE) of the asymptotic approximation is defined as $|gN^{-1/2} - \text{RMSE}^{(N)}|/\text{RMSE}^{(N)} \times 100\%$.

Parameters		Call and Put Options			
μ	$gN^{-1/2}$	$\text{RMSE}^{(N)}$	RE	$H(0, P_0)$	$\frac{\text{RMSE}^{(N)}}{H}$
0.0	0.0235	0.0226	4.3%	0.119	0.189
0.1	0.0236	0.0229	3.4%	0.119	0.192
0.2	0.0230	0.0226	1.7%	0.119	0.190
0.3	0.0218	0.0220	1.0%	0.119	0.184

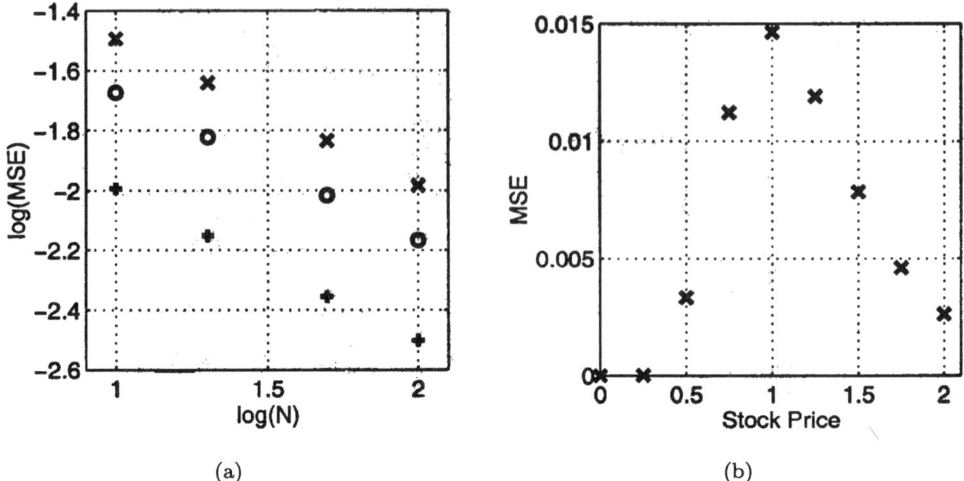

(a)   (b)

Fig. 2. (a) The logarithm of the root-mean-squared error $\log_{10}(\text{RMSE}^{(N)})$ is plotted as a function of the logarithm of the trading frequency $\log_{10}(N)$. The option is a European call (put) option with the strike price $K = 1$. The stock price process is given by (3.1) with parameters $\mu = 0.1$, $P_0 = 1.0$. The diffusion coefficient of the stock price process takes values $\sigma = 0.3$ (x's), $\sigma = 0.2$ (o's) and $\sigma = 0.1$ (+'s). (b) The root-mean-squared error RMSE is plotted as a function of the initial stock price P_0. The option is a European put option with the strike price $K = 1$. Parameters of the stock price process are $\mu = 0.1$, $\sigma = 0.3$.

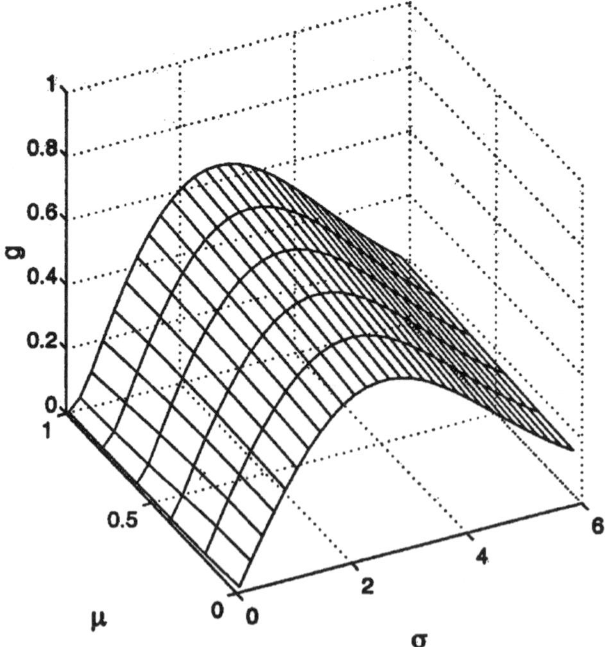

Fig. 3. The granularity g is plotted as a function of σ and μ. The option is a European call (put) option with strike price $K = 1$. The stock price process is geometric Brownian motion and initial stock price $P_0 = 1$.

Figure 2(b) plots the RMSE as a function of the initial stock price P_0. RMSE is a unimodal function of P_0/K (recall that the strike price has been normalized to $K = 1$ in all our calculations), achieving its maximum around 1 and decaying to zero as P_0/K approaches zero or infinity (see Table 1). This confirms the common intuition that close-to-the-money options are the most difficult to hedge—they exhibit the largest RMSE.

Finally, the relative importance of the RMSE can be measured by the ratio of the RMSE to the option price: $\text{RMSE}^{(N)}/H(0, P_0)$. This quantity is the root-mean-squared error per dollar invested in the option. Table 1 shows that this ratio is highest for out-of-the-money options, despite the fact that the RMSE is highest for close-to-the-money options. This is due to the fact that the option price decreases faster than the RMSE as the stock moves away from the strike.

Now consider the case of mean-reverting stock price dynamics (3.3). Recall that under these dynamics, the Black–Scholes formula still holds.[i] Nevertheless, the behavior of granularity and RMSE is quite different in this case. Figure 4 plots the granularity g of call and put options for the Ornstein–Uhlenbeck process (3.3) as a function of α and P_0. Figure 4(a) assumes a value of 0.1 for the mean-reversion parameter γ and Fig. 4(b) assumes a value of 3.0. It is clear from these two plots that the degree of mean reversion γ has an enormous impact on granularity. When γ is

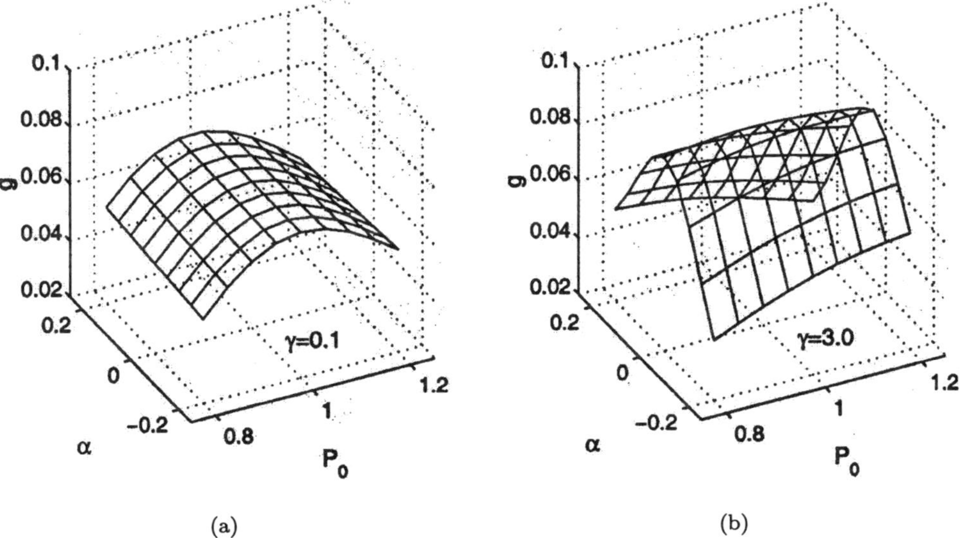

(a) (b)

Fig. 4. Granularity g is plotted as a function of P_0 and α. The option is a European call (put) option with the strike price $K = 1$. Parameters of the stock price process are $\sigma = 0.2$, $\mu = 0.05$. The stock price process is given by (3.4). Mean-reversion parameter γ takes two values: (a) $\gamma = 0.1$ and (b) $\gamma = 3.0$.

[i] However, the numerical value for σ may be different than that of a geometric Brownian motion because the presence of mean-reversion can affect conditional volatility, holding unconditional volatility fixed. See Lo and Wang (1995) for further discussion.

small, Fig. 4(a) shows that the RMSE is highest when P_0 is close to the strike price and is not sensitive to α. But when γ is large, Fig. 4(b) suggests that the RMSE is highest when $\exp(\alpha)$ is close to the strike price and is not sensitive to P_0.

The influence of γ on the granularity can be understood by recalling that granularity is closely related to the option's gamma (see Sec. 2.3). When γ is small, the stock price is more likely to spend time in the neighborhood of the strike price — the region with the highest "gamma" or $\partial^2 H(t, P_t)/\partial P_t^2$ — when P_0 is close to K. However, when γ is large, the stock price is more likely to spend time in a the neighborhood of $\exp(\alpha)$, thus g is highest when $\exp(\alpha)$ is close to K.

5. Extensions and Generalizations

The analysis of Sec. 2 can be extended in a number of directions, and we briefly outline four of the most important of these extensions here. In Sec. 5.1, we show that the normalized tracking error converges in a much stronger sense than simply in distribution, and that this stronger "sample-path" notion of convergence — called, ironically, "weak" convergence — can be used to analyze the tracking error of American-style derivative securities. In Sec. 5.2, we characterize the asymptotic joint distributions of the normalized tracking error and asset prices, a particularly important extension for investigating the tracking error of delta hedging a portfolio of derivatives. In Sec. 5.3, we provide another characterization of the tracking error, one that relies on PDE's, that offers important computational advantages. In Sec. 5.4, we consider alternatives to mean-squared-error loss functions and show that for quite general loss functions, the behavior of the expected loss of the tracking error is characterized by the same stochastic integral (2.17) as in the mean-squared-error case.

5.1. *Sample-path properties of tracking errors*

Recall that the normalized tracking error *process* is defined as:

$$\sqrt{N}\, \epsilon_t^{(N)} = \sqrt{N}\, (H(t, P_t) - V_t^{(N)}), \quad t \in [0, 1].$$

It can be shown that $\sqrt{N}\epsilon_t^{(N)}$ converges weakly to the stochastic process G_t, characterized by the stochastic integral in (2.12) as a function of its upper limit:[j]

$$G_t = \frac{1}{\sqrt{2}} \int_0^t \sigma^2(t, P_s) P_s^2 \frac{\partial^2 H(s, P_s)}{\partial P_s^2}\, dW_s'.$$

[j]The proof of this result consists of two steps. The first step is to establish that the sequence of measures induced by $\sqrt{N}\, \epsilon_t^{(N)}$ is tight (relatively compact). This can be done by verifying local inequalities for the moments of processes $\sqrt{N}\, \epsilon_t^{(N)}$ using the machinery developed in the proof of Theorem 2 (we must use Burkholder's inequality instead of the isometric property and Hölder's inequality instead of Schwarz's inequality throughout — see Bertsimas, Kogan, and Lo [1998] for further details). The second step is to characterize the limiting process. Such a characterization follows from the proof in Appendix A.2 and the fact that the results in Duffie and Protter (1992) guarantee weak convergence of stochastic processes, not just convergence of their one-dimensional marginal distributions.

This stronger notion of convergence yields stronger versions of Theorem 1 and 2 that can be used to analyze a number of sample-path properties of the tracking error by appealing to the Continuous Mapping Theorem (see Billingsley [1968]). This well-known result shows that the asymptotic distribution of any continuous functional $\xi(\cdot)$ of the normalized tracking error is given by $\xi(G_t)$. For example, the maximum of the normalized tracking error over the entire life of the derivative security, $\max_t \sqrt{N}\epsilon_t^{(N)}$, is distributed as $\max_t G_t$ asymptotically.

These results can be applied to the normalized tracking errors of American-style derivatives in a straightforward manner. Such derivatives differ from European derivatives in one respect: they can be exercised prematurely. Therefore, the valuation of these derivatives consists of both computing the derivative price function $H(t, P_t)$ and the optimal exercise schedule, which can be represented as a stopping time τ. Then the tracking error at the moment when the derivative is exercised behaves asymptotically as $G_\tau/\sqrt{N}$.[k] The tracking error, conditional on the derivative not being exercised prematurely, is distributed asymptotically as $(G_1/\sqrt{N} \,|\, \tau = 1)$.

5.2. *Joint distributions of tracking errors and prices*

Theorems 1 and 2 provide a complete characterization of the tracking error and RMSE for individual derivatives, but what is often of more practical interest is the behavior of a portfolio of derivatives. Delta-hedging a portfolio of derivatives is typically easier because of the effects of diversification — as long as tracking errors are not perfectly correlated across derivatives, the portfolio tracking error will be less volatile than the tracking error of individual derivatives.

To address portfolio issues, we require the joint distribution of tracking errors for multiple stocks, as well as the joint distribution of tracking errors and prices. Consider another stock with price $P_t^{(2)}$ governed by the diffusion equation

$$\frac{dP_t^{(2)}}{P_t^{(2)}} = \mu^{(2)}(t, P_t^{(2)})\, dt + \sigma^{(2)}(t, P_t^{(2)})\, dW_t^{(2)} \tag{5.1}$$

where $W_t^{(2)}$ can be correlated with W_t. According to the proof of Theorem 1(b) (see Appendix A.2), since the random variables $(W_{t_{i+1}} - W_{t_i})^2 - (t_{i+1} - t_i)$ and $W_{t_{i+1}}^{(2)} - W_{t_i}^{(2)}$ are uncorrelated,[l] the Wiener processes W_t' and $W_t^{(2)}$ are independent. Therefore, as N increases without bound the pair of random variables $(\sqrt{N}\,\epsilon_1^{(N)}, P_1^{(2)})$ converges in distribution to:

$$\left(\sqrt{N}\,\epsilon_1^{(N)}\,,\ P_1^{(2)}\right) \Rightarrow \left(\frac{1}{\sqrt{2}} \int_0^1 \sigma^2(t, P_t) P_t^2 \frac{\partial^2 H(t, P_t)}{\partial P_t^2}\, dW_t', P_1^{(2)}\right) \tag{5.2}$$

where W_t' is independent of W_t and $W_t^{(2)}$.

---

[k] Some technical regularity conditions, e.g., the smoothness of the exercise boundary, are required to ensure convergence. See, for example, Kushner and Dupuis (1992).

[l] This follows from the fact that, for every pair of standard normal random variables X and Y with correlation ρ, $X = \rho Y + \sqrt{1 + \rho^2}\, Z$, where Z is a standard normal random variable, independent of Y. Thus X and $Y^2 - 1$ are uncorrelated.

An immediate corollary of this result is that the normalized tracking error is uncorrelated with any asset in the economy. This follows easily from (5.2) since, conditional on the realization of P_t and $P_t^{(2)}$, $t \in [0,1]$, the normalized tracking error has zero expected value asymptotically. However, this does not imply that the asymptotic joint distribution of $(\sqrt{N}\,\epsilon_1^{(N)}, P_1^{(2)})$ does not depend on the correlation between W_t and $W_t^{(2)}$ — it does, since this correlation determines the joint distribution of P_t and $P_t^{(2)}$.

The above argument applies without change when the price of the second stock follows a diffusion process different from (5.1), and can also easily be extended to the case of multiple stocks.

To derive the joint distribution of the normalized tracking errors for multiple stocks, we consider the case of two stocks since the generalization to multiple stocks is obvious. Let W_t and $W_t^{(2)}$ have mutual variation $dW_t\,dW_t^{(2)} = \rho(t, P_t, P_t^{(2)})\,dt$, where $\rho(\cdot)$ is a continuously differentiable function with bounded first-order partial derivatives. We have already established that the asymptotic distribution of the tracking error is characterized by the stochastic integral (2.12). To describe the asymptotic joint distribution of two normalized tracking errors, it is sufficient to find the mutual variation of the Wiener processes in the corresponding stochastic integrals. According to the proof of Theorem 1(b) (Appendix A.2), this amounts to computing the expected value of the product

$$\left((W_{t_{i+1}} - W_{t_i})^2 - (t_{i+1} - t_i) \right) \left((W_{t_{i+1}}^{(2)} - W_{t_i}^{(2)})^2 - (t_{i+1} - t_i) \right).$$

Using Itô's formula, it is easy to show that the expected value of the above expression is equal to

$$\mathbf{E}_0 \left[2\rho^2(t, P_{t_i}, P_{t_i}^{(2)}) \right] (\Delta t)^2 + \mathbf{O}\left((\Delta t)^{\frac{5}{2}} \right).$$

This implies that $\rho^2(t, P_t, P_t^{(2)})$ is the mutual variation of the two Wiener processes in the stochastic integrals (2.12) that describe the asymptotic distributions of the normalized tracking errors of the two stocks. Together with Theorem 1(b), this completely determines the asymptotic joint distribution of the two normalized tracking errors.[m]

Note that the correlation of two Wiener processes describing the asymptotic behavior of two normalized tracking errors is always nonnegative, regardless of the sign of the mutual variation of the original Wiener processes W_t and $W_t^{(2)}$. In particular, when two derivatives have convex price functions, this means that even if the returns on the two stocks are negatively correlated, the tracking errors resulting from delta hedging derivatives on these stocks are asymptotically positively correlated.

---

[m]This result generalizes the findings of Boyle and Emanuel (1980).

5.3. *A PDE characterization of the tracking error*

It is possible to derive an alternative characterization of the tracking error using the intimate relationship between diffusion processes and PDE's. Although this may seem superfluous given the analytical results of Theorems 1 and 2, the numerical implementation of a PDE representation is often computationally more efficient.

To illustrate our approach, we begin with the RMSE. According to Theorem 1(c), the RMSE can be completely characterized asymptotically if g is known (see (2.14)). Using the Feynman–Kac representation of the solutions of PDE's (see Karatzas and Shreve [1991, Proposition 4.2.]), we conclude that $g^2 = u(0, P_0)$, where $u(t, x)$ solves the following:

$$\left[\frac{\partial}{\partial t} + \mu(t, x)x\frac{\partial}{\partial x} + \frac{1}{2}\sigma^2(t, x)x^2\frac{\partial^2}{\partial x^2} \right] u(t, x) + \frac{1}{2}\left(\sigma^2(t, x)x^2\frac{\partial^2 H(t, x)}{\partial x^2} \right)^2 = 0$$

(5.3)

$$u(1, x) = 0 \ , \ \forall\, x \ .$$

(5.4)

The PDE (5.3)–(5.4) is of the same degree of difficulty as the fundamental PDE (2.2)–(2.3) that must be solved to obtain the derivative-pricing function $H(t, P_t)$. This new representation of the RMSE can be used to implement an efficient numerical procedure for calculating RMSE without resorting to Monte Carlo simulation.[n]

Summary measures of the tracking error with general loss functions can also be computed numerically along the same lines, using the Kolmogorov backward equation. The probability density function of the normalized tracking error $\sqrt{N}\epsilon_1^{(N)}$ can be determined numerically as a solution of the Kolmogorov forward equation (see, for example, Karatzas and Shreve [1991, pp. 368–369]).

5.4. *Alternative measures of the tracking error*

As we observed in Sec. 2.2, the root-mean-squared error is only one of many possible summary measures of the tracking error. An obvious alternative is the L_p-norm:

$$\mathbf{E}_0\left[\left| \epsilon_1^{(N)} \right|^p \right]^{\frac{1}{p}}$$

(5.5)

where p is chosen so that the expectation is finite (otherwise the measure will not be particularly informative). More generally, the tracking error can be summarized by

$$\mathbf{E}_0\left[U(\epsilon_1^{(N)}) \right]$$

(5.6)

where $U(\cdot)$ is an arbitrary loss function.

---

[n] Results of some preliminary numerical experiments provide encouraging evidence of the practical value of this new representation.

Consider the set of measures (5.5) first and assume for simplicity that $p \in [1, 2]$. From (2.17), it follows that

$$\mathbf{E}_0 \left[\left| \epsilon_1^{(N)} \right|^p \right]^{\frac{1}{p}} \sim N^{-1/2} \mathbf{E}_0 \left[\left| \frac{1}{\sqrt{2}} \int_0^1 \sigma^2(t, P_t) P_t^2 \frac{\partial^2 H(t, P_t)}{\partial P_t^2} dW_t' \right|^p \right]^{\frac{1}{p}}. \tag{5.7}$$

Hence the moments of the stochastic integral in (2.17) describe the asymptotic behavior of the moments of the tracking error. Conditional on the realization of $\{P_t\}$, $t \in [0, 1]$, the stochastic integral on the right side of (5.7) is normally distributed with zero mean and variance

$$\int_0^1 \left(\sigma^2(t, P_t) P_t^2 \frac{\partial^2 H(t, P_t)}{\partial P_t^2} \right)^2 dt$$

which follows from Hull and White (1987). The intuition is that, conditional on the realization of the integrand, the stochastic integral behaves as an integral of a deterministic function with respect to the Wiener process which is a normal random variable. Now let m_p denote an L_p-norm of the standard normal random variable.[o] Then (5.7) can be rewritten as:

$$\mathbf{E}_0 \left[\left| \epsilon_1^{(N)} \right|^p \right]^{\frac{1}{p}} \sim \frac{m_p}{\sqrt{N}} \mathbf{E}_0 \left[\mathcal{R}^{\frac{p}{2}} \right]^{\frac{1}{p}} \tag{5.8}$$

where $\mathcal{R}$ is given by (2.15).

As in the case of a quadratic loss function, $\mathcal{R}$ plays a fundamental role here in describing the behavior of the tracking error. When $p = 2$, $\mathcal{R}$ enters (5.8) linearly and closed-form expressions can be derived for special cases. However, even when $p \neq 2$, the qualitative impact of $\mathcal{R}$ on the tracking error is the same as for $p = 2$ and our discussion of the qualitative behavior of the tracking error applies to this case as well.

For general loss functions $U(\cdot)$ that satisfy certain growth conditions and are sufficiently smooth near the origin, the delta-method can be applied and we obtain:

$$\mathbf{E}_0 \left[U \left(\epsilon_1^{(N)} \right) \right] \sim \frac{1}{N} |U''(0)| g^2 = \frac{1}{N} |U''(0)| \mathbf{E}_0[\mathcal{R}]. \tag{5.9}$$

When $U(\cdot)$ is not differentiable at 0, the delta method cannot be used. However, we can use the same strategy as in our analysis of L_p-norms to tackle this case. Suppose that $U(\cdot)$ is dominated by a quadratic function. Then

$$\mathbf{E}_0 \left[U \left(\epsilon_1^{(N)} \right) \right] \approx \mathbf{E}_0 \left[U \left(\frac{1}{\sqrt{2N}} \int_0^1 \sigma^2(t, P_t) P_t^2 \frac{\partial^2 H(t, P_t)}{\partial P_t^2} dW_t' \right) \right]. \tag{5.10}$$

Now let

$$m_U(x) = \mathbf{E}\left[U(x\eta) \right], \quad \eta \sim \mathcal{N}(0, 1).$$

[o] If X is a standard normal random variable, then $m_p = \mathbf{E}[|X|^p]^{1/p}$.

Then

$$\mathbf{E}_0 \left[U \left(\epsilon_1^{(N)} \right) \right] \approx \mathbf{E}_0 \left[m_U \left(\sqrt{\mathcal{R}/N} \right) \right] . \tag{5.11}$$

When the loss function $U(\cdot)$ is convex, $m_U(\cdot)$ is an increasing function (by second-order stochastic dominance). Therefore, the qualitative behavior of the measure (5.6) is also determined by $\mathcal{R}$ and is the same as that of the RMSE.

6. Conclusions

We have argued that continuous-time models are meant to be approximations to physical phenomena, and as such, their approximation errors should be better understood. In the specific context of continuous-time models of derivative securities, we have quantified the approximation error through our definition of temporal granularity. The combination of a specific derivative security and a stochastic process for the underlying asset's price dynamics can be associated with a measure of how "grainy" the passage of time is. This measure is related to the ability to replicate the derivative security through a delta-hedging strategy implemented in discrete time. Time is said to be very granular if the replication strategy does not work well — in such cases, time is not continuous. If, however, the replication strategy is very effective, time is said to be very smooth or continuous.

Under the assumption of general Markov diffusion price dynamics, we show that the tracking errors for derivatives with sufficiently smooth or continuous piecewise linear payoff functions behave asymptotically (in distribution) as $G/\sqrt{N}$. We characterize the distribution of the random variable G as a stochastic integral, and also obtain the joint distribution of G with prices of other assets and with other tracking errors. We demonstrate that the root-mean-squared error behaves asymptotically as $g/\sqrt{N}$, where the constant g is what we call the coefficient of *temporal granularity*. For two special cases — call or put options on geometric Brownian motion and on an Ornstein–Uhlenbeck process — we are able to evaluate the coefficient of granularity explicitly.

We also consider a number of extensions of our analysis, including an extension to alternative loss functions, a demonstration of the weak convergence of the tracking error process, a derivation of the joint distribution of tracking errors and prices, and an alternative characterization of the tracking error in terms of PDE's that can be used for efficient numerical implementation.

Because these results depend so heavily on continuous-record asymptotics, we perform Monte Carlo simulations to check the quality of our asymptotics. For the case of European puts and calls with geometric Brownian motion price dynamics, our asymptotic approximations are excellent, providing extremely accurate inferences over the range of empirically relevant parameter values, even with a small number of trading periods.

Of course, our definition of granularity is not invariant to the derivative security, the underlying asset's price dynamics, and other variables. But we regard this as a positive feature of our approach, not a drawback. After all, any plausible definition of granularity must be a relative one, balancing the coarseness of changes in the time domain against the coarseness of changes in the "space" or price domain. Although the title of this paper suggests that time is the main focus of our analysis, it is really the relation between time and price that determines whether or not continuous-time models are good approximations to physical phenomena. It is our hope that the definition of granularity proposed in this paper is one useful way of tackling this very complex issue.

Appendix A

The essence of these proofs involves the relation between the delta-hedging strategy and mean-square approximations of solutions of systems of stochastic differential equations described in Milstein (1974, 1987, 1995). Readers interested in additional details and intuition should consult these references directly. We present the proof of Theorem 1 only, and refer readers to Bertsimas, Kogan, and Lo (1998) for the others.

A.1. *Proof of Theorem 1(a)*

First we observe that the regularity conditions (2.10) imply the existence of a positive constant K_1 such that

$$\left| \frac{\partial^{\beta+\gamma}}{\partial \tau^\beta \partial x^\gamma} H(\tau, x) \right| \leq K_1 \tag{A.1}$$

for $(\tau, x) \in [0,1] \times [0, \infty)$, $0 \leq \beta \leq 1$, $1 \leq \gamma \leq 4$, and all partial derivatives are continuous.[P] Next, by Itô's formula,

$$H(1, P_1) = H(0, P_0) + \int_0^1 \left(\frac{\partial H(t, P_t)}{\partial t} + \frac{1}{2}\sigma^2(t, P_t) P_t^2 \frac{\partial^2 H(t, P_t)}{\partial P_t^2} \right) dt$$

$$+ \int_0^1 \frac{\partial H(t, P_t)}{\partial P_t} dP_t . \tag{A.2}$$

According to (2.2), the first integral on the right-hand side of (A.2) is equal to zero. Thus,

$$H(1, P_1) = H(0, P_0) + \int_0^1 \frac{\partial H(t, P_t)}{\partial P_t} dP_t \tag{A.3}$$

[P]Since the price of the derivative $H(\tau, x)$ is defined as a solution of (2.2), it is equal to the expectation of $F(P_1)$ with respect to the equivalent martingale measure (see Duffie [1996]), i.e.,

$$H(\tau, x) = \mathbf{E}_{(t=\tau, P_t^*=x)}[F(P_1^*)]$$

where

$$\frac{dP_t^*}{P_t^*} = \sigma(t, P_t^*)\, dW_t^* .$$

and W_t^* is a Brownian motion under the equivalent martingale measure. Equation (A.1) now follows from Friedman (1975; Theorems 5.4 and 5.5, p. 122). The same line of reasoning is followed in He (1989, p. 68). Of course, one could derive (A.1) using purely analytic methods, e.g., Friedman (1964; Theorem 10, p. 72, Theorem 11, p. 24; and Theorem 12, p. 25).

which implies that $H(t, P_t)$ can be characterized as a solution of the system of stochastic differential equations

$$\begin{cases} dX_t = \dfrac{\partial H(t, P_t)}{\partial P_t} \mu(t, P_t) P_t \, dt + \dfrac{\partial H(t, P_t)}{\partial P_t} \sigma(t, P_t) P_t \, dW_t \\[2mm] dP_t = \mu(t, P_t) P_t \, dt + \sigma(t, P_t) P_t \, dW_t \end{cases} . \qquad (A.4)$$

At the same time, $V_1^{(N)}$ is given by

$$V_1^{(N)} = H(0, P_0) + \sum_{i=0}^{N-1} \frac{\partial H(t, P_t)_{t=t_i}}{\partial P_t} (P_{t_{i+1}} - P_{t_i}), \qquad (A.5)$$

which can be interpreted as a solution of the following approximation scheme of (A.4) (as defined in Milstein [1987]):

$$\begin{cases} \bar{X}_{t_{i+1}} - X_{t_i} = \dfrac{\partial H(t, P_t)_{t=t_i}}{\partial P_t} (P_{t_{i+1}} - P_{t_i}) \\[2mm] \bar{P}_{t_{i+1}} - P_{t_i} = P_{t_{i+1}} - P_{t_i}, \end{cases} \qquad (A.6)$$

where $\bar{X}$ and $\bar{P}$ denote approximations to X and P, respectively. We now compare (A.6) to the Euler approximation scheme in Milstein (1995)

$$\begin{cases} \bar{X}_{t_{i+1}} - X_{t_i} = \dfrac{\partial H(t, P_t)_{t=t_i}}{\partial P_t} \mu(t_i, P_{t_i}) P_{t_i} (t_{i+1} - t_i) \\[2mm] \qquad\qquad + \dfrac{\partial H(t, P_t)_{t=t_i}}{\partial P_t} \sigma(t_i, P_{t_i}) P_{t_i} (W_{t_{i+1}} - W_{t_i}) \\[3mm] \bar{P}_{t_{i+1}} - P_{t_i} = \mu(t_i, P_{t_i})(t_{i+1} - t_i) \\[2mm] \qquad\qquad + \sigma(t_i, P_{t_i})(W_{t_{i+1}} - W_{t_i}) \end{cases} . \qquad (A.7)$$

Regularity conditions (2.10) and (A.1) allow us to conclude (see Milstein [1995, Theorem 2.1]) that a one-step version of the approximation scheme (A.7) has order-of-accuracy 2 in expected deviation and order-of-accuracy 1 in mean-square deviation (see Milstein (1987), Milstein (1995) for definitions and discussion). It is easy to check that the approximation scheme (A.6) exhibits this same property. Milstein (1995, Theorem 1.1) relates the one-step order-of-accuracy of the approximation scheme to its order-of-accuracy on the whole interval (see also Milstein [1987]). We use this theorem to conclude that (A.6) has mean-square order-of-accuracy $1/2$, i.e.,

$$\sqrt{\mathbf{E}_0 \left[(X(1, P_1) - \bar{X}(1, P_1))^2 \right]} = \mathbf{O}\left(\frac{1}{\sqrt{N}}\right). \qquad (A.8)$$

We now recall that $X(t, P_t) = H(t, P_t)$ and $\bar{X}(1, P_1) = V_1^{(N)}$ and conclude that

$$\sqrt{\mathbf{E}_0 \left[\left(H(1, P_1) - V_1^{(N)}\right)^2 \right]} = \mathbf{O}\left(\frac{1}{\sqrt{N}}\right) \qquad (A.9)$$

which completes the proof. $\qquad\square$

A.2. *Proof of Theorem 1(b)*

We follow the same line of reasoning as in the proof of Theorem 1(a), but we use the Milstein approximation scheme for (A.4) instead of the Euler scheme:

$$
\begin{cases}
\bar{X}_{t_{i+1}} - X_{t_i} = \dfrac{\partial H(t, P_t)_{t=t_i}}{\partial P_t} \mu(t_i, P_{t_i}) P_{t_i}(t_{i+1} - t_i) \\[2mm]
\quad + \dfrac{\partial H(t, P_t)_{t=t_i}}{\partial P_t} \sigma(t_i, P_{t_i}) P_{t_i}(W_{t_{i+1}} - W_{t_i}) \\[2mm]
\quad + \left(\dfrac{\partial^2 H(t, P_t)_{t=t_i}}{\partial P_t^2} \sigma(t_i, P_{t_i}) P_{t_i} + \dfrac{\partial H(t, P_t)_{t=t_i}}{\partial P_t} \dfrac{\partial(\sigma(t, P_t) P_t)_{t=t_i}}{\partial P_t} \right) \\[2mm]
\quad \times \dfrac{1}{2} \sigma(t_i, P_{t_i}) P_{t_i} \left((W_{t_{i+1}} - W_{t_i})^2 - (t_{i+1} - t_i) \right) \\[4mm]
\bar{P}_{t_{i+1}} - P_{t_i} = \mu(t_i, P_{t_i}) P_{t_i}(t_{i+1} - t_i) + \sigma(t_i, P_{t_i}) P_{t_i}(W_{t_{i+1}} - W_{t_i}) \\[2mm]
\quad + \dfrac{1}{2} \sigma(t_i, P_{t_i}) P_{t_i} \dfrac{\partial(\sigma(t, P_t) P_t)_{t=t_i}}{\partial P_t} \left((W_{t_{i+1}} - W_{t_i})^2 - (t_{i+1} - t_i) \right).
\end{cases}
\tag{A.10}
$$

According to Milstein (1974) (see also Milstein [1995, Theorem 2.1]), this one-step scheme has order-of-accuracy 2 in expected deviation and 1.5 in mean-square deviation. It is easy to check by comparison that the scheme

$$
\begin{cases}
\bar{X}_{t_{i+1}} - X_{t_i} = \dfrac{\partial H(t, P_t)_{t=t_i}}{\partial P_t} (P_{t_{i+1}} - P_{t_i}) + \dfrac{1}{2} \sigma(t_i, P_{t_i})^2 P_{t_i}^2 \dfrac{\partial^2 H(t, P_t)_{t=t_i}}{\partial P_t^2} \\[2mm]
\quad \times \left((W_{t_{i+1}} - W_{t_i})^2 - (t_{i+1} - t_i) \right) \\[4mm]
\bar{P}_{t_{i+1}} - P_{t_i} = P_{t_{i+1}} - P_{t_i}
\end{cases}
\tag{A.11}
$$

has the same property. We now use Milstein (1995, Theorem 1.1) to conclude that

$$
H(1, P_1) - V_1^{(N)} = \sum_{i=0}^{N-1} \frac{1}{2} \sigma(t_i, P_{t_i})^2 P_{t_i}^2 \frac{\partial^2 H(t, P_t)_{t=t_i}}{\partial P_t^2}
$$

$$
\times \left((W_{t_{i+1}} - W_{t_i})^2 - (t_{i+1} - t_i) \right) + \mathbf{O}\left(\frac{1}{N} \right)
\tag{A.12}
$$

where $f = \mathbf{O}(\frac{1}{N})$ means that $\lim_{N \to \infty} N \sqrt{\mathbf{E}_{t=0}[f^2]} < \infty$. By Slutsky's theorem, we can ignore the $\mathbf{O}(\frac{1}{N})$ term in considering the convergence in distribution of $\sqrt{N}\,(H(1, P_1) - V_1^{(N)})$, since $\sqrt{N}\,\mathbf{O}(\frac{1}{N})$ converges to zero in mean-squared and, therefore, also in probability. Observe now that, since $W_{t_{i+1}} - W_{t_i}$ and $W_{t_{j+1}} - W_{t_j}$ are independent for $i \neq j$, $(W_{t_{i+1}} - W_{t_i})^2$ and $W_{t_{i+1}} - W_{t_i}$ are uncorrelated, $\mathbf{E}_0[(W_{t_{i+1}} - W_{t_i})^2 - (t_{i+1} - t_i)] = 0$ and $\mathbf{E}_0[((W_{t_{i+1}} - W_{t_i})^2 - (t_{i+1} - t_i))^2] = 2/(t_{i+1} - t_i)^2$, by the functional central limit theorem (see Ethier and Kurtz [1986]), a piecewise constant martingale

$$\sqrt{N/2} \sum_{i=0}^{[Nt]-1} \left((W_{t_{i+1}} - W_{t_i})^2 - (t_{i+1} - t_i) \right) \qquad (A.13)$$

converges weakly on $[0,1]$ to a standard Brownian motion W_t', which is independent of W_t.[q] We complete the proof by applying Duffie and Protter (1992, Lemma 5.1 and Corollary 5.1). $\qquad\Box$

A.3. *Proof of Theorem 1(c)*

Equation (2.13) follows immediately from Theorem 1(a) and the proof of Theorem 1(b).[r] Combined with Theorem 1(b), (2.13) implies that

$$g = \sqrt{\frac{1}{2}\mathbf{E}_0 \left[\left(\int_0^1 \sigma^2(t, P_t) P_t^2 \frac{\partial^2 H(t, P_t)}{\partial P_t^2} \, dW_t' \right)^2 \right]} \, . \qquad (A.14)$$

Equation (2.14) follows from (A.14) using the isometric property of stochastic integrals. $\qquad\Box$

References

Avellaneda, M. and A. Paras, "Dynamic hedging portfolios for derivative securities in the presence of large transactions costs", *Applied Mathematical Finance* **1** (1994) 165–193.

Bachelier, L., "Theory of speculation", reprinted in *The Random Character of Stock Market Prices*, P. Cootner, ed., (Cambridge, MA, MIT Press, 1964).

Bertsimas, D., Lo, A. and L. Kogan, "Pricing and Hedging Derivative Securities in Incomplete Markets: An ϵ-Arbitrage Approach", Working Paper No. LFE–1027–97, MIT Laboratory for Financial Engineering, 1997.

Bertsimas, D., Lo, A. and L. Kogan, "When Is Time Continuous?", Working Paper No. LFE–1033–98, MIT Laboratory for Financial Engineering, 1998.

Bensaid, B., Lesne, J., Pages, H. and J. Scheinkman, 1992, "Derivative asset pricing with transaction costs", *Mathematical Finance* **2** (1992) 63–86.

Billingsley, P., *Probability and Measure*, Second Edition, John Wiley & Sons, New York, 1986.

Black, F., M. Scholes, "Pricing of options and corporate liabilities", *Journal of Political Economy* **81** (1973) 637–654.

Boyle, P., D. Emanuel, "Discretely adjusted option hedges", *Journal of Financial Economics* **8** (1980) 259–282.

Boyle, P. and T. Vorst, "Option replication in discrete time with transaction costs", *Journal of Finance* **47** (1992) 271–294.

Breeden, D., "An intertemporal asset pricing model with stochastic consumption and investment opportunities", *Journal of Financial Economics* **7** (1979) 265–196.

Cox, J., Ingersoll, J. and S. Ross, "An intertemporal general equilibrium model of asset prices", *Econometrica* **36** (1985) 363–384.

---

[q]The notation $[Nt]$ denotes the integer part of Nt and we use the convention $\sum_0^{-1} = 0$.
[r]Relation (A.12), established as a part of the proof of Theorem 1(b), guarantees that convergence in (2.12) occurs not only in distribution, but also in mean-squared.

Cox, J., Ross, S. and M. Rubinstein, "Option pricing: A simplified approach", *Journal of Financial Economics* **7** (1979) 229–263.

Duffie, D., "Stochastic equilibria with incomplete financial markets", *Journal of Economic Theory* **41** (1987) 405–416.

Duffie, D., *Dynamic Asset Pricing Theory*, Second Edition. Princeton University Press, Princeton, NJ, 1996.

Duffie, D. and C. Huang, "Implementing Arrow–Debreu equilibria by continuous trading of few long-lived securities", *Econometrica* **53** (1985) 1337–1356.

Duffie, D. and M. Jackson, "Optimal hedging and equilibrium in a dynamic futures market", *Journal of Economic Dynamics and Control* **14** (1990) 21–33.

Duffie, D. and P. Protter, "From discrete- to continuous-time finance: Weak convergence of the financial gain process", *Mathematical Finance* **2** (1992) 1–15.

Duffie, D. and W. Shafer, "Equilibrium in incomplete markets II: Generic existence in stochastic economies", *Journal of Mathematical Economics* **15** (1986) 199–216.

Eberlein, E., "On modeling questions in security valuation", *Mathematical Finance* **2** (1992) 17–32.

Edirisinghe, C., Naik, V. and R. Uppal, "Optimal replication of options with transaction costs and trading restrictions", *Journal of Financial and Quantitative Analysis* **28** (1993) 117–138.

Ethier, S. and T. Kurtz, *Markov Processes: Characterization and Convergence*, John Wiley & Sons, New York, 1986.

Figlewski, S., "Options arbitrage in imperfect markets", *Journal of Finance* **44** (1989) 1289–1311.

Foley, D., "On two specifications of asset equilibrium in macroeconomic models", *Journal of Political Economy* **83** (1975) 303–324.

Föllmer, H. and D. Sonderman, "Hedging of non-redundant contingent-claims", *Contributions to Mathematical Economics, in Honor of Gérard Debreu* in W. Hildebrand and A. Mas-Colell, eds., (North-Holland, Amsterdam , 1986).

Friedman, A., *Partial Differential Equations of Parabolic Type* Prentice-Hall, Englewood Cliffs, NJ, 1964.

Friedman, A., *Stochastic Differential Equations and Applications*, Volume 1, Academic Press, New York, 1975.

Garman, M., "A General Theory of Asset Valuation under Diffusion State Processes", Working Paper No. 50, University of California Berkeley, 1976.

Grannan, E. and G. Swindle, "Minimizing transaction costs of option hedging strategies", *Mathematical Finance* **6** (1996) 341–364.

Harrison, J. M., D. M. Kreps, "Martingales and arbitrage in multiperiod securities markets", *Journal of Economic Theory* **20** (1979) 381–408.

He, H., "Essays in Dynamic Portfolio Optimization and Diffusion Estimations", PhD thesis, MIT, 1989.

He, H., "Convergence from discrete- to continuous-time contingent claim prices", *Review of Financial Studies* **3** (1990) 523–546.

He, H., "Optimal consumption-portfolio policies: A convergence from discrete to continuous time models", *Journal of Economic Theory* **55** (1991) 340–363.

He, H. and N. Pearson, "Consumption and portfolio policies with incomplete markets and short-sale constraints: The infinite-dimensional case", *Journal of Economic Theory* **54** (1991) 259–304.

Henrotte, P., "Transactions Costs and Duplication Strategies", unpublished working paper, Stanford University, 1993.

Hodges, S. and A. Neuberger, "Optimal Replication of Contingent Claims Under Transactions Costs", unpublished paper, Financial Options Research Center, University of Warwick, 1989.

Huang, C., "An intertemporal general equilibrium asset pricing model: The case of diffusion information", *Econometrica* **55** (1987) 117–142.

Hull, J., A. White, "The pricing of options on assets with stochastic volatilities", *Journal of Finance* **42** (1987) 281–300.

Hutchinson, J., Lo, A. and T. Poggio, "A non-parametric approach to the pricing and hedging of derivative securities via learning networks", *Journal of Finance* **49** (1994) 851–889.

Ingersoll, J., *Theory of Financial Decision Making*, Rowman & Littlefield, Totowa, NJ, 1987.

Itô, K., "On stochastic differential equations", *Memoirs of the American Mathematical Society* **4** (1951) 1–51.

Jerison, D., Singer, I. and D. Stroock, eds., *The Legacy of Norbert Wiener: A Centennial Symposium*, American Mathematical Society, Providence, RI, 1997.

Karatzas, I. and S. Shreve, *Brownian Motion and Stochastic Calculus*, Springer-Verlag, New York, 1991.

Kushner, H. and P. Dupuis, *Numerical Methods for Stochastic Control Problems in Continuous Time* Springer Verlag, New York, 1992.

Leland, H., "Option pricing and replication with transaction costs", *Journal of Finance* **40** (1985) 1283–1301.

Lo, A. and J. Wang, "Implementing option pricing models when asset returns are predictable", *Journal of Finance* **50** (1995) 87–129.

Merton, R., "Lifetime portfolio selection under uncertainty: The continuous-time case", *Review of Economics and Statistics* **51** (1969) 247–257.

Merton, R., "Theory of rational option pricing", *Bell Journal of Economics and Management Science* **4** (1973) 141–183.

Merton, R., "Theory of finance from the perspective of continuous time", *Journal of Financial and Quantitative Analysis* **10** (1975) 659–674.

Merton, R., "On the microeconomic theory of investment under uncertainty", in *Handbook of Mathematical Economics*, Volume II, K. Arrow and M. Intriligator, eds., (North-Holland Publishers, Amsterdam, 1982).

Merton, R., "On the mathematics and economic assumptions of continuous-time models", in *Financial Economics: Essays in Honor of Paul Cootner*, W. Sharpe and C. Cootner, eds., (Prentice-Hall, Englewood Cliffs, NJ, 1982).

Milstein, G. N., *Numerical Integration of Stochastic Differential Equations*, Kluwer Academic Publishers, Boston, MA, 1995.

Milstein, G. N., "Approximate integration of stochastic differential equations", *Theory Prob. Appl.* **19** (1974) 557–563.

Milstein, G. N., "A theorem on the order of convergence of mean-square approximations of solutions of systems of stochastic differential equations", *Theory Prob. Appl.* **32** (1987) 738–741.

Neuberger, A., "Option replication with transaction costs: An exact solution for the pure jump process", in D. Chance and R. Trippi, eds., *Advances in Futures and Options Research*, Volume 7, D. Chance and R. Trippi, eds., (JAI Press, Greenwich, CT, 1994).

Platen, E., "An approximation method for a class of Itô processes", *Lit. Mat. Sb.* **21** (1981) 121–133 (In Russian).

Rubinstein, M., "Implied binomial trees", *Journal of Finance* **49** (1994) 771–818.

Stroock, D., *Probability Theory: An Analytic View*, Cambridge University Press, Cambridge, UK, 1993.

Toft, K., "On the mean-variance tradeoff in option replication with transactions costs", *Journal of Financial and Quantitative Analysis* **31** (1996) 233–263.

Wiener, N., "Differential space", *Journal of Mathematical Physics* **2** (1923) 131–174.

Wiggins, J., "Option values under stochastic volatility: Theory and empirical estimates", *Journal of Financial Economics* **5** (1987) 351-372.

Willinger, W. and Taqqu M. S., "Towards a convergence theory for continuous stochastic securities market models", *Mathematical Finance* **1** (1991) 55–99.

ASSET PRICES ARE BROWNIAN MOTION: ONLY IN BUSINESS TIME*

HELYETTE GEMAN

University Paris IX Dauphine and ESSEC

DILIP B. MADAN

University of Maryland

MARC YOR

University Paris VI - Laboratoire de Probabilités

Received 27 October 1998

This paper argues that asset price processes arising from market clearing conditions should be modeled as pure jump processes, with no continuous martingale component. However, we show that continuity and normality can always be obtained after a time change. We study various examples of time changes and show that in all cases they are related to measures of economic activity. For the most general class of processes, the time change is a size-weighted sum of order arrivals. The paper provides a number of new processes for modeling prices. Characteristic functions for these processes are also given in closed form.

Keywords: Lévy processes, characteristic functions, Brownian excursions, stochastic volatility.

1. Introduction

Continuity of price processes has served economic theory as a convenient and powerful assumption, delivering market completeness and unique pricing of contingent claims by arbitrage. Such assumptions are critical to the validity of the Black and Scholes (1973) and Merton (1973) option pricing theories and their associated dynamic hedging strategies, and they are also fundamental to the Cox and Huang (1989) approach to solving the Merton (1971) intertemporal consumption and investment allocation problem. The assumption of continuity justifies the Cox, Ross and Rubinstein (1979) binomial approximation of the process, using up and down price shocks tending to zero with the size of the time grid. We grant that such price processes can accurately represent market prices in economies that instantaneously and continuously equilibrate to information flows driven by *diffusions or Ito*

---

*This paper was completed during the time Dilip Madan was visiting the Centre de Recherches pour l'Enseignement de la Gestion, at the University of Paris IX Dauphine.

processes. Examples of such equilibrium models in the literature include Duffie and Huang (1985), Dumas (1989), He and Leland (1993), and Detemple and Murthy (1994).

We question the validity of diffusions as an appropriate model for the underlying uncertainties. Instead, we represent the price process as instantaneously and continuously adjusting to exogenous demand and supply shocks. We view the underlying uncertainties (represented by cumulated shocks) as consisting of increasing random processes. As such, our price processes turn out to be the difference of two increasing random processes, representing respectively the up and down moves of the market. In contrast with Brownian motion, such processes are of finite variation. Furthermore, we demonstrate that in the presence of local uncertainty such processes are generally purely discontinuous.[a]

The implications of our theory for economic analysis and risk management are profound. At variance with the Black–Scholes–Merton setting, options can no longer be replicated by trading in the stock and money market account. In fact, the primary motivation for the use of continuous processes as representing price movements was not its accuracy, but rather the dynamic hedging argument, valid in this context, that made options redundant assets and led to the Hakansson (1979) paradox. In our economy, not only do we obtain potentially greater accuracy in describing stock price changes, but options become primary market completing assets useful in hedging jump risks while option prices constitute a rich source of information to be employed in designing optimal risk exposures.

The possibility of discontinuities or jumps in asset prices has a long history in the economics literature. Merton (1976) considered the addition of a jump component to the classical geometric Brownian motion model for the pricing of options on stocks. In explaining the distribution of returns such models were used as early as Press (1965). A number of authors have since considered the issue of market completeness in this context (see for instance Jones (1984), Jarrow and Madan (1995)), while others have assessed and argued for their necessity in explaining implied volatility smiles in low maturity options (see for instance Bates (1996), Bakshi, Cao and Chen (1997)). However, these models all contain a diffusion component in addition to a low or finite activity jump part: the diffusion component accounts for high activity in price fluctuations while the jump component is used to account for rare and extreme movements.

By contrast, we account for the small, high activity and rare large moves of the price process in both a unified and connected manner. For the processes we consider, price jumps are the rule and all motion occurs via jumps. High activity is accounted for by a large (in fact infinite) number of small jumps. The activity at various jump sizes is analytically connected by the requirement that smaller jumps occur at a higher rate than larger jumps.

The property of an infinite number of small moves is shared with the diffusion-based models, with the additionally attractive feature that the sum of absolute

---

[a]An exception is provided by the local times of Brownian motion.

changes in price is finite for our processes while for diffusions this is infinite (for diffusions, the price changes must be squared before they sum to a finite value). This makes it possible for us to create and price contracts based on the instantaneous upward, downward or total variability (positive, negative, or absolute price jump size) of underlying asset prices, in addition to the more traditional contracts with payoffs that are functionally related to the level of the underlying price. Processes similar to ours have recently been used in option pricing models by Heston (1993), Madan and Milne (1991), and include the α-stable increments for $\alpha < 1$ that were studied by Mandelbrot (1963) and McCulloch (1978).

Though the processes we advocate are pure jump processes and of finite total variation, we generalize the results of Clark (1973) and show that they may always be viewed as continuous processes in an economic measure of time. Clark (1973) considered subordinated processes where prices were represented by a geometric Brownian motion and time was given by another independent geometric Brownian motion. The processes we propose can also be written as Brownian motion evaluated at a random time change (or a stochastic clock), but in general the process of the clock need not be independent of the price process. We study the relationship between the stochastic clock and the demand and supply shocks driving the price process. As in Ané and Geman (1997) who show empirically that a quasi-perfect normality of returns on a high frequency data base of FTSE100 futures prices can be recovered using a stochastic clock driven by the number of trades, the time change we consider in our different examples is related to a measure of economic activity. We observe that price continuity does arise, but in an activity-based measure of time, as opposed to calendar time directly; hence the title of the paper.

The literature relating price changes to measures of activity (see for examples, Tauchen and Pitts (1983), Karpoff (1987), Gallant, Rossi and Tauchen (1992), and Jones, Kaul and Lipson (1994)) has considered as relevant measures of activity either the number of trades or the volume. Our analytical results on the nature of the time change show that both the number and size of orders enter the time calculation in specific ways. The precise time changes relevant for various example economies suggests a measure of time based on a size weighted total orders. Furthermore, the time change generally belongs to the *same class* of processes as the process for the underlying activity.

The analysis of our paper provides a useful reduced form synthesis of continuous price equilibrium processes. First we note that equilibrium processes are free of arbitrage opportunities. A rich literature has studied the consequences of the no arbitrage assumption (Harrison and Kreps(1979), Kreps (1981), Harrison and Pliska (1981)) and has concluded that such processes are semimartingales (Delbaen and Schachermayer (1994)). Monroe (1978) characterized all semimartingales and showed that every semimartingale may be represented as a time-changed Brownian motion. A consequence of these results taken jointly is that equilibrium price processes may be identified by focusing attention on the Brownian motion and its time change. We note that time changes are increasing processes and argue that when they are locally uncertain, they must be pure jump processes. We relate the time change to the information provided by demand and supply shocks in the market.[b]

We first consider examples where the time change is independent of the Brownian motion (which was the assumption made in Clark (1973)). The identification of the time change in this context is easier. However, we also study more generally, examples where this independence assumption is relaxed (we note that the theorem of Monroe (1978) does not involve this property). In the case where there is dependence in our example economies, we show that the time change is cumulated volatility, where the latter depends on the Brownian motion. For a large class of processes both representations are valid: they can be written as a Brownian motion evaluated at an independent time, and also as another Brownian motion evaluated at another, now dependent time. We consider both representations as they provide a richer source of interesting and tractable models.

The two formulations are similar, in that in both price changes are related to excess demand. In the examples involving independence, excess demand coincides with orders as we assume no coincidence of demand and supply, and the price change is linearly related to the excess demand. In the second situation excess demand is modeled by a Brownian motion and the price response may be non-linear in the excess demand. The model primitives for the first formulation are the arrival rates of buy and sell orders, while for the second formulation it is the function relating price responses to excess demand.

In many cases we derive closed-form expressions for the characteristic function of the log price relative. It is well-known how one may then extract the density or the distribution function via Fourier inversion from this characteristic function. Bakshi and Madan (1997) show how one may use this characteristic function to obtain option prices.[c] Hence, our analysis provides a wide class of operational models for continuous-time price processes in economics. With the advent of price discovery in many previously regulated markets, and the opening of derivative markets for a much greater range of underlying processes, it is imperative that we expand the class of processes for which estimation and derivative pricing are feasible. This paper makes substantial contributions in this direction.

The outline of the paper is as follows: Sec. 2 presents a general economic model for the process of price discovery in market economies. It is observed that this process is purely discontinuous and of finite variation. Section 3 shows how in general one may recover continuity of prices in a time given by a stochastic clock.

---

[b]The resulting jump process need not be one of bounded variation and it may not be capable of being decomposed into the difference of two increasing processes if the infinite activity of small changes is too large; however, we view the hypothesis of infinite variation as unrealistic and consider mainly processes of bounded variation as good pricing models. In any case, the infinite variation jump processes can always be approximated by the difference of two increasing processes, even if they are not identical to it.

[c]The difficulty addressed in Bakshi and Madan (1997), is that of deriving the characteristic function of a probability element constructed to correspond to the term $N(d_1)$ in the Black–Scholes formula. For many cases this probability element is not in the same parametric class as the equivalent of $N(d_2)$.

Section 4 considers specific examples, where we relate the stochastic time to the uncertainties driving the price process and observe that this time is as an activity based measure. In Sec. 5 we consider some time changes that have been proposed in the literature, and are not related to the process of price discovery described in Sec. 2, but can be approximated by such processes. Section 6 concludes.

2. The Price Discovery Process

Consider an economy over the time interval $[0, T]$ in which a single commodity is traded continuously in time and has a price process, denoted $p(t)$, $0 \leq t \leq T$. The uncertainties driving the price process are identified as exogenous events representing demand or supply shocks respectively. Let $U(t)$ be the process of cumulated demand shocks, given by an increasing pure jump process. At any time t,

$$u_t = \Delta U(t) = U(t) - U(t_-) \geq 0 \,,$$

represents the number of units of the asset that are demanded by agents in the economy consequent upon the occurrence of a liquidity or information based event. The quantity u_t is the amount demanded at the prevailing price of $p(t_-)$ and in the absence of any price response to the processing of the demand order. We think of u_t as the amount some economic agent would like to buy at the price $p(t_-)$. The considerations determining u_t are either liquidity considerations, wealth or cash balance accumulations of individuals or information considerations reflecting beliefs of individuals that the asset is underpriced at $p(t_-)$. $U(t)$ is the cumulated level of such demand shocks that are motivated by liquidity or information shocks. Similarly, let $V(t)$ be the cumulated level of such supply shocks or amounts that agents would like to sell at the price $p(t_-)$. The supply shock is

$$v_t = \Delta V(t) = V(t) - V(t_-) \,.$$

Assumption 1. Non-coincidence of demand and supply shocks in continuous time.

We suppose that cumulated demand shocks, $U(t)$, and cumulated supply shocks, $V(t)$, are increasing pure jump processes with no coincidence of jumps in continuous time.

The primary sources for these shocks are the arrival of information in the economy that is transformed into market buy and sell orders and we view these primary sources of uncertainty as exogenous. We recognize that differences in beliefs among market participants can lead to simultaneous buy and sell orders triggered by a single information event, but suppose that there are sufficient differences in the access to the information, and the subsequent execution decisions, for the actual orders to materialize at different instants of continuous time. Under this assumption the markets are dealing, at all instants, with either one or the other type of order.

2.1. *Modeling the price increases*

All market participants realize that orders won't be executed at the price $p(t_-)$ and expect a price response to the execution of orders.[d] We suppose that the actual quantity demanded in response to a demand shock u_t, q_t^{du} at time t, is given by a demand function

$$q_t^{du} = q^{du}(p(t)/p(t_-), u_t, t) \tag{1}$$

that reflects a falling off of demand in the face of a price response. The quantity u_t is, as stated earlier, the amount an economic agent would like to buy at the price $p(t_-)$, and (1) gives the agent's demand response to price increases. We suppose that

$$\frac{\partial q^{du}}{\partial p(t)} < 0\,.$$

Furthermore, we have that u_t is the demand at no price response or that:

$$q^{du}(1, u_t, t) = u_t\,.$$

In addition to the demand function, there is a supply function with suppliers being aware of the liquidity or information event and the extent of this exogenous demand shock. The supply is given by

$$q_t^{su} = q^{su}(p(t)/p(t_-), u_t, t) \tag{2}$$

where we suppose that

$$\frac{\partial q^{su}}{\partial p(t)} > 0$$

and furthermore we assume no supply at a zero price response or that:

$$q^{su}(1, u_t, t) = 0\,.$$

Under these conditions all supply occurs at a price response.[e] Equation (2) may also be viewed as reflecting the price response in a limit order sell book on the market, though more generally it is the supply function of the economy that is relevant.

The market price $p(t)$ and quantity transacted q_t^u are simultaneously determined in equilibrium by the market clearing condition

$$q_t^u = q_t^{du} = q_t^{su}\,. \tag{3}$$

---

[d]We recognize that small orders are not likely to experience an adverse price impact and accommodate this by elasticity assumptions on demand and supply for low quantities. We also recognize that incentives for order splitting in the presence of low price responses to small orders can lead to strategies of submitting an infinite number of infinitesimal orders to complete finite transactions, but rely on the exogeneous costs associated with such splitting, to rule out these possibilities.
[e]The supply curve q^{su} represents supply at prices higher than $p(t_-)$. Supply at $p(t_-)$ comes forth as a market sell order, at instants that differ from those at which buy orders are coming to market.

We suppose that the market clearing condition (3) may be solved to determine simultaneously the price response

$$\ln\left(\frac{p(t)}{p(t_-)}\right) = \Phi^u(u_t, t) > 0 \tag{4}$$

and the quantity transacted

$$q_t^u = \Psi^u(u_t, t) > 0. \tag{5}$$

We may accommodate very small price responses to small shocks by supposing that the supply elasticity is high for small quantities.

It is useful to contrast this formulation with the more standard general equilibrium formulation where at each instant, all markets are simultaneously cleared to determine all asset prices. In our partial equilibrium model, the market clearing takes place in one market at a time. Our demand and supply curves are contingent on the arrival of a demand order for this asset at time t, in the quantity u_t if the price stays at $p(t_-)$.

2.2. *Modeling the price decreases*

For price decreases we follow a symmetric approach to that taken in modeling price increases. In the presence of a supply shock v_t at time t , to sell v_t if the price stays at $p(t_-)$, the supply function

$$q_t^{sv} = q^{sv}(p(t)/p(t_-), v_t, t) \tag{6}$$

reflects the curtailment of supply in the presence of a downward price response and we suppose that

$$\frac{\partial q^{sv}}{\partial p(t)} > 0$$

while the supply is v_t at no price response, or that:

$$q^{sv}(1, v_t, t) = v_t.$$

The demand function in the face of an exogenous supply shock is given by

$$q_t^{dv} = q^{dv}(p(t)/p(t_-), v_t, t) \tag{7}$$

and reflects the limit order buy book and more generally the economy wide demand response. We suppose that

$$\frac{\partial q^{dv}}{\partial p(t)} < 0$$

and furthermore that there is no demand at a zero price response,

$$q^{dv}(1, v_t, t) = 0.$$

Once again, the partial equilibrium condition in the presence of a supply shock determines the quantity supplied q_t^v and the condition is

$$q_t^v = q_t^{sv} = q_t^{dv}.$$ (8)

We suppose that the market clearing conditions may be solved for the price responses and the quantity transacted to yield

$$\ln\left(\frac{p(t)}{p(t_-)}\right) = -\Phi^v(v_t, t) < 0$$ (9)

and

$$q_t^v = \Psi^v(v_t, t).$$ (10)

Note that $\Phi^v(v_t, t)$ is the absolute value of the log price response and will later be used to cumulate the total price decreases. We may accommodate small price responses for small shocks by supposing that the economy wide demand elasticity is high for small quantities.

2.3. *The price process*

In this section, we put together the results for the price increases and decreases and derive the complete price process. Under Assumption 1, the demand and supply shocks never arise at the same instance of time and hence the price process is given by

$$\ln(p(t)) = \ln(p(0)) + \sum_{s \le t} \Phi^u(\Delta U(s), s) - \sum_{s \le t} \Phi^v(\Delta V(s), s)$$ (11)

while the process for the total quantity transacted to time t, is given by $Q(t)$, where

$$Q(t) = \sum_{s \le t} \Psi^u(\Delta U(s), s) + \sum_{s \le t} \Psi^v(\Delta V(s), s).$$ (12)

We observe from Eq. (12) that the process for the volume of transactions to time t is the sum of two increasing pure jump processes

$$Y_1(t) = \sum_{s \le t} \Psi^u(\Delta U(s), s)$$ (13)

and

$$Y_2(t) = \sum_{s \le t} \Psi^v(\Delta V(s), s).$$ (14)

It follows that $Q(t)$ is a pure jump process of finite variation and the volume transacted between two time points $t_1 < t_2$, $V(t_1, t_2)$ may be recovered as

$$V(t_1, t_2) = Q(t_2) - Q(t_1).$$

Similarly, we observe from Eq. (11) that the price process is the difference of two increasing pure jump processes

$$X_1(t) = \sum_{s \leq t} \Phi^u(\Delta U(s), s) \tag{15}$$

and

$$X_2(t) = \sum_{s \leq t} \Phi^v(\Delta V(s), s). \tag{16}$$

It follows that $\ln(p(t))$ *is a pure jump process of finite variation.*[f]

This formulation of the price process is in sharp contrast to traditional assumptions about such processes in the finance literature. Typically it is assumed that asset prices follow a diffusion process, and as a result, are continuous processes of infinite variation. Duffie and Huang (1985) provide a general existence theorem for economies in which equilibrium price processes are diffusions. He and Leland (1993) consider conditions on price processes in the diffusion context, that ensure their consistency with an economic equilibrium.

The motivation for our departure from these more traditional formulations lies in the modeling of the underlying uncertainties. We view the underlying uncertainties as consisting of increasing random processes. These could be processes for total prevailing price buy orders, and total prevailing price sell orders, viewed separately. Net orders, being the difference, are then by construction a process of bounded variation and cannot be accurately modeled by a continuous diffusion, which is of infinite variation. Many authors assume that the underlying uncertainty is a continuous diffusion and this then implies the same for the price process via the market equilibrium conditions.

2.4. *Parameterizing the price process*

We now introduce specific functional forms for the demand and supply functions that permit a closed form parametrization of the price process.

Assumption 2. Parametric responses to demand shocks
Suppose that the demand function q_t^{du} is given by

$$q_t^{du} = u_t - \delta_t \ln\left(\frac{p(t)}{p(t_-)}\right) \tag{17}$$

while the supply function in the face of a demand shock is

$$q_t^{su} = a_t u_t^{-\gamma_t} \ln\left(\frac{p(t)}{p(t_-)}\right) \tag{18}$$

reflecting a lower supply if the demand shock is large for positive γ_t.

---

[f]A process $x(t)$ is said to be of finite variation over the interval $[0, T]$ if for any path, there exists a constant A, such that for all partitions $0 = t_0 < t_1 < \cdots < t_n = T$ we have $\sum_{i=0}^{n-1} |x(t_{i+1}) - x(t_i)| < A$. A process of finite variation can be written as the difference of two increasing processes. The converse is also true.

Note that the response functions (17) and (18) meet the requirements that at $p(t) = p(t_-)$, demand is u_t while supply is 0, as is required by the interpretation of u_t.

Assumption 3. Parametric responses to supply shocks
We suppose that in the presence of a supply shock we have

$$q_t^{sv} = v_t + \eta_t \ln\left(\frac{p(t)}{p(t_-)}\right) \tag{19}$$

while

$$q_t^{sd} = -b_t v_t^{-\lambda_t} \ln\left(\frac{p(t)}{p(t_-)}\right) \tag{20}$$

and the demand is curtailed for large supply shocks for positive λ_t.

Solving for the equilibrium using Eqs. (17) and (18) we obtain that

$$\Phi^u(u_t, t) = \frac{u_t}{\delta_t + a_t u_t^{-\gamma_t}} \tag{21}$$

while

$$\Psi^u(u_t, t) = \frac{a_t u_t^{1-\gamma_t}}{\delta_t + a_t u_t^{-\gamma_t}}. \tag{22}$$

Similarly, the equilibrium solutions using Eqs. (19) and (20) are

$$\Phi^v(v_t, t) = \frac{v_t}{\eta_t + b_t v_t^{-\lambda_t}} \tag{23}$$

and

$$\Psi^v(v_t, t) = \frac{b_t v_t^{1-\lambda_t}}{\eta_t + b_t v_t^{-\lambda_t}}. \tag{24}$$

Under these parametric assumptions one may write the price process as

$$\ln p(t) = \ln p(0) + \sum_{s \leq t} \frac{\Delta U(s)}{\delta_s + a_s(\Delta U(s))^{-\gamma_s}} - \sum_{s \leq t} \frac{\Delta V(s)}{\eta_s + b_s(\Delta V(s))^{-\lambda_s}} \tag{25}$$

and the process for cumulated transactions is

$$Q(t) = \sum_{s \leq t} \frac{a_s \Delta U(s)}{a_s + \delta_s(\Delta U(s))^{\gamma_s}} + \sum_{s \leq t} \frac{b_s \Delta V(s)}{b_s + \eta_s(\Delta V(s))^{\lambda_s}}. \tag{26}$$

Assumption 4. Stationary limit order books
The special case of stationary limit order books occurs when the supply responses to market order demand shocks and the demand responses to market sell shocks are not responsive to the size of these shocks. In this case γ_s, λ_s are zero and the coefficients of the demand and supply functions are constant through time, $\delta_t = \delta, a_t = a, \eta_t = \eta$, and $b_t = b$.

Under Assumptions 1 through 4, we may write the price process as

$$\ln p(t) = \frac{1}{\delta + a} U(t) - \frac{1}{\eta + b} V(t) + \ln p(0) \tag{27}$$

while

$$Q(t) = \frac{a}{\delta + a} U(t) + \frac{b}{\eta + b} V(t). \tag{28}$$

Under these assumptions, the price process is the difference of two increasing pure jump processes while the transacted quantity is related to the sum of these processes. Furthermore, we have linearity of the price response in the order flow. Later we do consider models that allow for non-linearities in the price response to excess demand.

3. Stochastic Time Changes and Prices

It was argued in Sec. 2 that in general the process for the price of an asset is the difference of two increasing pure jump processes and is a process of finite variation. In this section, we observe interestingly, that continuity of the price process can always be recovered by the use of a stochastic time change. In effect, there is always a sense of economic time, related to market activity, such that prices are continuous when time is measured in units of business activity. In the next section, we begin our study of the relationship between this time change and the original price process.

The price process of Sec. 2, by virtue of being of bounded variation, is a semimartingale. It is also known that every price process that is consistent with the absence of arbitrage opportunities, must be a semimartingale (Delbaen and Schachermayer (1994)). Hence all equilibrium price processes meet this condition, whether they are diffusions or not. The next question is, "What semimartingales are most appropriate as models for price processes?" For this we turn to Monroe (1978).

A remarkable result of Monroe (1978)[g] shows that every semimartingale can be written as a Brownian motion (possibly defined on some adequately extended probability space) evaluated at a random time.[h] By this result there exists a Brownian

---

[g]We wish to thank Joe Horowitz for bringing this result to our attention.

[h]It is instructive to note that one may even write calendar time as time changed Brownian motion, whereby

$$t = W(T(t))$$

where $T(t)$ is defined by

$$T(t) = \inf\{s | W(s) \geq t\}.$$

One clearly observes the dependence between the time change and Brownian motion in this case. We also observe that for any increasing process $A(t)$

$$A(t) = W(T(A(t))) = W(T^*(t)),$$

where $T^*(t)$ is another time change.

motion $W(t)$ and a random time change $T(t)$, where $T(t)$ is an increasing stochastic process,[i] such that

$$\ln(p(t)) = \ln(p(0)) + W(T(t)). \tag{29}$$

Equation (29) implies that the study of price processes for market economies may be reduced to the study of time changes for Brownian motion. This is a powerful reduced form representation of a complex phenomenon involving multi-dimensional considerations — those of modeling demand, supply and their interaction through market clearing — to a single entity: the correct unit of time for the economy, with respect to which we have a Brownian motion.

We note that the price process will be continuous in calendar time only if the time change is continuous. This observation has some restrictive implications. If the time change is continuous it must essentially be a stochastic integral with respect to Brownian motion (Revuz and Yor (1994), page 190 Theorem 3.9).[j] For a time change, it follows that the diffusion component must be zero: otherwise it will not be an increasing process and so cannot be a time change. We therefore have the implication that continuous time changes are locally deterministic. We summarize this discussion in the following proposition.

Proposition 1. *In the absence of arbitrage opportunities, continuous-time price processes of market economies are time-changed Brownian motions. Furthermore, if there is local uncertainty in the time change then the price process is not a continuous process in any interval of time.*

In addition, Proposition 1 suggests that return distributions should be normal, not when measured in calendar time, but when measured per unit of what may be termed, economic time. The search for such a formulation began in earnest with Clark (1973), and has been pursued more recently by Ané and Geman (1997) and Madan and Chang (1998). In fact, it is shown in Ané and Geman (1997) that one minute, FTSE100, calendar time returns are highly non-normal, while returns measured per unit trade are normally distributed. Madan and Chang (1998) show that Brownian motion, when time changed by a gamma process, provides a significantly better description of historical asset returns and the risk neutral return distribution embedded in option prices. Our investigations here will focus on theoretically identifying and interpreting $T(t)$ from a knowledge of the process $\ln(p(t))$ as a process of bounded variation.[k] In particular we wish to explore the relationship between time changes and general measures of economic activity.

[i]More formally, it is required that there exists a filtration with respect to which the process $W(t)$ is adapted, and that $T(t)$ is an increasing sequence of stopping times adapted to this filtration.

[j]The qualification required is that its quadratic characteristic is absolutely continuous with respect to Lebesgue measure or equivalently that it has a well-defined sense of a variance rate. As noted earlier, the local time of Brownian motion is an example of a continuous increasing random process which has intervals of time when the clock does not move at all, and is hence not appropriate for our consideration. We restrict attention to processes with a well-defined sense of a realized variance rate.

[k]We recognize that semimartingales may be of infinite variation and so the price process may not be of bounded variation, but note that even in this case, an approximation by bounded variation

We first consider representations of the type given by Eq. (29) in which the Brownian motion is independent of the time change. Furthermore, to allow for additional generality we consider stochastic time changes applied to continuous Ito processes. From this perspective, let $x(t)$ be an Ito process of the general form

$$x(t) = x(0) + \int_0^t \theta(u)du + \int_0^t \sigma(u)dW(u)\,. \tag{30}$$

We consider the representation of the price process as $x(t)$ evaluated at a random time $T(t)$ or

$$\ln(p(t)) = \ln(p(0)) + x(T(t)) \tag{31}$$

where $T(t)$ is independent of $(x(u), u \geq 0)$.

Some general properties of the time change $T(t)$ are inherited from the price process. Firstly, as already noted, since the price process is pure jump, and $x(t)$ is continuous, it follows that $T(t)$ is a pure jump process. Second, we say that the price process represents a high level of activity if there is no interval of time in which prices are constant throughout the time interval. Processes of high activity levels have an infinite arrival rate of demand and supply shocks, though of necessity the arrival rate is finite for all shocks with a magnitude strictly bounded away from zero. If the price process is such a high activity process, then this property is also inherited by the time change. In the next section, we explore examples where the time change, as a process, reflects quite exactly the two processes for the demand and supply shocks.

4. Time Changes Related to Demand and Supply Shocks

This section studies the relationship of stochastic time changes to the process of demand and supply shocks, in some simple and tractable contexts. We begin with processes that have a finite arrival rate and then consider high activity processes with infinite arrival rates. We study these relationships between time changes and demand and supply shocks in the context of various examples for the price process.

We first consider two pure jump cases with finite jump arrival rates and jump sizes distributed as (i) a reflected normal distribution, as is the case with the Merton (1976) jump diffusion model, and (ii) for exponentially distributed jump sizes arriving at Poisson times, as this class of processes is a base model for the construction of a wide class of increasing random processes and time changes. In the latter case, we have jump arrival rates exponentially related to the jump size. For these cases we show that in (i) the number of arrivals, independent of the size, constitutes the time change, but in (ii) arrivals of different sizes have differing impacts on the stochastic clock. We show that the relevant time measure is a size weighted cumulation of order arrivals.

processes is admissible. Note for example that the usual Binomial approximation to Brownian motion is a process of bounded variation.

Our next example considers the case of the gamma process that turns out to be a fundamental building block for a wide class of infinite arrival rate candidate processes. We show that for the gamma process a proxy for the time change over an interval is related interestingly to the level of excess demand. This is reasonable as demand matched by supply does not contribute to price pressure.

The result for gamma processes is next generalized to a wide class of increasing random processes that have a desirable structure on the arrival rate of jump sizes, which is that larger jumps have a smaller arrival rate. The relationship between the level of economic activity and the time change generalizes the result obtained for the compound Poisson process with exponentially distributed jump sizes.

We next investigate the similarity of the time change and the order arrival processes. In particular we observe that when the order arrival processes are in a given subclass of processes, the time change also belongs to the same subclass. An important example of a subclass is given by processes for which larger jump, size weighted arrival rates, are smaller. This is the class of processes that can be generated from the gamma process by convolution, termed the class of generalized gamma convolutions. Once again, the time change is also a generalized gamma convolution. A useful generalization of the variance gamma model in the generalized gamma convolution family of processes is also provided.

The next subsection considers this question of similarity of order arrival and time change for the case of the stable processes. It is observed that for α-stable arrival rate processes the time change is a stable process with index $\alpha/2$. We then ask when the time change is a simple adjustment of scale and speed of the process for the arrival of shocks. This class consists of the gamma process and processes constructed from the gamma process by integrating the tail of the gamma arrival rates, as is made precise later. Interestingly, the gamma process itself is a consequence of applying this procedure to the compound Poisson process with exponentially distributed jump sizes.

For processes with completely monotone arrival rates as a function of the jump size,[1] we present an alternative economic representation of the process. We model excess demand by a Brownian motion and define a *force function* that expresses the derivative of prices as a function of this excess demand. The equilibrium price process is given by restricting attention to the zero excess demand situations or evaluating prices at the inverse local time of Brownian motion, where by construction Brownian motion is zero and we have equilibrium. We establish the connections between the force function and the arrival rates of price moves, providing in addition various examples relating these two constructs.

All the processes for the log price relative $X(t) = \ln(p(t)/p(0))$ considered in this section are pure jump processes with independent and identically distributed

---

[1]A real valued function of one variable is monotone if its derivative does not change sign. It is completely monotone, if all its derivatives have this property. With respect to arrival rates we require that their derivatives with respect to the jump size alternate in sign, begining with negative for positive jumps and positive for negative jumps.

increments over non-overlapping intervals of regular length. We recognize the considerable econometric evidence in support of time inhomogeneity in both the statistical and risk neutral price processes, but focus attention here primarily on the correct characterization of the local motion. For this purpose, time homogeneous processes, like Brownian motion itself, are an appropriate starting point for the discussion. Generalizations incorporating time inhomogeneity can be accommodated later as appropriate for the application at hand. The processes we consider are completely characterized by their Lévy densities $k(x)-$ defining the arrival rates of price jumps of size $x-$ that are identified by the unique decomposition of their characteristic functions provided by the Lévy Khintchine theorem that asserts that[m]

$$E[\exp(iuX(t))] = \exp\left(t \int_{-\infty}^{\infty} (e^{iux} - 1)k(x)dx\right). \tag{32}$$

In each case we seek to rewrite the process $X(t)$ as a time-changed Brownian motion. If we consider Brownian motion with drift

$$B(t) = \theta t + \sigma W(t) \tag{33}$$

and evaluate this at an independent random time $T(t)$, with characteristic function given by the Lévy measure $\tilde{k}(x)$, where

$$E[\exp(iuT(t))] = \exp\left(t \int_{0}^{\infty} (e^{iux} - 1)\tilde{k}(x)dx\right). \tag{34}$$

We seek to discover the relationship between k and $\tilde{k}$ under the condition that $X(t)$ may be written as Brownian motion time changed by $T(t)$, or that

$$X(t) = B(T(t)).$$

From the independence of the Brownian motion and the time change one may infer that

$$\begin{aligned}
E[\exp(iuB(T(t)))] &= E[\exp(iu\theta T(t) + \sigma W(T(t)))] \\
&= E[\exp((iu\theta - \sigma^2 u^2/2)T(t))] \\
&= \exp\left(t \int_{0}^{\infty} (e^{(iu\theta - \sigma^2 u^2/2)x} - 1)\tilde{k}(x)dx\right) \\
&= \exp\left(t \int_{-\infty}^{\infty} (e^{iux} - 1)k(x)dx\right). \tag{35}
\end{aligned}$$

The final equality in (35) follows from (32) on noting that $B(T(t))$ is also $X(t)$. In each case of independence of the Brownian motion and the time change, the Lévy

---

[m]We suppose both the absence of a continuous martingale component and a deterministic drift rate. A deterministic drift rate may easily be added in applications, we are here concerned with just the stochastic component.

measure for the time change $\tilde{k}$ is discovered given the arrival rate Lévy measure by solving for $\tilde{k}$ in the last equality of (35). In the absence of independence the construction is more complicated and we consider this in subsection 4.8.

4.1. *Compound Poisson demand and supply shocks*

This subsection considers two finite arrival rate processes with arrivals occurring at Poisson times. The first case is one that mirrors the Merton (1976) jump diffusion model and considers a reflected normal distribution for the jump sizes. In the second case, we consider exponentially distributed jump sizes. We shall note later that the first case is not completely monotone, while the second is the building block for this family of processes.

4.1.1. *Reflected normal prevailing price order sizes*

Let the prevailing price buy and sell orders be given by independent copies of the increasing compound Poisson process

$$X(t) = \sum_{i=1}^{N(t)} Y_i \tag{36}$$

where $N(t)$ is a Poisson process with arrival rate λt and the sequence of the magnitude of demand and supply shocks Y_i are independently and identically distributed with a reflected normal density,

$$f(y) = \frac{\sqrt{2}\exp\left(-\dfrac{y^2}{2\sigma^2}\right)}{\sigma\sqrt{\pi}}\,, \quad \text{for } y > 0\,. \tag{37}$$

In this example, the time change to be applied to Brownian motion to recover the price process turns out to be the total number of prevailing price orders, buy or sell.

Suppose that $\ln(p(t))$ is given by Eq. (27) with $(\delta + a)^{-1} = (\eta + b)^{-1} = \gamma$. In this case, we may write

$$\ln(p(t)/p(0)) = \gamma(X_1(t) - X_2(t)) \tag{38}$$

where $X_1(t), X_2(t)$ are two independent copies of the process satisfying (36).

Let $\phi_Y(u)$ be the characteristic function of Y, so that

$$\phi_Y(u) = \int_0^\infty \exp(iuy)\frac{\sqrt{2}\exp\left(-\dfrac{y^2}{2\sigma^2}\right)}{\sigma\sqrt{\pi}}dy\,. \tag{39}$$

Since the order size is positive or negative with equal probability, the characteristic function of the order size is given by

$$\text{Re}(\phi_Y(u)) = \exp\left(-\frac{\sigma^2 u^2}{2}\right).$$

It follows by direct computation of characteristic functions for compound Poisson processes (see for example, Karlin and Taylor (1981)) that

$$\phi_{\ln(p(t)/p(0))}(u) = \exp\left(2\lambda t \left(\exp\left(-\frac{\sigma^2\gamma^2 u^2}{2}\right) - 1\right)\right). \tag{40}$$

Equation (40), however, is also the characteristic function of $\sigma\gamma W(N_1(t) + N_2(t))$, where $W(t)$ is a standard Brownian motion and $N_1(t), N_2(t)$ are two independent copies of a Poisson process with arrival rate λt.

In this example, the time change just counts the number of all the demand and supply shocks, ignoring the magnitude of the shocks, which are accounted for by the distribution of the Brownian motion between successive arrivals. The volatility is of course scaled by γ to reflect the price sensitivity of the shocks. It is interesting to note that the time change of this example is akin to the number of trades, the time change observed to be relevant by Ané and Geman (1997) in their empirical study of high frequency returns on the FTSE100 futures index.

4.1.2. *Exponential jump sizes*

Suppose now that the prevailing price order sizes have an exponential density

$$f(y) = a\exp(-ay), \quad \text{for } y > 0 \tag{41}$$

with mean order size given by $1/a$. Further suppose that the Poisson arrival rate of jumps is $1/a$, so that the Lévy measure for the buy or sell orders is just the exponential function

$$k(x) = \exp(-ax). \tag{42}$$

Let $X_1(t)$ and $X_2(t)$ be two independent copies of this exponential compound Poisson process and suppose that log prices evolve in accordance with (38). The characteristic function for the log price relative is now easily computed as

$$\phi_{\ln(p(t)/p(0))}(u) = \exp\left(\frac{2t}{a}\left(\frac{a^2}{a^2 + u^2\gamma^2} - 1\right)\right). \tag{43}$$

To observe this process as a time-changed Brownian motion consider the time change given by $T(t)$, an increasing process with Lévy measure

$$\tilde{k}(y) = 2a\exp(-a^2 y), \quad \text{for } y > 0. \tag{44}$$

Now consider the process $\gamma\sqrt{2}W(T(t))$ and evaluate its characteristic function using (34) as

$$E[\exp(iu\gamma\sqrt{2}W(T(t)))] = E[\exp(-\gamma^2 u^2 T(t))]$$

$$= \exp\left(t\int_0^\infty (e^{-\gamma^2 u^2 y} - 1)2ae^{-a^2 y}dy\right)$$

$$= \exp\left(2ta\left(\frac{1}{u^2\gamma^2 + a^2} - \frac{1}{a^2}\right)\right)$$

$$= \exp\left(\frac{2t}{a}\left(\frac{a^2}{a^2 + u^2\gamma^2} - 1\right)\right). \tag{45}$$

We now investigate and comment on the relationship obtained in (45) between the time change and the order arrival process. An order of size y, be it buy or sell, arrives at the rate $2\exp(-ay)$. To construct the time change we have to weight the orders of different sizes differently; in fact, orders of size y amount to $a\exp(-a(a-1)y)$ units of time change. For $a = 1$ each order amounts to a unit of time change, as in the case of the reflected normal process, but for $a > 1$ (when most of the orders are small), the small orders count for a greater time change. The opposite holds for $a < 1$, in which case large orders count for greater time changes. The break even point of order sizes that count for a unit time change is given in both cases by $\ln(a)/(a(a-1))$.

Our next example considers two high activity processes for the prevailing price buy and sell order processes. In particular, we consider for $U(t)$ and $V(t)$, two independent gamma processes with a common coefficient of variation.

4.2. *Gamma process demand and supply shocks*

Suppose that the prevailing price buy order process, $U(t)$, is a gamma process with mean and variance rates μ_1, ν_1, respectively, while the prevailing price sell order process, $V(t)$, is also an independent gamma process of mean and variance rates μ_2, ν_2 respectively. It is useful to write these processes in terms of the standard gamma process with unit mean and variance rates.

Let $\gamma(t)$ denote the standard gamma process of unit mean and variance rate. The standard gamma process is a pure jump increasing random process with characteristic function

$$\phi_{\gamma(t)}(u) = \left(\frac{1}{1-iu}\right)^t \tag{46}$$

and Lévy measure

$$K_\gamma(x)dx = \frac{\exp(-x)}{x}dx \quad \text{for } x > 0 \tag{47}$$

that integrates to infinity, and hence we have a process that jumps infinitely often in any interval (or a high activity process), in sharp contrast to the processes considered in the last subsection.

One may write the prevailing price buy and sell order processes, $U(t)$ and $V(t)$, in terms of the standard gamma process by

$$U(t) = \frac{\nu_1}{\mu_1}\gamma\left(\frac{\mu_1^2}{\nu_1}t\right) \, , \text{ while } V(t) = \frac{\nu_2}{\mu_2}\gamma\left(\frac{\mu_2^2}{\nu_2}t\right) \, .$$

We suppose, as in Madan and Chang (1998), that the two processes for the prevailing price buy and sell orders share the same coefficient of variation $\kappa = \frac{\mu_1^2}{\nu_1} = \frac{\mu_2^2}{\nu_2}$. Under this assumption the price process of Eq. (27) may now be written, letting $\alpha_1 = (\delta + a)^{-1}\nu_1/\mu_1$ and $\alpha_2 = (\eta + b)^{-1}\nu_2/\mu_2$, as

$$\ln(p(t)/p(0)) = \gamma_1(t) - \gamma_2(t) = \alpha_1\gamma(\kappa t) - \alpha_2\gamma(\kappa t) \, . \tag{48}$$

The characteristic function of $\ln(p(t)/p(0))$ is then easily evaluated, using (46) as

$$\phi_{\ln(p(t)/p(0))}(u) = \left(\frac{1}{1 - i\alpha_1 u}\right)^{\kappa t}\left(\frac{1}{1 + i\alpha_2 u}\right)^{\kappa t}$$

$$= \left(\frac{1}{1 - i(\alpha_1 - \alpha_2)u + \alpha_1\alpha_2 u^2}\right)^{\kappa t} \, . \tag{49}$$

To represent this process as a time-changed Brownian motion, consider a Brownian motion with drift θ and volatility σ evaluated at the gamma time, $\gamma_3(t) = \gamma(\kappa t)$. Hence, let

$$Y(t) = \theta\gamma_3(t) + \sigma W(\gamma_3(t))$$

where $W(t)$ is a standard Brownian motion. The characteristic function of Y conditional on the gamma time is

$$\phi_{Y(t)|\gamma_3(t)}(u) = \exp\left(iu\theta\gamma_3(t) - \frac{\sigma^2 u^2}{2}\gamma_3(t)\right) \tag{50}$$

and the characteristic function of Y is

$$\phi_Y(u) = \left(\frac{1}{1 - iu\theta + \frac{\sigma^2 u^2}{2}}\right)^{\kappa t} \, . \tag{51}$$

A comparison of (51) and (49) shows that with $\theta = \alpha_1 - \alpha_2$ and $\sigma = \sqrt{2\alpha_1\alpha_2}$, the price process may be expressed in probability law as a Brownian motion with drift evaluated at the gamma time, $\gamma(\kappa t)$.

The time change $\gamma_3(t)$ is easily related to the individual gamma processes that were differenced and we observe that

$$\gamma_3(t) = (1/\alpha_i)\gamma_i(t) \, , \quad \text{for } i = 1, 2 \, . \tag{52}$$

which is a simple change of scale of the original order arrival rate processes. Later we inquire (in subsection 4.5) as to when we may expect the time change to be related to the original order arrival time by a simple change of scale and speed.

A different view of the relationship between the prevailing price buy and sell order arrival processes and the time change may be obtained by focusing attention on the prices at time $t = (2\kappa)^{-1}$, in which case

$$\ln\left(\frac{p\left(\frac{1}{2\kappa}\right)}{p(0)}\right) = \alpha_1\gamma(1/2) - \alpha_2\gamma(1/2)\,. \tag{53}$$

It is easily verified that the probability law of $2\gamma(1/2)$ is that of the square of a standard normal variate and hence that

$$2\ln\left(\frac{p\left(\frac{1}{2\kappa}\right)}{p(0)}\right) = \alpha_1 B - \alpha_2 S \tag{54}$$

where $B = N_d^2, S = N_s^2$, and N_d, N_s are two independent normal random variables of zero mean and unit variance. It is shown in the appendix that the time change at this same point is

$$\gamma_3(1/2) = \left(\sqrt{B} - \sqrt{S}\right)^2\,. \tag{55}$$

We thus observe that the time change is related to the level of excess demand. This is a reasonable property in that buy orders matched by supply orders do not result in price pressures.

4.3. *Demand and supply shocks as monotone Lévy processes*

An important structural property of Lévy densities is that of monotonicity. One expects that jumps of larger sizes have lower arrival rates than jumps of smaller sizes. This property amounts to asserting for differentiable densities that the derivative is negative for positive jump sizes and positive for negative jump sizes. Modeling the negative jumps symmetrically with the positive ones, we restrict the discussion to the Lévy density for the positive jumps.

The property of monotonicity may be strengthened to complete monotonicity by requiring derivatives of the same order to have the same sign. The mathematical definition, however, further requires the signs to be alternating. A completely monotone Lévy density is decreasing and convex, its derivative is increasing and concave and so on. Structural restrictions of this sort are useful in modeling discontinuous phenomena, given the wide class of choices that are otherwise available to model the Lévy density, which basically is any positive function that integrates the minimum of x^2 and 1. Complete monotonicity also has the interesting property of linking analytically the arrival rates of large jumps to that of small ones by requiring the latter to be larger than the former. The presence of such links makes it possible to learn about larger jumps from observing smaller ones.

In this regard, we note that the jump diffusion model based on the reflected normal distribution for the jump sizes is not completely monotone as is easily observed by noting that the normal density shifts from being a concave function near zero to a convex function near infinity. On the other hand, the exponentially distributed jump size is the foundation for all completely monotone Lévy densities. By Bernstein's theorem all completely monotone Lévy densities are given by the Laplace transforms of positive measures on the positive half line, or that there exists a measure $\rho(da)$ such that

$$k(y) = \int_0^\infty e^{-ay} \rho(da) \,. \tag{56}$$

Such Lévy densities have the useful interpretation that the economy is populated by individuals who submit prevailing price buy or sell orders with an exponential distribution with mean order rate $(1/a)$: the measure $\rho(da)$ is a measure of the number of orders per unit time of this mean level, with exponential size distribution. All completely monotone Lévy densities come from such an economy.

We consider now prevailing price buy orders given by a strictly increasing pure jump random process with completely monotone Lévy density $k(x)$, satisfying (56). The gamma process of the Sec. 4.2 is the special case when $k(x) = e^{-x}/x$ as defined in Eq. (47), and this is a completely monotone density. As in the gamma process case, we suppose the prevailing price sell orders $V(t)$ are given by an independent copy of the same Lévy process. The log price process of Eq. (27) now has the form

$$\ln(p(t)/p(0)) = \alpha_1 U(t) - \alpha_2 V(t) \tag{57}$$

with $\alpha_1 = (\delta + a)^{-1}, \alpha_2 = (\eta + b)^{-1}$. It is shown in the appendix, following the analysis of Leblanc (1997), Knight (1981), Kotani and Watanabe (1982), for the special case of $\alpha_1 = \alpha_2 = \alpha$, that the Lévy measure of the time change is $\tilde{k}(y) = k_3(\alpha y)$, where k_3 is given by

$$k_3(y) = \int_0^\infty \rho(da) 2a e^{-a^2 y} \,. \tag{58}$$

The time change Lévy density given explicitly by Eq. (58) generalizes to the completely monotone class of densities the result we demonstrated earlier for the exponential Lévy measure. We observe interestingly that the time change aggregates the counting of the arrival of orders with the same weighting function to provide the time change. An order by an individual with a mean order arrival rate of $(1/a)$ of size y amounts to $a \exp(-a(a-1)y)$ ticks of the stochastic clock.

The literature relating price changes to economic activity (Tauchen and Pitts (1983), Karpoff (1987), Gallant, Rossi and Tauchen (1992), and Jones, Kaul and Lipson (1994)) has focused on volume and the number of trades as the relevant measure of economic activity. The analysis of this section indicates that the time measure incorporates both the number and size of orders. For $a < 1$, the time or activity measure is positively and exponentially related to the size of orders.

We now consider two important subclasses of the class of completely monotone Lévy densities, and these are the class of generalized gamma convolution densities, and the stable processes with index $\alpha < 1$, and show in each case that the time change lies in the same subclass as well.

4.4. *Demand and supply shocks as generalized gamma convolutions*

An important subclass of completely monotone Lévy measures is the class of generalized gamma convolutions (Bondesson (1992)). A Lévy measure $k(x)$ is in the generalized gamma convolution class if the size weighted Lévy density $xk(x)$, is completely monotone. When we weight the arrival rate or probability by size, we get the expected impact on the process and this hypothesis requires that large jumps have a sufficiently small arrival rate so that their impact is below the impact of smaller jump sizes. Under this hypothesis, small moves dominate large moves in impact and in this section we ask that if the arrival of orders has this property of completely monotone impact, then is this property inherited by the time change. Do large clock ticks also have a lower expected impact on the passage of time than the smaller clock ticks? We show here that indeed this is the case.

For Lévy densities in the class of generalized gamma convolutions, the process is a mixture of gamma processes with the characteristic function for the cumulated prevailing price buy orders $U(t)$ now having the explicit form

$$\phi_{U(t)}(u) = \exp\left(iatu + t\int \log\left(\frac{c}{c-iu}\right) K(dc)\right) \tag{59}$$

where $K(dc)$ is a non-negative measure on the positive half line.[n]

When K is the Dirac delta measure at $c = 1$, $U(t)$ is the standard gamma process $\gamma(t)$ discussed in Sec. 4.2. The generalized gamma convolution class is very wide and includes as special cases the stable processes and the inverse Gaussian processes. It is an infinite parameter class, effectively being parameterized by the measures $K(dc)$. Clearly, the gamma process is the fundamental building block for processes in this class.

Suppose now that $U(t)$ and $V(t)$ are two independent copies of a generalized gamma convolution class of processes with drift a and mixing measure K. Consider the general symmetric price process of Eq. (57), and evaluate the characteristic function of the log price relative as

---

[n]It is required that $K(dc)$ satisfy the integrability conditions:

$$\int_0^1 |\log(c)|\, K(dc) < \infty \quad \text{and} \quad \int_1^\infty \frac{1}{c} K(dc) < \infty.$$

$$\phi_{\ln(p(t)/p(0))}(u) = \exp\left(t\int\left(\begin{array}{c}\log\left(\dfrac{c}{c-iu\alpha}\right)-\\[2ex]\log\left(\dfrac{c}{c+iu\alpha}\right)\end{array}\right)K(dc)\right)$$

$$= \exp\left(t\int\log\left(\frac{c^2}{c^2+u^2}\right)K(dc)\right). \tag{60}$$

We wish to determine whether this process can be written as Brownian motion evaluated at a generalized gamma convolution. Suppose for this purpose that $T(t)$ is an increasing process that is a generalized gamma convolution time change with zero drift and mixing measure $\tilde{K}$. The characteristic function for $T(t)$ has the form

$$\phi_{T(t)}(u) = \exp\left(t\int\log\left(\frac{c}{c-iu}\right)\tilde{K}(dc)\right). \tag{61}$$

Consider now the characteristic function of Brownian motion evaluated at this generalized gamma convolution time, $W(T(t))$. This is simply obtained by evaluating (61) at $-u^2/(2i)$ and is

$$\phi_{W(T(t))}(u) = \exp\left(t\int\log\left(\frac{\tilde{c}}{\tilde{c}+u^2/2}\right)\tilde{K}(d\tilde{c})\right). \tag{62}$$

The two characteristic functions (60) and (62) are equated on making the change of variable $2\tilde{c} = c^2$ in (62) and defining

$$\tilde{K}(d\tilde{c}) = \frac{1}{c}K(dc). \tag{63}$$

Specifically we have, in the case $K(dc) = g(c)dc$, that the characteristic function for the time change is

$$\phi_{T(t)}(u) = \exp\left(t\int\log\left(\frac{c}{c-iu}\right)\frac{g(\sqrt{2c})}{\sqrt{2c}}dc\right). \tag{64}$$

Equation (64) provides the exact representation of the time change as a generalized gamma convolution.[o]

The weighting that was given to the gamma process with mean $(1/\sqrt{2c})$ in the prevailing price order arrival process is given in the time change to the gamma

---

[o]In this case, the function g must meet the following four integrability conditions:

$$\int_0^1 |\log(c)|\, g(c)dc < \infty; \quad \int_0^1 |\log(c)|\, \frac{g(\sqrt{2c})}{\sqrt{2c}}dc < \infty$$

$$\int_1^\infty \frac{1}{c}g(c)dc < \infty; \quad \int_1^\infty \frac{1}{c}\frac{g(\sqrt{2c})}{\sqrt{2c}}dc < \infty.$$

process with mean $(1/c)$ after multiplication by $(1/\sqrt{2c})$. We now present a useful generalization of the variance gamma model within the generalized gamma convolution family of processes.

4.4.1. *Generalizing the variance gamma model to control arrival rates of different sizes and signs*

A potentially useful generalization of the variance gamma model studied in Madan and Chang (1998), within the class of generalized gamma convolutions is given by the Lévy measure, $k_{CGYM}(x)$,[P] for an increasing process (the log price process being obtained by differencing two such processes with possibly different parameters) where

$$k_{CGYM}(x) = \frac{\exp(-Mx)}{Cx^{Y+1}(1+x)^G}. \tag{65}$$

The special case $G = 0, Y = 0$, gives the symmetric variance gamma process when we difference two identical but independent copies of this process. When we take just $G = 0$, we get the Lévy measure studied by Vershik and Yor (1995), for the purpose of combining the gamma and stable laws. We note that the setting $Y = 1$ constitutes the boundary between processes of finite and infinite variation. Further, from the perspective of financial modeling, this Lévy measure, unlike the variance gamma model, allows one to control for the behavior near zero and at infinity separately via the parameters Y and G respectively, and this too separately, for the positive and negative moves, by differentiating these parameters on the two sides. We have here a fairly robust eight parameter stochastic process. It is amenable to statistical work as the characteristic function is available in closed form in terms of the special functions of mathematics. It is shown in the appendix that for the increasing process $X_{CGYM}(t)$,

$$E[\exp(iuX_{CGYM}(t)] = \exp \left(\begin{array}{l} t\dfrac{\Gamma(1-Y)}{CY} [M^{Y+G}U(G, Y+G+1, M) \\ -(M-iu)^{Y+G}U(G, Y+G+1, M-iu)] \end{array} \right) \tag{66}$$

where $U(a, b, z) = z_2^{-a}F_0(a, 1+a-b, -1/z)$, and $_2F_0$ is one of the confluent hypergeometric functions.

4.5. *Stable processes as time-changed Brownian motion*

The class of increasing stable processes of index $\alpha < 1$ could also be considered as a candidate model for the prevailing price buy and sell order processes. The difference of two independent stable processes of index $\alpha < 1$ is also a time-changed

---

[P]We name this Lévy measure after Peter Carr and ourselves, in recognition of Peter Carr's request to develop an extension of the variance gamma model in this direction.

Brownian motion and we enquire into the nature of the time change. For an increasing stable process of index α the Lévy measure is

$$\nu(dx) = \frac{1}{x^{\alpha+1}}dx \text{ for } x > 0, \tag{67}$$

and the difference, $X(t)$, of two independent copies of such a process is the symmetric stable process of index α with characteristic function

$$E[\exp(iuX(t))] = \exp(-tc|u|^\alpha), \tag{68}$$

for a positive constant c. We derive from the characteristic functions that the time change is not an increasing stable α-process.

The characteristic function of an independent Brownian motion evaluated at an independent increasing stable process of index $\alpha, T(t)$ is given by

$$E[\exp(iuW(T(t)))] = E[\exp(-u^2 T(t)/2)]$$

$$= \exp(-t(c/2)|u|^{2\alpha}),$$

or a symmetric stable process of index 2α.

It follows from this observation that the difference of two increasing stable α processes for $\alpha < 1$, is Brownian motion evaluated at an increasing stable $\alpha/2$ process.

4.6. *Scale and speed adjusted time changes*

We now consider, within the class of completely monotone Lévy densities when the time change Lévy process is simply related to the original order arrival process, in that it may involve speeding up or slowing down accompanied by a scaling of the original process. Specifically, we ask when $\gamma_3(t)$ has the form $a\gamma_1(bt)$ for scaling and speed adjustment coefficients a and b. For example, when we relate the time change to the volume of transactions, then we are in the case $b = 1$ and $a = 2$. For the scale and speed adjustment to be valid, one must have

$$E[\exp(iu\gamma_3(t))] = E[\exp(iua\gamma_1(bt))]$$

$$= \exp\left(bt \int_0^\infty (e^{iuax} - 1)k(x)dx\right) \tag{69}$$

or that

$$\exp\left(t \int_0^\infty (e^{iuy} - 1)k_3(y)dy\right) = \exp\left(t \int_0^\infty (e^{iuy} - 1)k\left(\frac{y}{a}\right)\frac{b}{a}dy\right). \tag{70}$$

It follows that we must have

$$k_3(y) = \frac{b}{a}k\left(\frac{y}{a}\right). \tag{71}$$

Suppose that $k(y) = \int f(a)e^{-ay}da$ and that $k_3(y) = \int f_3(a)e^{-ay}da$ then (71) implies that

$$\int f_3(u)e^{-uy}du = \frac{b}{a}\int f(u)e^{-uy/a}du$$

$$= \int bf(av)e^{-vy}dv.$$

It follows that one must have the equality

$$f_3(u) = bf(au) \tag{72}$$

among the transforms.

But by Eq. (58) on changing variables and writing $\rho(da) = f(a)da$ we note that we must have $f_3(x) = f(\sqrt{x})$. It follows that for a simple scale and speed adjustment in the time change we must have

$$f(\sqrt{x}) = bf(ax). \tag{73}$$

We know the gamma process has this property, and for the gamma process we have

$$\frac{\exp(-x)}{x} = \int_1^\infty e^{-xu}du.$$

Hence for this case $f(u) = 1_{u\geq 1}$. Equation (73) may be expressed as $f(x) = bf(ax^2)$ and for $f(x) = 1_{x\geq 1}$, this condition is satisfied for $a = b = 1$.

More generally, if we define $\tilde{f}(x) = f(cx)$, then we must have

$$\tilde{f}(x) = f(cx) = bf(ac^2x^2) = b\tilde{f}(x^2),$$

if we choose $a = 1/c$. Hence all possible solutions require $\tilde{f}(x)$ to be proportional to $\tilde{f}(x^2)$. A wide class of solutions is given by

$$\tilde{f}(x) = c_1|\ln(x)|^\alpha 1_{[0,1]}(x) + c_2[\ln(x)]^\alpha 1_{[1,\infty)}(x). \tag{74}$$

We note that the wider class of processes for which a scale and speed adjustment is valid contains the gamma Lévy measure and somewhat more generally, the solutions of (74) associated with α, $0 < \alpha < 1$ for $c_1 = 0, c_2 = 1$.[q] An interesting special case occurs when we consider the case $c_1 = 0, c_2 = 1$ and $\alpha = 1$ in (74).

---

[q]We must have $c_1 = 0$ for even if $\alpha = 0$, and $c_2 = 0$ we have

$$k(y) = \int_0^1 e^{-\xi y}d\xi = \frac{1 - e^{-y}}{y}$$

but then $\int_0^\infty (1 \wedge y)k(y)dy = \infty$.

This yields[r]

$$k(y) = \int_1^\infty \ln(\xi)e^{-\xi y}d\xi \tag{75}$$

$$= \frac{1}{y}\int_y^\infty \frac{e^{-u}}{u}du\,. \tag{76}$$

Equation (76) represents the tail of the Lévy measure of the gamma process divided by the lower limit of integration. This is an interesting operation on Lévy measures. In fact if $\rho(dx)$ is a Lévy measure, then defining

$$\tilde{\rho}(dx) = \frac{dx}{x}\int_x^\infty \rho(dy) \tag{77}$$

we also obtain a Lévy measure since $\int_0^\infty (1 \wedge x)\tilde{\rho}(dx) < \infty$ holds.[s] For many Lévy measures, and certainly for the gamma process, Eq. (77) gives new Lévy measures. We also observe that applying (77) twice amounts to taking $c_1 = 0, c_2 = 1$, and $\alpha = 2$ in (74).[t]

[r]Substituting $\xi = x/y$ we have

$$k(y) = \frac{1}{y}\int_y^\infty (\ln(x) - \ln(y))e^{-x}dx$$

$$= \frac{1}{y}\int_y^\infty \int_y^x \frac{du}{u}e^{-x}dx$$

$$= \frac{1}{y}\int_y^\infty \int_u^\infty \frac{e^{-x}}{u}dxdu$$

$$= \frac{1}{y}\int_y^\infty \frac{e^{-u}}{u}du$$

[s]To verify this, we observe

$$\int_0^\infty (1 \wedge x)\tilde{\rho}(dx) = \int_0^1 dx \int_x^\infty \rho(dy) + \int_1^\infty \frac{1}{x}\int_x^\infty \rho(dy)dx$$

$$= \int_0^\infty (1 \wedge u)\rho(du) + \int_1^\infty \int_1^u \frac{dx}{x}\rho(du)$$

$$= \int_0^\infty (1 \wedge u)\rho(du) + \int_1^\infty \ln(u)\rho(du)\,.$$

The first integral on the right hand side is finite as ρ is a Lévy measure, and the second is finite provided ρ is sufficiently damped at infinity.

[t]To see this, consider

$$k(y) = \int_1^\infty \ln(\xi)^2 e^{-\xi y}d\xi\,.$$

4.7. *Brownian excursions and equilibrium*

We now consider another and equivalent representation of the asset price process that is valid for price processes with completely monotone Lévy densities, a consequence of Krein's theory (Kotani and Watanabe (1982)). In this equivalent formulation, excess demand is modeled as a Brownian motion and price changes are related to the excess demand by a possibly non-linear response function that we term the *force function*. We show by examples that simple force functions are associated with fairly complex Lévy densities, while the force function for the gamma process is at this writing, to our knowledge, still unknown. Hence the two approaches complement each other and provide a wider class of interesting models than would be possible if we restricted attention to just one of these two equivalent formulations.

Unlike the examples of the earlier sections, the time change and the Brownian motion are no longer independent in the examples of this section. Here the time change represents cumulated volatility, where the latter depends on the Brownian motion itself.

We view the Brownian motion $W(t)$ as a measure of market departure from equilibrium, and view positive values of $W(t)$ as an excess demand in the market driving prices up, while negative values of $W(t)$ are indicative of excess supply driving prices downward. The extent of the price response to the disequilibrium is given by the force function $f(x)$. The price process is

$$S(u) = \int_0^u f(W(s))ds\,. \tag{78}$$

The process $S(u)$ is a continuous process that increases when Brownian motion is positive at rate $f(W(s))ds$ and decreases when $W(s)$ is negative. Hence if a trader takes a long position during a positive excursion of Brownian motion and

We may again set $\xi = x/y$ and write that

$$k(y) = \frac{1}{y} \int_y^\infty (\ln(x) - \ln(y))^2 e^{-x} dx$$

$$= \frac{1}{y} \int_y^\infty \int_y^x 2\frac{du}{u} \int_y^v \frac{dv}{v} e^{-x} dx$$

$$= \frac{1}{y} \int_y^\infty 2\frac{dv}{v} \int_v^\infty \frac{du}{u} \int_u^\infty e^{-x} dx$$

$$= \frac{2}{y} \int_y^\infty \frac{dv}{v} \int_v^\infty \frac{e^{-u}}{u} du$$

which is the application of (77) twice to the gamma Lévy measure.

successfully reverses his position before a return to zero, there is a pure arbitrage profit to be made. The same holds true for short positions in negative excursions of Brownian motion.

We avoid arbitrage opportunities by restricting trading to occur in equilibrium, i.e., on the zero set of Brownian motion. The period of the Brownian excursion is then akin to a tatonnement period during which the discovery of the equilibrium price takes place. The time domain of Eq. (78) includes disequilibrium situations when $W(s) \neq 0$ and the market is in the phase of equilibration. We now restrict the price process to equilibrium states when Brownian motion is zero. To perform this restriction to the equilibrium domain, we define $\sigma(t)$ to be the inverse local time of Brownian motion at zero. Our price process is the bounded variation process (see the integrability condition (80) below)

$$\ln(p(t)/p(0)) = \int_0^{\sigma(t)} ds\, f(W(s)) \,. \tag{79}$$

This is a pure jump process as it is $S(\sigma(t))$ and it inherits the jump properties from the inverse local time.

Inverse local time is a Lévy process with Lévy measure $L(x)$,

$$L(x) = \frac{1}{\sqrt{2\pi} x^{3/2}}, \ x > 0 \,.$$

Hence, this is a high activity jump process that jumps at a zero of Brownian motion to another zero of Brownian motion, with the size of the jump being the length of the Brownian excursion. Each jump time of inverse local time is an equilibrium trading time and there are infinitely many trading opportunities in any interval of time.

The price process (79) includes Lévy processes of infinite variation but finite quadratic variation. To obtain processes of bounded variation one must restrict the class of force functions to those that meet the integrability condition

$$\int_{-K}^{K} dx\, \mid f(x) \mid\, < \infty \tag{80}$$

for all K. Under this integrability condition (that essentially prevents an infinite force at zero), the two components of positive and negative moves may be constructed as separate processes. One may write the price process as

$$\ln(p(t)/p(0)) = \int_0^{\sigma(t)} ds\, f^+(W(s)) - \int_0^{\sigma(t)} ds\, f^-(W(s)) \tag{81}$$

where $f^+(x) = f(x)\mathbf{1}_{(x \geq 0)}$; $f^-(x) = f(x)\mathbf{1}_{(x \leq 0)}$.[u] We now model the term $(\delta + a)^{-1}U(t)$ of Eq. (27) by $\int_0^{\sigma(t)} ds f^+(W(s))$ while $(\eta + b)^{-1}V(t)$ is given by $\int_0^{\sigma(t)} ds f^- (W(s))$.

It is interesting to note that during an excursion of the Brownian motion away from 0, the market is equilibrating and the clock $\sigma(t)$ stops and does not count time. The only times we count and the only firm trading prices we permit are those when we have equilibrium, and $W(\sigma(t)) = 0$. We measure time by time spent in equilibrium and we note that $\sigma(t)$ is strictly increasing in t.

Some interesting facts about the Lévy measure for the price process may be inferred directly from the structure of the force function. For example, one may show that if the force function is zero in an interval around zero, then only excursions of a certain significant departure from zero contribute to a price movement, and one may infer that the arrival rate of the process is finite.[v] Similarly one may show that if the force function is strictly positive in absolute value in an open interval around zero then the arrival rate of the Lévy measure is infinite. Hence, we can conclude that economies that are dormant around the equilibrium point in that they have no force, have finite arrival rate Lévy measures. Furthermore, if the force function of one economy dominates the force function of another, then the tail of the Lévy measure of the economy with the dominating force function is heavier. Greater force generally means greater activity in terms of the arrival rate of price changes.

The price process of Eq. (79) may once again be expressed as time-changed Brownian motion. Define $\psi(y)$ by $\psi'(y) = f(y)$ and $\psi(0) = 0$, let

$$F(x) = \int_0^x \psi(y)dy \, . \tag{82}$$

Ito's lemma may now be applied to $F(W(t))$ in the following way:

$$0 = F(W(\sigma(t))) = \int_0^{\sigma(t)} \psi(W(s))dW(s) + \frac{1}{2}\int_0^{\sigma(t)} f(W(s))ds \, . \tag{83}$$

It follows from Eq. (79) substituted in (83) that

$$\ln(p(t)/p(0)) = -2\int_0^{\sigma(t)} \psi(W(s))dW(s) \, . \tag{84}$$

Hence the probability law of $\ln(p(t)/p(0))$ is that of Brownian motion evaluated at $4\int_0^{\sigma(t)} \psi^2(W(s))ds$. However, here the Brownian motion and the time change are

---

[u]It is interesting to note that when the integrability condition is not met and there is an infinite activity near the origin that only sums up when we square the moves, one may approximate the process by considering the difference of increasing processes that truncate the force function at values of x below ε in absolute value and obtain the infinite variation process as the limit as we let ε tend to zero, provided as a sufficient condition that $\int_0 x^\alpha f(x)dx < \infty$ for some value $\alpha < 1/2$. For the cases we consider this condition is met. In this sense our decomposition into the difference of two increasing processes is quite fundamental.

[v]This property can be formally proved by integrating the force functions with respect to the Ito measure for the height m of the excursion (see Yor (1995) page 68) which is $(1/m^2)dm$.

not independent processes. The time change is in fact constructed from the original Brownian motion as:[w]

$$T(t) = 4 \int_0^{\sigma(t)} \psi(W(s))^2 ds \,.$$

We note that the time change cumulates the instantaneous volatility of the price process, and observe that this volatility depends on the Brownian motion.

An interesting special case of such equilibrium processes derived from Brownian excursions is obtained by considering the case of

$$f(x) = a(x^+)^m + b(x^-)^m \,. \tag{85}$$

For $a > 0$ and $b = -a$, the log price process (79) is a symmetric stable process of index $\alpha = (m+1/2)^{-1}$. This process can be written as Brownian motion evaluated at a dependent time constructed from the price process using (84) as the quadratic variation of the price process. The Brownian motion may also be constructed out of the price process, as the price process evaluated at the inverse of the quadratic variation. We have here an alternative representation of stable α, $\alpha < 1$, processes as time-changed Brownian motions (as opposed to the representation of Sec. 4.5), that do not invoke independence between the Brownian and the time change.

It is instructive to consider the relationship between the Lévy measures of the stable processes and the force functions of the representations (79). We observe that when m is large, the force is high at large values of x and the Lévy measure is accordingly lower at these high values of x. Similarly, for low values of x, the force is higher for low values of m, and once again these low values of m have the lower Lévy measure. It therefore appears that tails for the Lévy measure that are less heavy than the stable laws would require a force function that dominates the polynomials.

4.8. *Characteristic functions for price processes based on Brownian excursions*

For the econometric evaluation and identification of price processes from data on financial prices, the characteristic function of the log price relative is a very useful and fundamental construct. For all the cases considered in Secs. 4.1 through 4.6, we have explicit characteristic functions for the log price relative. In this section, we present an algorithm to evaluate the characteristic function for the Brownian excursion model presented in Sec. 4.7, based on the results of Revuz and Yor (1994).

---

[w]It is possible for a continuous martingale to be written as a Brownian motion evaluated at an independent time change given by its quadratic characteristic. However, as shown by Ocone (1993), Dubins, Émery, Yor (1993), a necessary and sufficient condition for this to be possible is that the probability law of stochastic integrals with respect to the martingale coincide for all integrands that have absolute value unity. For the force functions f, and the associated function ψ, this is unlikely to be the case.

We are interested in the characteristic function

$$\phi_{\ln(p(t)/p(0))}(u) = E\left[\exp\left(iu\int_0^{\sigma(t)} f(W(s))ds\right)\right], \tag{86}$$

where $\sigma(t)$ is the inverse local time of the Brownian motion $W(t)$. Writing the function $f = f^+ - f^-$, noting that f is positive for positive arguments and negative for negative arguments, we may rewrite (86) as

$$\phi_{\ln(p(t)/p(0))}(u) = E\left[\exp\left(iu\int_0^{\sigma(t)} f^+(W(s))ds - iu\int_0^{\sigma(t)} f^-(W(s))ds\right)\right]. \tag{87}$$

By the Ray–Knight theorem, the two components are independent and the result follows on multiplication of the two expectations. Hence,

$$\phi_{\ln(p(t)/p(0))}(u) = E\left[\exp\left(iu\int_0^{\sigma(t)} f^+(W(s))ds\right)\right]$$
$$\times E\left[\exp\left(-iu\int_0^{\sigma(t)} f^-(W(s))ds\right)\right]. \tag{88}$$

The result follows on evaluating each of the expectations. It is shown in Revuz and Yor (1994), that the Laplace transform is given by

$$E\left[\exp\left(-\lambda\int_0^{\sigma(t)} f^+(W(s))ds\right)\right] = \exp(t\psi'(0^+)/2) \tag{89}$$

where $\psi(x)$ is the unique positive, decreasing solution to the Sturm–Liouville equation

$$\frac{1}{2}\psi''(x) = \lambda f^+(x)\psi(x)$$

subject to the boundary conditions $\psi(0) = 1$, $\psi(\infty) = 0$, when the function f meets the condition $\int_0^\infty xf^+(x)dx = \infty$. Here we present an analysis based on the Ray–Knight theorem that uses methods more commonly employed in the economics literature, and initiated by Cox, Ingersoll and Ross (1985).

By the Ray–Knight theorem one may write the process for the price increases as

$$\int_0^{\sigma(t)} f^+(W(s))ds = \int_0^\infty dx f^+(x)L^x_{\sigma(t)}(W), \tag{90}$$

where $L^x_{\sigma(t)}(W) = Z(x)$ is the local time of the Brownian motion W at x between 0 and $\sigma(t)$, the inverse local time at zero of W. The process $Z(x)$ viewed as a process in the space variable, for fixed t, is a Feller diffusion and is in fact a squared Bessel process of dimension 0, starting at t, or in finance terms a CIR process (see Geman and Yor 1993), that satisfies the stochastic differential equation

$$dZ(x) = 2\sqrt{Z(x)}dB(x)\,; \quad Z(0) = t \tag{91}$$

for a standard Brownian motion $B(x)$ in the space variable. We are therefore equivalently interested in the Laplace transform of $\int_0^\infty dx f^+(x)Z(x)$. We proceed by considering this problem in the familiar way of analyzing term structure models and define

$$G(y, Z) = E\left[\exp\left(-\lambda \int_y^\infty f^+(x)Z(x)dx\right) \mid Z(y) = Z\right]. \tag{92}$$

The partial differential equation for G may be derived as

$$G_y + 2ZG_{ZZ} - \lambda f(y)Z = 0\,, \tag{93}$$

which must be solved for the boundary conditions $G(y, 0) = 1$, and $G(\infty, Z) = 0$.[x] It is well known (see e.g., Revuz and Yor (1994) page 424 Theorem 4.7) that the solution in Z is of the form

$$G(y, Z) = \exp(b(y)Z) \tag{94}$$

where the function b satisfies the Ricatti differential equation (also classically obtained in finance for CIR type models of interest rates or stochastic volatility)

$$b' + 2b^2 = \lambda f^+\,.$$

The Sturm–Liouville equation follows on making the substitution $b = \frac{1}{2}\frac{\psi'}{\psi}$, and the result follows on evaluating $G(0, Z(0)) = G(0, t)$. We now consider some explicit examples where the characteristic function of the log price relative may be explicitly evaluated, beyond the polynomial case for the force function f for which we have already observed that we obtain a stable law for the price process.

We consider two further cases, one with a decreasing force function, and the other with an exponentially increasing force function. The second example has a force function that dominates the polynomials yielding the stable laws that we have already reported on.

4.8.1. *Example with diminishing force function*

Suppose that

$$f^+(x) = \frac{1}{(kx + l)^2}\,.$$

---

[x]The condition at infinity is related to the requirement that $f^+(x)x$ integrates to infinity. When this is finite, the condition at infinity is different, but we shall only be concerned with cases where this condition is met.

Let Z_x be the Feller diffusion associated with the local times at x evaluated at the inverse local time at t. As noted earlier, $dZ_x = 2\sqrt{Z_x}dB(x)$ for a standard Brownian motion B, and $Z_0 = t$. Define $\varphi(x) = (kx + l)^{-1}$ and consider $Y_x = \varphi(x)Z_x$. By construction $Y_0 = t/l$, and an application of Ito's lemma shows that

$$dY_x = -\frac{k}{(kx + l)^2}Z_x dx + 2\varphi(x)\sqrt{Z_x}dB(x)\,.$$

Writing the martingale $\int_0^x 2\varphi(y)\sqrt{Z_y}dB(y)$ in its Dubins-Schwarz form: $w(A_x)$, where $(w(u), u \geq 0)$ is a Brownian motion, and $A_x = 4\int_0^x dy\varphi^2(y)Z_y$, we may represent $\{Y_x\}$ as: $\gamma(A_x)$, where

$$\gamma(u) = \frac{t}{l} - \frac{k}{4}u + w(u)\,.$$

It follows that $\gamma(4\int_0^\infty dy(ky+l)^{-2}Z_y)$ is Y_∞, which by construction is zero. Hence,

$$4\int_0^{\sigma(t)} f^+(W(s))ds = 4\int_0^\infty dx(kx + l)^{-2}Z_x = T_0(\gamma)$$

the time at which the Brownian motion γ with drift $-k/4$ and initial value t/l reaches zero. This is the same as the first passage time distribution of Brownian motion starting at zero and with drift $k/4$ reaching t/l. The density of $T_0(\gamma)$ is given by

$$P(T_0(\gamma)\epsilon ds) = \frac{t/l}{\sqrt{2\pi s^3}} \exp\left(-\frac{(t/l)^2}{2s}\right) \exp\left(\frac{tk}{4l} - \frac{k^2 s}{32}\right) ds\,.$$

The Laplace transform is given by

$$E[\exp(-\lambda T_0(\gamma))] = \exp\left(-\frac{t}{l}\left(\sqrt{\frac{k^2}{16} + 2\lambda} - \frac{k}{4}\right)\right)$$

and the Lévy measure is

$$k(x) = \frac{1}{l}\frac{\exp\left(-\frac{k^2}{32}x\right)}{\sqrt{2\pi x^3}}\,.$$

We observe in this example that the force function and the Lévy measure are both inversely related to the parameters k and l, and hence are positively related to each other.

4.8.2. *Example with exponentially increasing force function*

Suppose that

$$f^+(x) = \theta \exp(\alpha x)\,, \quad \alpha, \theta > 0\,. \tag{95}$$

For this case, following Jeanblanc, Pitman and Yor (1997), Example 6, we observe that the solution to the Sturm–Liouville equation with the boundary condition $\psi(0) = 1$ and $\psi(\infty) = 0$, is

$$\psi(x) = \frac{K_0\left(\dfrac{2\sqrt{2\lambda\theta}}{\alpha}\exp\left(\dfrac{\alpha}{2}x\right)\right)}{K_0\left(\dfrac{2\sqrt{2\lambda\theta}}{\alpha}\right)}$$

where K_0 is the modified Bessel function of second kind of order zero. It follows on differentiation, noting that $K_1 = -K_0'$ and substituting into (89) that

$$E\left[\exp\left(-\lambda\int_0^{\sigma(t)} f^+(W(s))ds\right)\right] = \exp\left(-\frac{t}{2}\frac{\sqrt{2\lambda\theta}K_1\left(\dfrac{2\sqrt{2\lambda\theta}}{\alpha}\right)}{K_0\left(\dfrac{2\sqrt{2\lambda\theta}}{\alpha}\right)}\right). \tag{96}$$

For the Lévy measure associated with this Laplace transform we note from Donati-Martin and Yor (1997) page 1055 that

$$\sqrt{\xi}\frac{K_1(\sqrt{\xi})}{K_0(\sqrt{\xi})} = \xi\int_0^\infty \exp(-\xi y)H_{-1}(y)dy = -\int_0^\infty(1-\exp(-\xi y))\frac{\partial}{\partial y}H_{-1}(y)dy.$$

Substituting $2\sqrt{2\lambda\theta}/\alpha$ for $\sqrt{\xi}$ and making the change of variable $8\theta y/\alpha^2 = z$ we obtain that the Lévy measure for this process is

$$k(x) = \frac{\alpha^3}{16\theta}\frac{1}{\pi^2}\int_0^\infty dz\exp\left(-z\frac{\alpha^2 x}{8\theta}\right)\frac{1}{J_0^2(\sqrt{z}) + Y_0^2(\sqrt{z})}$$

where we have differentiated H_{-1} using the definition provided in Donati-Martin and Yor (1997) Eq. (6.6).

5. Time Changes Related to Demand and Supply Shocks in the Limit

Recently, Barndorff-Nielsen (1997) has proposed the normal inverse Gaussian distribution as a possible model for the stock price process. This process may also be represented as a time-changed Brownian motion, where the time change $T(t)$ is the first passage time of another independent Brownian motion with drift to the level t. The time change is therefore an inverse Gaussian process and as one evaluates a Brownian motion at this time, this suggests the nomenclature normal inverse Gaussian. An interesting question from the perspective of this paper is whether such a process may be represented as the difference of two increasing processes that constitute the price responses to demand and supply shocks, respectively. However, this is not possible as the normal inverse Gaussian process is one of infinite variation.

We note that the inverse Gaussian process is a homogeneous Lévy process that is in fact a stable process of index $\alpha = 1/2$. We observed in Sec. 4.5, that if $2\alpha < 1$, then time changing Brownian motion with such a process leads to the symmetric stable process of index $\alpha < 1$. For $\alpha = 1/2$, we observe below that the process is of infinite variation.

In general, for $W(T(t))$ to be a process of bounded variation we must have that

$$\int (1 \wedge |x|)\tilde{\nu}(dx) < \infty, \tag{97}$$

where $\tilde{\nu}$ is the Lévy measure of the time-changed Brownian motion. We may relate $\tilde{\nu}$ to the Lévy measure ν, of the time change by

$$\tilde{\nu}(dx) = dx \int \nu(dy) \frac{\exp\left(-\dfrac{x^2}{2y}\right)}{\sqrt{2\pi y}}. \tag{98}$$

The time changed process is of bounded variation just if (see the Appendix for a proof),

$$\int_0 \nu(dy)\sqrt{y} < \infty. \tag{99}$$

For the inverse Gaussian time change, we see from (67) for $\alpha = 1/2$ that (99) is infinite. It follows that the normal inverse Gaussian process is a time-changed Brownian motion of infinite variation and therefore it cannot be expressed as the difference of two increasing processes. It may, however, be approximated by such processes by ignoring jumps of absolute size below ε and then letting ε tend to zero.

From the perspective of Brownian excursions, we offer the example of a force function that diverges to infinity at 0, and results in an infinite variation Lévy process. The specific force function is studied in Donati-Martin and Yor (1997) and is given by

$$f(x) = \text{sign}(x)\frac{1}{\exp(2\theta \mid x \mid) - 1}.$$

It is shown in Donati-Martin and Yor (1997), that the Laplace transform of the positive moves

$$E\left[\exp\left(-\lambda \int_0^{\sigma(t)} f^+(W(s))ds\right)\right] = \exp\left(-\frac{t\lambda}{2\theta}\left\{\begin{array}{c} -2\gamma - \Psi\left(1 + i\sqrt{\dfrac{\lambda}{2}}\right) \\ -\Psi\left(1 - i\sqrt{\dfrac{\lambda}{2}}\right) \end{array}\right\}\right) \tag{100}$$

where γ is Euler's constant and $\Psi(x) = \Gamma'(x)/\Gamma(x)$. It is shown in the Appendix that the characteristic function of the log price process is given by

$$\phi_{\ln(p(t)/p(0))}(u) = \exp\left(-\frac{t}{\theta}\sum_{n=1}^{\infty}\frac{4u^2 n}{u^2 + 4n}\right), \tag{101}$$

and the Lévy measure is

$$k(x) = \frac{\partial^2}{\partial x^2} \frac{\exp(-4x)}{1 - \exp(-4x)} \ . \tag{102}$$

6. Conclusion

We argue in this paper that price processes, representing market responses to underlying uncertainties given by the increasing random processes of cumulated demand and supply shocks are bounded variation semimartingales. Furthermore, being semimartingales (an implication of the no arbitrage condition) they are time-changed Brownian motions-as shown by Monroe (1978). The focus of attention in modeling the price process then shifts to modeling the time change. We show, by various examples, that one may generally relate this time change to a measure of economic activity and hence deduce that it is an increasing process with local uncertainty. This has the implication that the time change is a pure jump process and hence, so is the price process. Specifically, we conclude that the process for the price of a traded asset should not be modeled as possessing a continuous martingale or diffusion component. This result is contrary to the fundamental paradigm of modern pricing theory and the dominant practice of the past 25 years. So far our investigation has concentrated on homogeneous time changes that are independent of the price path, and we recognize that it may be necessary to develop methodologies that allow for time inhomogeneous and price path dependent stochastic time changes. We anticipate that future research will address these issues.

The specific time changes considered in this paper are compound Poisson processes, gamma processes, general Lévy processes with completely monotone Lévy densities, generalized gamma convolutions and the inverse local time of Brownian motion at zero. In each case we exhibit the price process as a finite variation process that is the difference of two increasing processes, one recording the price increases and the other the price decreases. In each case we show how the price process may be viewed as Brownian motion evaluated at a random time that is related to the sum of the processes being differenced to get the price process. We interpret the time change in all cases as a measure of economic activity. For the widest class of processes considered, we show that time is to be measured as a size weighted cumulation of orders. In this sense, the correct analytical measure turns out to be a combination of the number of trades and volume proxies often considered in the empirical literature. In addition we provide a wide class of operational models, with associated test procedures, for the price processes of market economies in continuous time.

Appendix

Derivation of Eq. (55).

We may write the right hand side of (54) as

$$\alpha_1 N_d^2 - \alpha_2 N_s^2 = (\sqrt{\alpha_1} N_d - \sqrt{\alpha_2} N_s)(\sqrt{\alpha_1} N_d + \sqrt{\alpha_2} N_s)$$

$$= \sqrt{\alpha_1 + \alpha_2} M (\sqrt{\alpha_1} N_d + \sqrt{\alpha_2} N_s) \tag{A.1}$$

where $M = (\sqrt{\alpha_1} N_d - \sqrt{\alpha_2} N_s)/\sqrt{\alpha_1 + \alpha_2}$ is another standard normal variate. Now project $\sqrt{\alpha_1} N_d + \sqrt{\alpha_2} N_s$ on M and write

$$\sqrt{\alpha_1} N_d + \sqrt{\alpha_2} N_s = \frac{\alpha_1 - \alpha_2}{\sqrt{\alpha_1 + \alpha_2}} M + \frac{2\sqrt{\alpha_1 \alpha_2}}{\sqrt{\alpha_1 + \alpha_2}} \tilde{M} \tag{A.2}$$

where $\tilde{M}$ is a standard normal variate independent of M. Substitution of (A.2) into (A.1) shows that one may write

$$\alpha_1 N_d^2 - \alpha_2 N_s^2 = (\alpha_1 - \alpha_2) M^2 + 2\sqrt{\alpha_1 \alpha_2} |M| \tilde{M} . \tag{A.3}$$

If we define by $\gamma_3(1/2)$, by the relation $M^2 = 2\gamma_3(1/2)$, one may then write (A3) as

$$\alpha_1 N_d^2 - \alpha_2 N_s^2 = (\alpha_1 - \alpha_2) 2\gamma_3(1/2) + 2\sqrt{\alpha_1 \alpha_2}\sqrt{2\gamma_3(1/2)}\tilde{M} \tag{A.4}$$

and division of (A.4) by 2, noting (54) yields

$$\ln\left(\frac{p\left(\frac{1}{2\kappa}\right)}{p(0)}\right) = (\alpha_1 - \alpha_2)\gamma_3(1/2) + \sqrt{2\alpha_1 \alpha_2}\sqrt{\gamma_3(1/2)}\tilde{M} \tag{A.5}$$

or the result that the log price process is Brownian motion with drift $(\alpha_1 - \alpha_2)$ and volatility $\sqrt{2\alpha_1 \alpha_2}$ evaluated at $\gamma_3(t)$.

We see from (A.5) that the price process is basically the difference of squares of normals. The non-negative prevailing price buy and sell orders are essentially the squares of Gaussian variates, N_d, and N_s, respectively. The time change, M, is the square of the excess demand, $(\sqrt{B} - \sqrt{S})^2$, which is the activity measure in this case.

Derivation of Eq. (58).

First note that the characteristic function of $\ln(p(t)/p(0))$ may be written as

$$\phi_{\ln(p(t)/p(0))}(u) = \exp\left(-2t \int_0^\infty (1 - \cos(u\alpha x))k(x)dx\right) .$$

On the other hand the characteristic function of $\sigma W(t) = Y(t)$ evaluated at a time change $\gamma_3(t)$ conditional on the time change is

$$\phi_{Y(\gamma_3(t))|\gamma_3(t)}(u) = \exp\left(-\frac{\sigma^2 u^2 \gamma_3(t)}{2}\right) .$$

Suppose that $\gamma_3(t)$ is a Lévy process with Lévy measure $k_3(x)dx$, then the characteristic function of $Y(\gamma_3(t))$ is given by

$$\phi_{Y(\gamma_3(t))}(u) = \exp\left(-t \int_0^\infty (1 - e^{-\frac{\sigma^2 u^2}{2}x})k_3(x)dx\right) .$$

For the log price process to be a time-changed Brownian motion, using the Lévy process $\gamma_3(t)$ for the time change, one must have

$$2\int_0^\infty (1 - \cos(u\alpha x))k(x)dx = \int_0^\infty (1 - e^{-\frac{\sigma^2 u^2}{2}x})k_3(x)dx\,. \tag{A.6}$$

We may let $\sigma^2/2 = \alpha$, $\tilde{k}(y) = k(y/\alpha)$, $\tilde{k}_3(y) = k_3(2y/\sigma^2)$ and then write (A.6) as

$$2\int_0^\infty (1 - \cos(ux))\tilde{k}(x)dx = \int_0^\infty (1 - e^{-u^2 x})\tilde{k}_3(x)dx\,. \tag{A.7}$$

Differentiating (A.7) with respect to u yields

$$\int_0^\infty \frac{\sin(ux)}{u}x\tilde{k}(x)dx = \int_0^\infty e^{-u^2 x}x\tilde{k}_3(x)dx\,. \tag{A.8}$$

We now recall that

$$\frac{\sin(ux)}{u} = \frac{1}{2}\int_{-\infty}^\infty 1_{|y|<x}e^{iuy}dy \tag{A.9}$$

and

$$e^{-u^2 x} = \int_{-\infty}^\infty \frac{1}{\sqrt{4\pi x}}e^{-\frac{y^2}{4x}}e^{iuy}dy\,. \tag{A.10}$$

Substituting (A.10) and (A.9) into (A.8) and using the uniqueness of Fourier transforms, we deduce that for each y

$$\int_0^\infty x\tilde{k}(x)\frac{1}{2}1_{|y|<x}dx = \int_0^\infty x\tilde{k}_3(x)\frac{1}{2\sqrt{\pi}}\frac{1}{\sqrt{x}}e^{-\frac{y^2}{4x}}dx\,. \tag{A.11}$$

Differenting (A.11) with respect to $y \geq 0$ yields

$$\tilde{k}(y) = \frac{1}{2}\int_0^\infty \frac{\tilde{k}_3(x)}{\sqrt{\pi x}}e^{-\frac{y^2}{4x}}dx\,. \tag{A.12}$$

Equation (A.12) may be solved for $\tilde{k}_3(x)$ satisfying (58) when $\tilde{k}(y)$ is given as the Laplace Eq. (56). To observe this, we recall that

$$e^{-ay} = \int_0^\infty \frac{a}{\sqrt{2\pi t^3}}e^{-\frac{y^2 t}{2}}e^{-\frac{a^2}{2t}}dt\,. \tag{A.13}$$

Employing (A.13) we may write

$$\tilde{k}(y) = \int e^{-ay}\rho(da)$$

$$= \int \rho(da)\int_0^\infty \frac{dt}{\sqrt{2\pi t^3}}ae^{-\frac{y^2 t}{2}}e^{-\frac{a^2}{2t}}\,.$$

Making the change of variable $t = \frac{1}{2x}$ we obtain

$$\tilde{k}(y) = \int \rho(da)\int_0^\infty \frac{dx}{\sqrt{\pi x}}ae^{-\frac{y^2}{4x}}e^{-a^2 x}\,.$$

It follows that defining $\tilde{k}_3$ by Eq. (58) satisfies (A.11) as was to be shown.

Derivation of Eq. (66).

We wish to evaluate the integral

$$A = \int_0^\infty \frac{\exp(-Mx)}{Cx^{Y+1}(1+x)^G}(e^{iux} - 1)dx\,. \tag{A.14}$$

For this we first consider the general form for $\delta < 1$ (the case of finite variation)

$$\int_0^\infty (e^{iux} - 1)\frac{e^{-\gamma x}}{x^{\delta+1}}dx = -\frac{1}{\delta}x^{-\delta}(e^{-(\gamma-iu)x} - e^{-\gamma x})|_0^\infty$$

$$+ \int_0^\infty \frac{1}{\delta}x^{-\delta}\left(\gamma e^{-\gamma x} - (\gamma - iu)e^{-(\gamma-iu)x}\right)dx$$

$$= \frac{\Gamma(1-\delta)}{\delta}\left(\gamma^\delta - (\gamma - iu)^\delta\right)\,. \tag{A.15}$$

Coming now to the integration of (A.14) we first note that

$$\frac{1}{(1+x)^G} = \int_0^\infty \frac{1}{\Gamma(G)}y^{G-1}e^{-(1+x)y}dy\,. \tag{A.16}$$

Substitution of (A.16) into (A.14) and reversing the order of integration yields that

$$A = \int_0^\infty \frac{y^{G-1}e^{-y}}{\Gamma(G)C}\int_0^\infty (e^{iux} - 1)\frac{e^{-(M+y)x}}{x^{Y+1}}dx\,. \tag{A.17}$$

We now employ (A.15) in (A.17) to obtain that

$$A = \frac{\Gamma(1-y)}{C\Gamma(G)y}\int_0^\infty y^{G-1}e^{-y}((M + y)^y - (M - iu + y)^y)dy\,. \tag{103}$$

The result follows on noting that

$$\int_0^\infty y^{G-1}e^{-y}(M + y)^Y dy = M^{Y+G}\Gamma(G)U(G, Y + G + 1, M)$$

from the integral representation of the U function (Abramowitz and Stegun (1972) page 505, 13.2.5).

Proof of condition (99).

The condition for obtaining a process of bounded variation on time changing Brownian motion may be expressed in terms of ν by

$$\int_0^\infty dx \int \frac{\nu(dy)}{\sqrt{2\pi y}}(x \wedge 1)\exp\left(-\frac{x^2}{2y}\right) < \infty\,. \tag{A.18}$$

Partitioning the integration in (A.18) over x into the intervals below and above unity we require that

$$\int \frac{\nu(dy)}{\sqrt{2\pi y}}\left\{\int_0^1 dx\, x\exp\left(-\frac{x^2}{2y}\right) + \int_1^\infty dx\, \exp\left(-\frac{x^2}{2y}\right)\right\} < \infty\,.$$

Changing the variable of integration from x to $t = x/\sqrt{y}$ we write that

$$\int \frac{\nu(dy)}{\sqrt{2\pi y}} \left\{ \int_0^{1/\sqrt{y}} ydt \, t \exp(-t^2/2) + \int_{1/\sqrt{y}}^{\infty} \sqrt{y}dt \, \exp(-t^2/2) \right\} < \infty \, .$$

Performing the first inner integration we may write

$$\int \frac{\nu(dy)}{\sqrt{2\pi}} \sqrt{y} \left(1 - \exp\left(-\frac{1}{2y}\right) \right) + \int \frac{\nu(dy)}{\sqrt{2\pi}} \int_{1/\sqrt{y}}^{\infty} dt \, \exp(-t^2/2) < \infty \, . \qquad (\text{A.19})$$

Consider first the second integral in (A.19). This may be rewritten as

$$\int_0^{\infty} dt \, \exp(-t^2/2) \int \frac{\nu(dy)}{\sqrt{2\pi}} \mathbf{1}_{(\frac{1}{\sqrt{y}} < t)} = \int_0^{\infty} dt \, \exp(-t^2/2) \int \frac{\nu(dy)}{\sqrt{2\pi}} \mathbf{1}_{(\frac{1}{t^2} < y)}$$

$$= \int_0^{\infty} dt \, \exp(-t^2/2) \frac{\bar{\nu}(1/t^2)}{\sqrt{2\pi}} \, ,$$

where $\bar{\nu}(x) = \int_x^{\infty} \nu(dy)$. This integral is always finite. For the first integral we observe that as

$$\sqrt{y} \left(1 - \exp\left(-\frac{1}{2y}\right) \right) = O(1) \, ,$$

as y tends to infinity, this integral is finite near ∞. The condition follows from the behavior of the first integral for y near zero. $\qquad \square$

Proof of Eqs. (101) and (102).

We use the representation of the Ψ function given by (see Abramowitz and Stegun, page 259, 6.3.22)

$$\Psi(x) + \gamma = \int_0^{\infty} \frac{e^{-z} - e^{-zx}}{1 - e^{-z}} dz$$

to evaluate the Laplace transform (100) at $\lambda = -iu$ and $\lambda = iu$ and multiply the results to obtain that

$$\phi_{\ln(p(t)/p(0))}(u) = \exp\left(\begin{array}{c} -\dfrac{tu}{\theta} \displaystyle\int_0^{\infty} \dfrac{e^{-z}}{1 - e^{-z}} \sin\left(\dfrac{\sqrt{u}}{2}z\right) X \\[2ex] \left(\exp\left(\dfrac{\sqrt{u}}{2}z\right) - \exp\left(-\dfrac{\sqrt{u}}{2}z\right) \right) dz \end{array} \right) \, .$$

The result follows on writing $(1 - e^{-z})^{-1}$ as a power series and integrating each term separately and simplifying.

For the Lévy measure we note that

$$\frac{u^2 a}{u^2 + a} = \int_0^{\infty} dx a^2 e^{-ax}(1 - e^{-u^2 x})$$

so taking $a = 4n$, and summing over n we obtain:

$$\sum_{n=1}^{\infty} (4n)^2 \exp(-4nx) = \Phi''(x)$$

where

$$\Phi(x) = \sum_{n=1}^{\infty} \exp(-4nx) = \frac{\exp(-4x)}{1 - \exp(-4x)} \, . \qquad \square$$

References

[1] M. Abramowitz and I. A. Stegun, *Handbook of Mathematical Functions*, National Bureau of Standards, Applied Mathematics Series, 55, Washington, D.C., 1972.

[2] T. Ané and H. Geman, "Order Flow, Transactions Clock and Normality of Asset Returns", forthcoming *Journal of Finance*.

[3] G. Bakshi, C. Cao and Z. Chen, "Empirical performance of alternative option pricing models", forthcoming *Journal of Finance*, 1997.

[4] G. Bakshi and D. B. Madan, "Spanning and Derivative-security valuation", *Journal of Financial Economics* **55** (2000) 205–238.

[5] D. Bates, "Post-'87 Crash Fears in S&P500 Futures Options", Working paper, University of Iowa, Iowa City, Iowa 52242-1000, 1996.

[6] F. Black and M. Scholes, "The pricing of options and corporate liabilities", *Journal of Political Economy* **81** (1973) 637–654.

[7] L. Bondesson, *Generalized Gamma Convolutions and Related Classes of Distributions and Densities*, Lecture Notes in Statistics 76, Springer-Verlag, Heidelberg, 1992.

[8] P. K. Clark, "A subordinated stochastic process with finite variance for speculative prices", *Econometrica* **41** (1973) 135–155.

[9] J. C. Cox, S. A. Ross and M. Rubinstein, "Option pricing: A simplified approach", *Journal of Financial Economics* **7** (1979) 229–263.

[10] J. C. Cox and C. F. Huang, "Optimum consumption and investment portfolio policies when asset prices follow a diffusion process", *Journal of Economic Theory* **49** (1989) 33–83.

[11] J. Cox, J. Ingersoll and S. Ross, "A theory of the term structure of interest rates", *Econometrica* **53** (1985) 385–408.

[12] F. Delbaen and W. Schachermayer, "A general version of the fundamental theorem of asset pricing", *Mathematische Annalen* **300** (1994) 563–520.

[13] J. Detemple and S. Murthy, "Intertemporal asset pricing with heterogeneous beliefs", *Journal of Economic Theory* **62**(2) (1994) 294–000.

[14] C. Donati-Martin and Marc Yor, "Some Brownian functionals and their laws", *The Annals of Applied Probability* **25** (1997) 1011–1058.

[15] L. E. Dubins, M. Émery and M. Yor, "On the Lévy transformation of Brownian motions and continuous martingales", *Seminaire de Probabilité, XXVII, Lecture Notes in Mathematics* **1557** (1993) 122–132.

[16] D. Duffie and C. Huang, "Implementing arrow-debreu equilibria by continuous trading of few long-lived securities", *Econometrica* **53** (1985) 1337–1356.

[17] B. Dumas, "Two-person dynamic equilibrium in the capital market", *Review of Financial Studies* **2** (1989) 157–188.

[18] A. R. Gallant, P. E. Rossi and G. Tauchen, "Stock prices and volume", *Review of Financial Studies* **5** (1992) 199–242.

[19] H. Geman and M. Yor, "Bessel processes, Asian options and perpetuities", *Mathematical Finance* (1993) 349–375.

[20] N. H. Hakansson, "The fantastic world of finance: Progress and the free lunch", *Journal of Financial and Quantitative Analysis* **14** (1979) 717–734.

[21] M. Harrison and D. Kreps, "Martingales and arbitrage in multiperiod security markets", *Journal of Economic Theory* **20** (1979) 381–408.

[22] M. Harrison and S. Pliska, "Martingales and stochastic integrals in the theory of continuous trading", *Stochastic Processes and Their Applications* **11** (1981) 215–260.

[23] H. He and H. Leland, "On equilibrium asset pricep processes", *Review of Financial Studies* **6** (1993) 593–617.

[24] S. L. Heston, "Invisible parameters in option prices", *Journal of Finance* **48**(3) (1993) 933–947.

[25] R. A. Jarrow and D. B. Madan, "Option pricing using the term structure of interest rates to hedge systematic discontinuities in asset returns", *Mathematical Finance* **5**(4) (1995) 311–336.

[26] M. Jeanblanc, J. Pitman and M. Yor, "The Feynman-Kac formula and decomposition of Brownian paths", *Computational and Applied Mathematics, Special Issue on Stochastic Analysis* **16**(1) (1997) 27–52.

[27] E. P. Jones, "Option arbitrage and strategy with large price changes", *Journal of Financial Economics* **13** (1984) 91–113.

[28] C. Jones, G. Kaul and M. L. Lipson, "Transactions, volumes and volatility", *Review of Financial Studies* **7** (1994) 631–651.

[29] S. Karlin and H. Taylor, *A Second Course in Stochastic Processes*, Academic Press, London, 1981.

[30] Karpoff and M. Jonathan, "The relation between price changes and trading volume: A survey", *Journal of Financial and Quantitative Analysis* **22** (1987) 109–126.

[31] F. B. Knight, "Characterization of the Lévy measure of inverse local time for gap diffusions", *Seminar on Stochastic Processes*, Birkhauser, 1981, 53–78.

[32] D. Kreps, "Arbitrage and equilibrium in economies with infinitely many commodities", *Journal of Mathematical Economics* **8** (1981) 15–35.

[33] S. Kotani and S. Watanabe, "Krein's spectral theory of strings and generalized diffusion processes", *Functional Analysis in Markov Processes* in Fukushima, ed., Lecture Notes in Mathematics 923 (Springer-Verlag, 1982).

[34] B. Leblanc, "Modélisation de la Volatilité d'un Actif Financier et Applications, dissertation", University de Paris VII-Denis Diderot, 1997.

[35] D. B. Madan and F. Milne, "Option pricing with V. G. Martingale components", *Mathematical Finance* **1**(4) (1991) 39–55.

[36] D. B. Madan, P. Carr and E. Chang, "The variance gamma process and option pricing", *European Finance Review* **2** (1998) 79–105.

[37] B. Mandelbrot, "New methods in statistical economics", *Journal of Political Economy* **61** (1963) 421–440.

[38] J. H. McCulloch, "Continuous time processes with stable increments", *Journal of Business* **51**(4) (1978) 601–620.

[39] R. C. Merton, "Optimum consumption and portfolio rules in a continuous time model", *Journal of Economic Theory* **3** (1971) 373–413.

[40] R. C. Merton, "Theory of rational option pricing", *Bell Journal of Economics and Management Science* **4** (1973) 141–183.

[41] R. C. Merton, "Option pricing when underlying stock returns are discontinuous", *Journal of Financial Economics* **3** (1976) 125–144.

[42] I. Monroe, "Processes that can be embedded in Brownian motion", *The Annals of Probability* **6**(1) (1978) 42–56.

[43] D. L. Ocone, "A symmetry characterization of conditionally independent increment Martingales", *Barcelona Seminar on Stochastic Analysis, St. Felice de Guixols, 1991*, D. Nualart and M. Sanz eds. (Birkhauser, New York, 1993).

[44] J. S. Press, "A compound events model for security prices", *Journal of Business* **40** (1967) 317–335.

[45] D. Revuz and M. Yor, *Continuous Martingales and Brownian Motion* (Springer-Verlag, Berlin, 1994).

[46] G. Tauchen and M. Pitts, "The price-volume relationship on speculative markets", *Econometrica* **51** (1983) 485–505.

[47] A. Vershik and M. Yor, "Multiplicativité du processus gamma et étude asymptotique des lois stables d'indice α, lorsque α tend vers 0″, Prépublication N° 289 du Laboratoire de Probabilités de L'Université Paris VI, 1995.

[48] M. Yor, *Local Times and Excursions for Brownian Motion: A Concise Introduction*, Lecciones en Matemáticas, Número 1, Postgrado de Matemáticas, Facultdad de Ciencias, Universidad Central de Venezuela, 1995.

HEDGING UNDER STOCHASTIC VOLATILITY

K. RONNIE SIRCAR*

Received 24 November 1998

We present a family of hedging strategies for a European derivative security in a stochastic volatility environment. The strategies are robust to specification of the volatility process and do not need a parametric description of it or estimation of the volatility risk premium. They allow the hedger to control the probability of hedging success according to risk aversion. The formula exploits the separation between the time-scale of asset price fluctuation (ticks) and the longer time-scale over which volatility fluctuates, that is, the observed "persistence" of volatility. We run simulations that demonstrate the effectiveness of the strategies over the classical Black–Scholes strategy.

1. Introduction

In this article, we present a family of hedging strategies for a European derivative security that super-replicate the claim with a controllable success probability, in a stochastic volatility environment. The strategy has the following features:

- It is an approximate (asymptotic) solution to the problem, but as such, it is computable (we give an explicit formula).
- It is based on a nonparametric description of the random volatility process of the underlying asset. Thus it is robust to specific modelling of the volatility (under technical restrictions).
- It requires estimates of certain simple statistics of the volatility process that are easily obtained from historical asset price data.
- It does not need identification (or estimation) of the volatility risk premium or the market's (equivalent martingale) pricing measure.

The strategy is selected as follows: choose a minimum acceptable probability p_0 with which the strategy should dominate the perfect hedging strategy. Then, find the number of standard deviations ρ of a standard normal distribution whose confidence interval has this probability

*Department of Mathematics, University of Michigan, Ann Arbor, MI 48109-1109, *sircar@umich.edu*. The author is grateful for conversations on the subject with George Papanicolaou, to Hans Föllmer for suggesting the particular hedging problem considered here, and participants of the Courant Finance Seminar for helpful questions and comments.

$$\frac{1}{\sqrt{2\pi}} \int_{-\rho}^{\rho} e^{-z^2/2} dz = p_0 \, . \tag{1}$$

This is easily found in tables (for example, $\rho = 1$ corresponds to $p_0 = 67\%$, $\rho = 2$ to 95%). Then the hedging strategy is given by formulas (15) and (20). Note that p_0 is not the probability that the hedge is successful, but increasing p_0 increases that probability.

That the randomness of the volatility process (whose distribution is unspecified) translates into a family of hedging strategies that are distinguished simply by a *normally distributed* random variable comes from an application of the central limit theorem for Markov processes to the Black–Scholes derivative pricing PDE with a random volatility coefficient. That such a convenient characterization is a good approximation is due to the persistent nature (or burstiness) of volatility (in at least equity and F/X markets).

The next section briefly sketches the background and motivation for stochastic volatility models. In Sec.3, we explain how volatility persistence is modelled here and how uncertainty in volatility translates into uncertainty in derivative prices and hedging strategies. Section 4 presents the main result which is illustrated by the simulations of Sec. 5. The issue of estimation, crucial for the theory to be applicable, is explained in Sec. 6, followed by conclusions.

2. Why Stochastic Volatility?

Several excellent survey articles, for example, Ref. 7 outline the main features of stochastic volatility modelling for derivative pricing, starting from the work of Hull and White[8] in 1987. The stock price (or exchange rate) process $\{X_t, t \geq 0\}$ satisfies

$$dX_t = \mu X_t dt + \sigma_t X_t dW_t \, ,$$

where $\{\sigma_t, t \geq 0\}$ is the stochastic volatility process. An overview of the usual approach as it relates to the work here is given in Ref. 13, and the important points are:

- Empirical studies of stock price data strongly suggest volatility is *not* constant (as assumed by the Black–Scholes theory), but has a random component. ARCH/GARCH models, whose continuous-time diffusion limits are stochastic volatility models, provide much better descriptions of the data. See Ref. 2 for details.
- Empirical studies of implied volatility data, for example, Ref. 12 report frequent observation of the smile curve, a U-shaped variation of implied volatility with strike price for options with the same time-to-maturity. The minimum is at or near the current stock price.
- Any stochastic volatility model in which the volatility process is independent of the Brownian motion W_t results in predicted European option prices whose implied volatility curve smiles, with minimum at today's stock price adjusted

by compound interest earned from today to expiration. See Ref. 11 or Ref. 13 for a proof.

- Most analysis and estimation is parametric: σ_t is modelled as the solution to a particular Itô SDE.
- A stochastic volatility environment is a simple example of an incomplete market. As such there is no unique pricing (or equivalent martingale) measure and this indeterminacy can be characterized by an unknown process, labelled the volatility risk premium, playing a part in derivative pricing and hedging. Most usually this process is taken to be zero, sometimes a constant or a deterministic function of present volatility, the main reasons being feasibility of estimation or to preserve the Markovian structure. Here we shall not need to make such a choice, as explained in Sec. 3.

We mention briefly some recent work related to the hedging problem considered here. The problem of almost sure super-replication is considered by Cvitanić *et al.*[3] (and authors referenced therein) by stochastic control methods. Such strategies that guarantee a successful hedge are usually expensive, which motivates Föllmer and Leukert[4] (and others cited there) to allow some risk of shortfall and look for a strategy that maximizes the probability of a successful hedge given some initial cash input that the hedger is willing to spend. This is explicitly computed in the constant volatility framework, and when volatility can jump by a random amount at a known time. Avellaneda *et al.*[1] construct worst-case super-replicating strategies for complex portfolios of options given that volatility lies in a known band.

3. Separation of Time-Scales

Figure 1 shows two simulated realizations of possible volatility paths over the course of a year. In the first, volatility is low (4–8%) for a large part of that year (roughly $t = 0.1$ till $t = 0.8$ — over eight months) and then for the rest of the year it is at a higher level. In the second path, volatility fluctuates between periods of high and periods of low far more often — it seems to be low for a few weeks and then high for a few weeks, then low again, and so on. That is, often, when it is low, it stays low for a period, and similarly when it is high.

Empirical studies suggest that the latter realization is a much more typical yearly volatility pattern than the former — it exhibits volatility clustering, or the tendency of high volatility to come in bursts.

In fact, the sample paths of the second process can be obtained by simulating the first process for a much longer time (50 years) and squeezing the realization into one year. That is, the second process comes from speeding-up the first. Mathematically, if we call the first process $\{\sigma(t), t \geq 0\}$, it is convenient to denote the second process $\{\sigma(\frac{t}{\varepsilon}), t \geq 0\}$, where $\varepsilon > 0$ is a dimensionless parameter that represents the speeding-up. Because volatility clustering is really a distinguishable feature, we shall think of ε as being small.

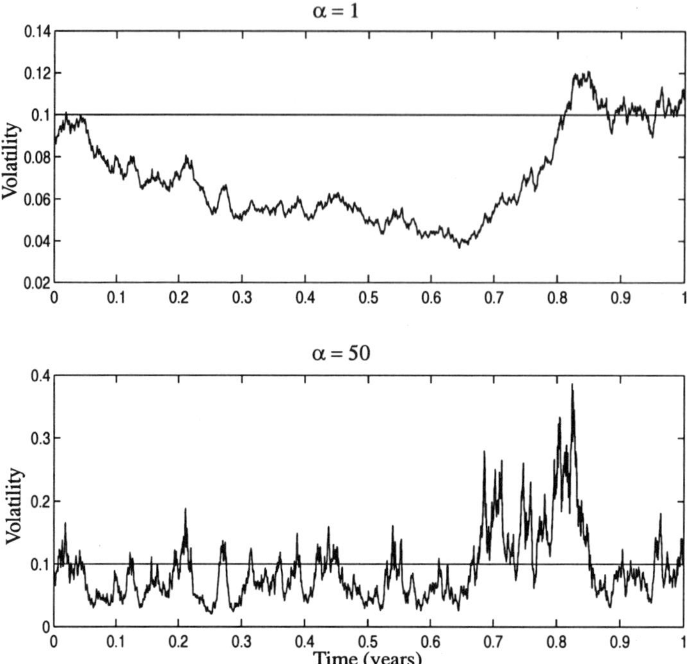

Fig. 1. The top figure shows a simulated path of $\sigma_t = e^{Y_t}$, Y_t a mean-reverting Ornstein–Uhlenbeck process defined by Eq. (2), with $\alpha = 1$, and the bottom one shows a path with $\alpha = 50$. Note how volatility "clusters" in the latter case.

3.1. *Examples*

1. Suppose $\sigma(t)$ is a two-state Markov chain representing a crude model of volatility taking a high or a low state: $\sigma(t) \in \{\sigma_1, \sigma_2\}$. Then if the generator matrix Q of the process has elements that are $\mathcal{O}(1)$ in size (neither big nor small), for example,

$$Q = \begin{pmatrix} -2 & 2 \\ 8 & -8 \end{pmatrix},$$

then a typical six-month sample path might look like the first graph of Fig. 2. However, if we consider such a process with a generator having large entries, for example,

$$Q' = \begin{pmatrix} -200 & 200 \\ 800 & -800 \end{pmatrix},$$

then a typical path, shown in the bottom graph, gives a much better description of high volatility coming in bursts of a few days or weeks, rather than months.

But notice that $Q' = \frac{1}{0.01} Q$, and that if $\sigma'(t) := \sigma(\frac{t}{\varepsilon})$, with $\varepsilon = 0.01$, then the generator of $\sigma'(t)$ is exactly this $Q' = \frac{1}{\varepsilon} Q$. Thus the speeding-up notation $\sigma(\frac{t}{\varepsilon})$ is a concise description of the fact that such a parametric model of volatility (and indeed

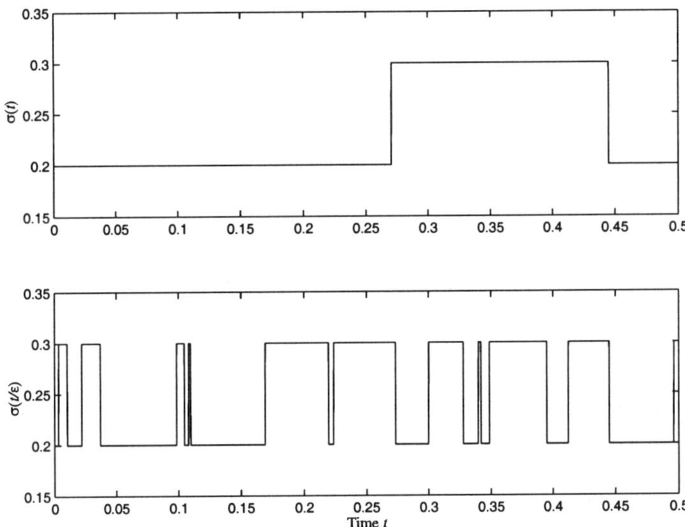

Fig. 2. The top figure shows a simulated path of $\sigma(t)$ a two-state Markov chain, and the bottom one shows a path of $\sigma\left(\frac{t}{\varepsilon}\right)$, with $\varepsilon = 0.05$.

any Markovian model) contains some parameters that are small, if the model is to reflect clustering periods of lengths that are usually observed. In this representation, the "smallness" is controlled by the ε.

2. Suppose $\sigma(t) = f(Y_t)$, where $f(\cdot)$ is a positive increasing function and Y_t is a mean-reverting Ornstein–Uhlenbeck process: it satisfies

$$dY_t = \alpha(m - Y_t)dt + \beta dZ_t \,. \tag{2}$$

The top graph of Fig. 1 shows $\sigma(t)$ for $f(y) = e^y$ (the expOU model), with parameters $\alpha = 1$, $\beta^2 = 0.5$ and m chosen so that the RMS volatility (or long-run mean level) is 10%. With these $\mathcal{O}(1)$ values, we do not see significant volatility persistence.

Now consider the speeded-up process $\sigma(\frac{t}{\varepsilon}) = f(Y_{t/\varepsilon})$. The generator of the original Y_t is

$$\mathcal{L} := \alpha(m - y)\frac{\partial}{\partial y} + \frac{1}{2}\beta^2\frac{\partial^2}{\partial^2 y} \,,$$

while the generator of $Y_{t/\varepsilon}$ is

$$\mathcal{L}^\varepsilon = \frac{1}{\varepsilon}\mathcal{L} = \frac{\alpha}{\varepsilon}(m - y)\frac{\partial}{\partial y} + \frac{1}{2}\frac{\beta^2}{\varepsilon}\frac{\partial^2}{\partial^2 y} \,.$$

From this, we see that speeding-up Y_t (and hence $\sigma(t)$) is analogous to replacing α by α/ε and β by $\beta/\sqrt{\varepsilon}$. That is, the rate of mean-reversion α is scaled by $1/\varepsilon$ with the noise factor β scaled correspondingly to keep $\beta^2/2\alpha$ constant. For example, if $\varepsilon = 1/50$, we can simulate $\sigma(\frac{t}{\varepsilon})$ by using $\alpha = 50$ and $\beta^2 = 25$ in (2). A realization of this process is shown in the bottom graph of Fig. 1, and again, it better captures volatility clustering.

In this context, it is then convenient to think of speeding-up as simply the presence of a *large* mean-reversion rate: there are always Itô fluctuations in the volatility from the Brownian motion Z_t and it reverts slowly to its mean-level when looked at relative to this time-scale. But it reverts fast to the mean when looked at over the time-scale of a year. In the path shown, it crosses the mean-level over 30 times during the year. This class of models is analyzed and estimated from market data in Refs. 5 and 6.

3.2. *Characterization in terms of time-scales*

In summary, there are three distinct time-scales in the modelling of the underlying asset price (or exchange rate) $\{X_t^\varepsilon, t \geq 0\}$, which satisfies

$$dX_t^\varepsilon = \mu X_t^\varepsilon dt + \sigma\left(\frac{t}{\varepsilon}\right) X_t^\varepsilon dW_t. \tag{3}$$

Firstly, there is the "infinitely small" scale of the "infinitely fast" fluctuations of the Brownian motion W_t. These model the tick-by-tick fluctuations of the price. The volatility process might also have a component fluctuating on this scale (for example, the Z_t in the second example above). However there is a longer time-scale representing volatility persistence which might be on the order of a few days or weeks (for example, the average mean-reversion time in the second example). These variations are slow in comparison to the tick-tick scale, but still fast when looked at over the lifetime of a derivative contract (many months) which is the long time-scale of our pricing or hedging problems.

Brownian		Volatility		Option
fluctuations	$\ll$	fluctuations	$\ll$	$[0, T]$
$\sim$ minutes		$\sim$ days		$\sim$ months .

Such a separation of scales is utilized by the asymptotic analysis of the next section.

4. Hedging

Suppose at time $t = 0$, we wish to hedge the risk of having written a European call option with strike price K and expiration date T on a stock by buying and selling only the underlying stock. The problem is to find a hedging strategy $H(t, x)$ that gives the number of units of the underlying held at time t (when its price is x) so that at time T, when we might have to sell the stock to the option holder for price K, we break even or make a profit with high probability.

In the complete market constant volatility case, a perfect (probability 1) break even hedging scheme is to hold $C_x^{BS}(t, x)$ units of stock (the delta), where C^{BS} denotes the Black–Scholes formula. In the stochastic volatility case, the additional source of randomness cannot be exactly hedged by trading in just the stock, and so we look for strategies with a high probability of success.

4.1. *Illustration of method*

To explain how we find such a strategy, consider first the simplified scenario in which the stock price has a deterministic volatility $\sigma_1(t)$

$$dX_t = \mu X_t dt + \sigma_1(t) X_t dW_t \,,$$

but we hedge with the wrong deterministic volatility function $\sigma_2(t)$. That is, we solve the (generalized) Black–Scholes PDE

$$\bar{C}_t + \frac{1}{2}\sigma_2(t)^2 x^2 \bar{C}_{xx} = 0 \,,$$
$$\bar{C}(T, x) = (x - K)^+ \,, \tag{4}$$

whose solution we denote $\bar{C}(t, x; [\sigma_2])$, and hold $\bar{C}_x(t, X_t; [\sigma_2])$ of the stock at time t. We take the interest rate $r = 0$ here, but our final formulas are given for the general case. At time $t = 0$, we put up the amount $\bar{C}(0, X_0; [\sigma_2])$, the cost of the hedge. The value of the hedging portfolio V_t is

$$V_t = \bar{C}(0, X_0; [\sigma_2]) + \int_0^t \bar{C}_x(s, X_s; [\sigma_2]) dX_s \,.$$

When the wrong volatility $\sigma_2(\cdot)$ is used, the strategy is not self-financing and V_t may be negative at certain times. We assume we are willing to put up extra cash after $t = 0$ if the strategy demands it, but our profit/loss bookkeeping is with respect to the initial cost of the hedge. The question of interest is how close is V_T to $(X_T - K)^+$, the payoff of the written claim?

Following Ref. 9, we know that

$$\bar{C}(t, X_t; [\sigma_2]) = \bar{C}(0, X_0; [\sigma_2]) + \int_0^t \bar{C}_x(s, X_s; [\sigma_2]) dX_s$$

$$+ \int_0^t \left(\bar{C}_t(s, X_s; [\sigma_2]) + \frac{1}{2}\sigma_1(s)^2 X_s^2 \bar{C}_{xx}(s, X_s; [\sigma_2]) \right) ds \,,$$

by Itô's lemma, and the last integral can be re-written as

$$\frac{1}{2} \int_0^t (\sigma_1(s)^2 - \sigma_2(s)^2) X_s^2 \bar{C}_{xx}(s, X_s; [\sigma_2]) ds \,, \tag{5}$$

using (4). This gives

$$V_T = (X_T - K)^+ + \frac{1}{2} \int_0^T (\sigma_2(s)^2 - \sigma_1(s)^2) X_s^2 \bar{C}_{xx}(s, X_s; [\sigma_2]) ds \,.$$

Clearly, taking $\sigma_2 \equiv \sigma_1$ hedges the claim perfectly. When $\sigma_1(t)$ is a random process, we want to find a $\sigma_2(\cdot)$ so that $V_T \geq (X_T - K)^+$ with a high probability. This reduces to finding $\sigma_2(\cdot)$ such that

$$\frac{1}{2} \int_0^T \sigma_2(s)^2 X_s^2 \bar{C}_{xx}(s, X_s; [\sigma_2]) ds \geq \frac{1}{2} \int_0^T \sigma_1(s)^2 X_s^2 \bar{C}_{xx}(s, X_s; [\sigma_2]) ds \tag{6}$$

in some "best" way. This is a stochastic control type problem to find a volatility path that maximizes a probability and this is the approach of Refs. 3 and 4. To our knowledge, the exact solution is not easily computable, so we will look for a dominating solution: find $\sigma_2(\cdot)$ such that (6) holds with high probability.

The remaining ingredient is to characterize possible hedging strategies in a convenient way that translates uncertainty in the volatility into uncertainty in the final value V_T. This is achieved by an asymptotic approximation that exploits the separation of time-scales.

4.2. *Asymptotic approximation*

We define the stochastic option price $C^\varepsilon(t,x)$ as the solution to the Black–Scholes PDE with the random speeded-up volatility coefficient $\sigma(\frac{t}{\varepsilon})$ from (3):

$$C_t^\varepsilon + \frac{1}{2}\sigma^2\left(\frac{t}{\varepsilon}\right)x^2 C_{xx}^\varepsilon = 0\,,$$

$$C^\varepsilon(T,x) = (x - K)^+\,. \tag{7}$$

This can be thought of as a conditional Black–Scholes PDE with each realization of the process $\{C^\varepsilon(t,x), 0 \le t \le T\}$ a call option pricing function given the path of the volatility. As $\sigma(\frac{t}{\varepsilon})$ and W_t are independent, the derivation of this equation exactly follows the derivation of the classical Black–Scholes PDE.

We are interested in the asymptotic behaviour of C^ε as $\varepsilon \downarrow 0$: this approximation will tell us how to deal with the risk from the randomness of the volatility. We shall make the following assumptions on the process $\{\sigma(t), t \ge 0\}$ (which also hold after speeding-up):

1. $\{\sigma^2(t), t \ge 0\}$ is wide-sense stationary: it has time-independent mean $\bar{\sigma}^2 := E\{\sigma^2(t)\}$ and its autocorrelation $E\{(\sigma^2(t) - \bar{\sigma}^2)(\sigma^2(s) - \bar{\sigma}^2)\}$ is a function of $|t - s|$ only.

2. It is ergodic and Markov.

As ε becomes smaller and smaller, the distinction between the time-scales disappears and $C^\varepsilon(t,x)$ looks more and more like the Black–Scholes formula with a constant *averaged* volatility. This is a standard averaging principle result following from the ergodic theorem. It says that $C^\varepsilon(t,x)$ converges in probability to $C^{BS}(t,x; \sqrt{\bar{\sigma}^2})$, the Black–Scholes formula with volatility $\sqrt{\bar{\sigma}^2}$, which satisfies the (averaged) PDE

$$C_t^{BS} + \frac{1}{2}\bar{\sigma}^2 x^2 C_{xx}^{BS} = 0\,,$$

$$C^{BS}(T,x) = (x - K)^+\,. \tag{8}$$

That is,

$$P\left(\sup_{0 \le t \le T}\sup_{x > 0}\left|C^\varepsilon(t,x) - C^{BS}(t,x; \sqrt{\bar{\sigma}^2})\right| > \delta\right) \longrightarrow 0$$

as $\varepsilon \downarrow 0$, for any $\delta > 0$. A proof is given in Ref. 13.

So far we have a crude approximation to possible prices under stochastic volatility by the Black–Scholes formula. What is of use is the next correction term, valid for small $\varepsilon > 0$, that quantifies volatility risk.

Let us write

$$C^\varepsilon(t, x) = C^{BS}(t, x; \sqrt{\bar{\sigma^2}}) + \sqrt{\varepsilon} Z^\varepsilon(t, x),$$

which defines the error term $Z^\varepsilon(t, x)$. Then subtracting (8) from (7), we find that $Z^\varepsilon(t, x)$ satisfies

$$Z_t^\varepsilon + \frac{1}{2}\sigma^2\left(\frac{t}{\varepsilon}\right) x^2 Z_{xx}^\varepsilon + \frac{1}{2}\left(\frac{\sigma^2\left(\frac{t}{\varepsilon}\right) - \bar{\sigma^2}}{\sqrt{\varepsilon}}\right) x^2 C_{xx}^{BS} = 0,$$

$$Z^\varepsilon(T, x) = 0.$$
(9)

We shall use two convergence results: the first is **weak averaging for stochastic differential equations** which says that as $\varepsilon \downarrow 0$, the solution to (3) with initial condition $X_0^\varepsilon = x$ converges weakly to the solution of the analogous SDE with the averaged volatility coefficient $\sqrt{\bar{\sigma^2}}$:

$$d\bar{X}_t = \mu \bar{X}_t dt + \sqrt{\bar{\sigma^2}} \bar{X}_t d\bar{W}_t,$$
(10)

with the same starting value $\bar{X}_0 = x$, where $\{\bar{W}_t, t \geq 0\}$ is a standard Brownian motion. See, for example, Ref. 14 for a proof.

This weak approximation of X^ε by $\bar{X}$ can then be combined with the **central limit theorem for Markov processes** to show that the scaled fluctuation of the volatility process behaves weakly like the increment of a Brownian motion $\{B_t, t \geq 0\}$:

$$\int_t^T g(s, X_s^\varepsilon)\left(\frac{\sigma^2\left(\frac{s}{\varepsilon}\right) - \bar{\sigma^2}}{\sqrt{\varepsilon}}\right) ds \longrightarrow \gamma \int_t^T g(s, \bar{X}_s) dB_s,$$
(11)

for bounded non-anticipating functions g; see Ref. 10, for example. The Brownian motion B_t is standard and γ contains the remaining trace of the original volatility process through the integral of its correlation function:

$$\gamma^2 = 2 \int_0^\infty E\{(\sigma^2(s) - \bar{\sigma^2})(\sigma^2(0) - \bar{\sigma^2})\} ds.$$
(12)

It is shown in Ref. 13 that $Z^\varepsilon(t, x)$ converges weakly to the Gaussian process $Z(t, x)$ that satisfies the linear stochastic PDE

$$dZ_t + \frac{1}{2}\bar{\sigma^2} x^2 Z_{xx} dt = -\frac{1}{2}\gamma x^2 C_{xx}^{BS} dB_t,$$
(13)

the limit equation of (9), with $Z(T, x) = 0$. Thus we have the approximation

$$C^\varepsilon(t, x) = C^{BS}(t, x; \sqrt{\bar{\sigma^2}}) + \sqrt{\varepsilon} Z(t, x) + \mathcal{O}(\varepsilon),$$
(14)

for small $\varepsilon > 0$: possible option prices are decomposed as the sum of a Black–Scholes price and a normally distributed random function $Z(t, x)$ no matter what the original volatility distribution. All that is left is γ^2 (known as the power spectral density at zero frequency of the volatility), a statistic whose estimation from data is considered in Sec. 6.

Similar approximations are derived in Ref. 13 in the more general situation that volatility is of the form $\sigma(t, x)$, a function-space-valued random process.

4.3. *Hedging strategy*

How now does this representation help with the hedging problem? Motivated by the simple form of (14), let us look for a strategy $H(t, x)$ that is the delta of a correction to the Black–Scholes delta with the averaged volatility coefficient:

$$H(t, x) = C_x^{BS}(t, x; \sqrt{\overline{\sigma^2}}) + \sqrt{\varepsilon} F_x(t, x). \tag{15}$$

To measure the performance of such strategies, it is convenient to define the *effective volatility function* $E^\varepsilon(t, x)$. We want to write $C^{BS}(t, x; \sqrt{\overline{\sigma^2}}) + \sqrt{\varepsilon} F(t, x)$ as $\bar{C}(t, x; [E^\varepsilon])$, the solution to (4) with $E^\varepsilon(t, x)$ instead of $\sigma_2(t)$. Expanding $E^\varepsilon(t, x)$ in the form $\sqrt{\overline{\sigma^2}}$ plus correction, substituting into (4) and comparing powers of ε gives

$$E^\varepsilon(t, x) = \sqrt{\overline{\sigma^2}} - \sqrt{\varepsilon} \frac{\mathcal{L}^{BS} F(t, x)}{x^2 \sqrt{\overline{\sigma^2}} C_{xx}^{BS}(t, x)} + \mathcal{O}(\varepsilon),$$

where

$$\mathcal{L}^{BS} := \frac{\partial}{\partial t} + \frac{1}{2} \overline{\sigma}^2 x^2 \frac{\partial^2}{\partial x^2}.$$

Now $H(t, x) = \bar{C}_x(t, x; [E^\varepsilon])$ and, given an initial cash input V_0 (to be determined), the value of our hedging portfolio is

$$V_t = V_0 + \int_0^t \bar{C}_x(s, X_s^\varepsilon; [E^\varepsilon]) dX_s^\varepsilon$$

$$= V_0 - \bar{C}(0, X_0^\varepsilon; [E^\varepsilon]) + \bar{C}(t, X_t^\varepsilon; [E^\varepsilon])$$

$$- \frac{1}{2} \int_0^t \left(\sigma^2 \left(\frac{s}{\varepsilon}\right) - E^\varepsilon(s, X_s^\varepsilon)^2 \right) (X_s^\varepsilon)^2 \bar{C}_{xx}(s, X_s^\varepsilon; [E^\varepsilon]) ds,$$

analogous to (5). Therefore, the final value is

$$V_T = V_0 - \left(C^{BS}(0, X_0^\varepsilon; \sqrt{\overline{\sigma^2}}) + \sqrt{\varepsilon} F(0, X_0^\varepsilon) \right) + (X_T^\varepsilon - K)^+$$

$$- \frac{1}{2} \int_0^T \left(\sigma^2 \left(\frac{s}{\varepsilon}\right) - \overline{\sigma}^2 + 2\sqrt{\varepsilon} \frac{\mathcal{L}^{BS} F(s, X_s^\varepsilon)}{(X_s^\varepsilon)^2 C_{xx}^{BS}(s, X_s^\varepsilon)} \right)$$

$$\times (X_s^\varepsilon)^2 \bar{C}_{xx}(s, X_s^\varepsilon) ds + \mathcal{O}(\varepsilon).$$

So the replication error, which determines whether the strategy yields a profit or a loss is weakly approximated by

$$V_T - (X_T^\varepsilon - K)^+ = V_0 - \left(C^{BS}(0, x; \sqrt{\bar{\sigma}^2}) + \sqrt{\varepsilon}F(0, x) \right)$$

$$- \sqrt{\varepsilon} \left(\int_0^T \mathcal{L}^{BS} F(s, \bar{X}_s) ds + \frac{1}{2}\gamma \int_0^T (\bar{X}_s)^2 C_{xx}^{BS}(s, \bar{X}_s) dB_s \right)$$

$$+ \mathcal{O}(\varepsilon), \tag{16}$$

where $x = X_0^\varepsilon$, the observed current stock price, and we have used (11) and the weak approximation of X^ε by $\bar{X}$, the solution of (10), and that $\bar{C}_{xx} = C_{xx}^{BS} + \mathcal{O}(\sqrt{\varepsilon})$ within our region of asymptoticity.

Let us choose $V_0 = C^{BS}(0, x; \sqrt{\bar{\sigma}^2}) + \sqrt{\varepsilon}F(0, x)$, defined to be the cost of the hedge, and find an F that, with respect to the probability measure defined by the Brownian motion $\{B_t, t \geq 0\}$ (on some abstract space to which we do not make specific reference) makes the combined last two terms positive, given the path of $\bar{X}$.

Suppose we knew the path of the average (Black–Scholes) stock price $\bar{X}_s, 0 \leq s \leq T$. Since

$$\int_0^T \mathcal{L}^{BS} F(s, \bar{X}_s) ds + \frac{1}{2}\gamma \int_0^T (\bar{X}_s)^2 C_{xx}^{BS}(s, \bar{X}_s) dB_s$$

is, given this path, a Gaussian random variable with mean $M := \int_0^T \mathcal{L}^{BS} F(s, \bar{X}_s) ds$ and variance

$$S^2 := \gamma^2 \int_0^T \frac{1}{4}(X_s^\varepsilon)^4 C_{xx}^{BS}(s, X_s^\varepsilon)^2 ds,$$

our choice of F is based on a quantity that makes M negative (so that the average profit in (16) is positive). We also want $-M$ to be a number ρ times the standard deviation S. We can then choose $\rho > 0$ depending on how much risk we are prepared to allow of the normal random variable exceeding that number of its standard deviations.

First we solve

$$\mathcal{L}^{BS} \zeta(t, x) = -\frac{1}{2}\gamma x^2 C_{xx}^{BS}(t, x) \tag{17}$$

$$\zeta(T, x) = 0,$$

and then taking (at time $t = 0$),

$$F(t, x) = \frac{\rho}{\sqrt{T}}\zeta(t, x) \tag{18}$$

for some $\rho > 0$ (the number of standard deviations we want), we have

$$\int_0^T \mathcal{L}^{BS} F(s, \bar{X}_s) ds = \frac{\rho\gamma}{2\sqrt{T}} \int_0^T \bar{X}_s^2 C_{xx}^{BS}(s, \bar{X}_s) ds$$

$$\geq \rho S. \tag{19}$$

The last inequality follows from the Cauchy–Schwarz inequality

$$\left(\int_0^T f(s)ds \right)^2 \geq T \int_0^T f(s)^2 ds$$

for non-negative functions $f(\cdot)$.

Note that from (17) and (18), $\mathcal{L}^{BS} F \leq 0$ so that $M \leq 0$ and the average replication error is positive. Thus, with this choice of F, the replication error is weakly approximated by a random variable that, conditional on the path $\bar{X}_s, 0 \leq s \leq T$, is normal with mean ρ times the standard deviation. Because the convergence is weak, this cannot be translated into a result along almost all paths of the Brownian motion $\{W_t, t \geq 0\}$. Thus we can say the chosen hedging strategy dominates the perfect hedging strategy with B-probability (probability with respect to the limiting volatility fluctuation measure defined by B_t) along almost all paths of the *average* stock price $\bar{X}$, but as the individual paths of $\bar{X}$ may not be close to the paths of X^ε, we cannot make a precise quantification in terms of the joint law of X^ε. The simulations of the next section will demonstrate the effectiveness of the strategy.

In practice, ρ controls the lower bound on the risk: taking more standard deviations increases the probability of a successful hedge. Choosing $\rho = 2$ means the hedging strategy dominates the perfect strategy with B-probability 95%. Again, we stress that this is probability on the space of paths of the limiting Brownian motion B_t. We expect that the hedge success probability will also be high although we cannot quantify it exactly. A higher ρ also increases the cost of the hedge, as seen from the choice of V_0 above.

It remains to compute $F(t, x)$ by solving (17). Using the Green's function for the Black–Scholes PDE (see, for example, Ref. 13 and Appendix C), we find

$$F(t, x) = \rho \gamma \sqrt{\frac{\tau}{2\pi \bar{\sigma}^2 T}} \frac{K^{a+1}}{x c^a} \exp\left(-\nu \tau - \frac{L^2}{2\bar{\sigma}^2 \tau} \right),$$

where $L = \log(x/K), \tau = T - t, a = r/\bar{\sigma}^2 - 1/2$. This is the formula incorporating a non-zero interest rate r which was omitted from the equations so far for simplicity of presentation.

The hedging strategy is given by (15), where

$$F_x = -\rho \gamma \sqrt{\frac{\tau}{2\pi \bar{\sigma}^2 T}} \left(\frac{K}{x} \right)^{a+1} \left(\frac{\log(x/K)}{\bar{\sigma}^2 \tau} + \frac{r}{\bar{\sigma}^2} - \frac{1}{2} \right) \exp\left(-\nu \tau - \frac{L^2}{2\bar{\sigma}^2 \tau} \right).$$

$$(20)$$

5. Simulations

In this section, we demonstrate the effectiveness of the hedging strategy derived above by simulating many stock price paths in a stochastic volatility environment, and presenting profit/loss histograms with respect to these realizations. We simulate

(3) in which $\sigma(t/\varepsilon)$ is a rapidly fluctuating two-state Markov chain volatility process. This is not proposed as a realistic model of volatility, but we use it here to illustrate the performance of the asymptotic theory.

We compare two alternative hedging strategies in the underlying stock, the first with the Black–Scholes strategy using the averaged volatility $\sqrt{\bar{\sigma^2}}$, in which the option writer holds $C_x^{BS}(t, X_t^\varepsilon)$ units of the stock at time t. The cost of the strategy is $C^{BS}(0, x)$, where x is the observed stock price at $t = 0$.

The second strategy incorporates the asymptotic correction for the randomly fluctuating volatility: it involves holding $H(t, X_t^\varepsilon)$ of the stock, where H is defined by (15), and we choose $\rho = 1$ in (18). That is, the strategy dominates the perfect hedging strategy with B-probability 67%. The cost of this hedge is larger than the Black–Scholes hedge: it is $C^{BS}(0, x) + \sqrt{\varepsilon}F(0, x)$.

In Fig. 3, we show the stock, volatility and hedging processes along a typical realization, and in Fig. 4 the profit-loss histograms from 3000 runs implementing the two strategies over the length of a 12-month $(T = 1)$ contract with 200 equally spaced re-hedgings. The profits/losses are with respect to the different costs of each strategy.

We see that the conservative second strategy yields profits much more often. On average, the profit is \$3.19 with respect to the initial cost of \$16.38. The Black–Scholes strategy produces an average profit of \$0.28 with respect to the lower cost of the strategy of \$13.32. An even more successful strategy would be to take $\rho = 2$, for example, though of course the cost will be much higher.

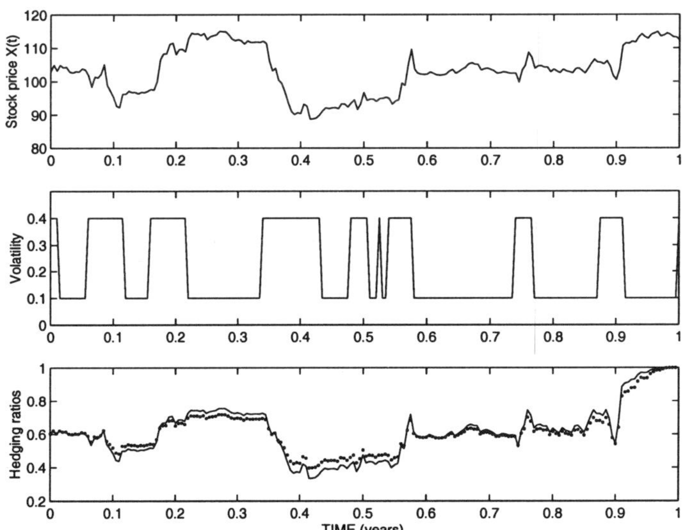

Fig. 3. Stock price X_t^ε, volatility $\sigma(t/\varepsilon)$ and hedging ratios along one path. The parameter values are $\varepsilon = 0.0005, \rho = 1, K = 100, T = 1$. Volatility is a two-state Markov chain with values 0.1 or 0.4 and $\gamma = 1$. In the bottom graph, the dotted line shows the asymptotics-adjusted hedging ratio $H(t, X_t^\varepsilon)$ and the solid line is the Black–Scholes strategy $C_x^{BS}(t, X_t)$.

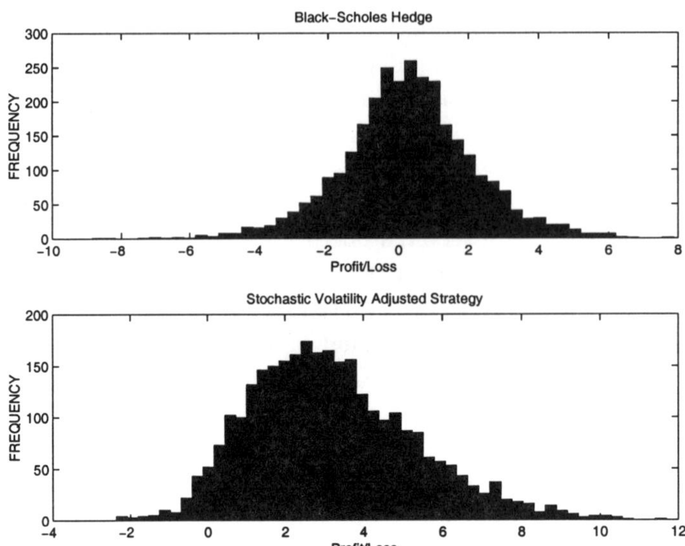

Fig. 4. Profit/loss statistics of the two strategies from 3000 simulations using the parameter values listed in Fig. 3. The adjusted hedge is sucessful more often. Note that this does not imply an arbitrage as the cost of the second hedge is greater.

6. Estimation of Parameters

We now present a simple algorithm to estimate $\bar{\sigma}^2$ and $\varepsilon\gamma^2$ using long-run historical stock price data. The method exploits the conditional lognormal distribution of X_t^ε in the model (3). These are the only parameters needed in the asymptotic theory.

Suppose we have discrete observations $X^\varepsilon(t_n)$ of the stock price at evenly-spaced times $t_n = n\Delta t$, $n = 0, \ldots, N$. Then, as $Y^\varepsilon(t) := \log X^\varepsilon(t)$ satisfies

$$dY^\varepsilon(t) = \left(\mu - \frac{1}{2}\sigma^2\left(\frac{t}{\varepsilon}\right)\right) dt + \sigma\left(\frac{t}{\varepsilon}\right) dW_t \,,$$

the discrete increments of the logs of the observations satisfy

$$D_n := Y_n - Y_{n-1} = \sigma\left(\frac{t_n}{\varepsilon}\right)\Delta W_n + \left(\mu - \frac{1}{2}\sigma^2\left(\frac{t_n}{\varepsilon}\right)\right)\Delta t \,,$$

where ΔW_n is a $\mathcal{N}(0, \Delta t)$ random variable.

Then the quantities

$$M_k := \frac{1}{N-k}\sum_{n=1}^{N-k}(D_n - D_{n+k})^2 \,,$$

for $k = 1, \ldots, N-1$, can be used to estimate $\bar{\sigma}^2$ because

$$E\{M_k\} = 2\bar{\sigma}^2\Delta t + \mathcal{O}\left(\Delta t^{3/2}\right) \,,$$

where we have used the stationarity of $\sigma^2(\cdot)$.

Similarly, the quantities

$$T_k := \frac{1}{N-k} \sum_{n=1}^{N-k} D_n^2 D_{n+k}^2$$

can be used to estimate the (non-centred) autocorrelation of σ because

$$E\{T_k\} = E\left\{\sigma^2\left(\frac{k\Delta t}{\varepsilon}\right)\sigma^2(0)\right\}\Delta t^2 + \mathcal{O}\left(\Delta t^{5/2}\right),$$

where we have used the fact that $E\{\sigma^2(t+h)\sigma^2(t)\}$ depends only on h (second-order stationarity).

From our observations, we calculate the empirical autocorrelation

$$R_k := \frac{1}{\Delta t^2}\left[T_k - \frac{1}{4}\left(\frac{1}{k}\sum_{l=1}^{k} M_l\right)^2\right].$$

The expected value of each R_k approximates the autocorrelation:

$$E\{R_k\} = E\left\{\sigma^2\left(\frac{k\Delta t}{\varepsilon}\right)\sigma^2(0)\right\} - (\bar{\sigma^2})^2 + \mathcal{O}\left(\Delta t^{1/2}\right)$$

$$= E\left\{\left(\sigma^2\left(\frac{k\Delta t}{\varepsilon}\right) - \bar{\sigma^2}\right)(\sigma^2(0) - \bar{\sigma^2})\right\} + \mathcal{O}\left(\Delta t^{1/2}\right).$$

Then as

$$\int_0^\infty E\left\{\left(\sigma^2\left(\frac{s}{\varepsilon}\right) - \bar{\sigma^2}\right)(\sigma^2(0) - \bar{\sigma^2})\right\}ds$$

$$= \varepsilon \int_0^\infty E\left\{\left(\sigma^2(s) - \bar{\sigma^2}\right)(\sigma^2(0) - \bar{\sigma^2})\right\}ds$$

by a change of variable, it follows that twice the area under the curve obtained by interpolating the empirical autocorrelation $\{R_k\}$ is an estimate of $\varepsilon\gamma^2$.

This procedure is tested on simulated data in Ref. 13 and practical issues (such as adaptation to non-evenly-spaced high-frequency data) are addressed there. We are presently working with real market data.

7. Conclusions

The hedging strategies computed here (distinguished by the number $\rho(p_0)$) dominate the perfect hedging strategy (which depends on the realized volatility path) with B-probability p_0. This approximate bound comes from two modelling features: separation of time-scales and the generality of studying the Black–Scholes pricing PDE with a random volatility coefficient, allowing the results to apply for a large class of possible stochastic volatility processes, with unspecified distribution. The result is restricted to uncorrelated volatility (no skew or leverage effect) which is realistic in F/X and some equity markets. The simulations demonstrate the effectiveness of the $\rho = 1$ strategy, although a precise analytical quantification of the success probability is not possible because the asymptotic approximations converge weakly and not pathwise.

Greater precision in the quantification of the risk of an unsuccessful hedge would require a more detailed (parametric) description of volatility and the market pricing measure. This is studied in Ref. 6. The separation of scales analysis then aids estimation of the parameters of the model from S&P 500 index data.

The major benefit of the asymptotic analysis is computability, and it is a topic of future work to apply it to the stochastic control problems that arise when one wants to maximize a probability over possible hedging strategies.

References

[1] M. Avellaneda, A. Levy and A. Paras, "Pricing and hedging derivative securities in markets with uncertain volatilies", *Applied Mathematical Finance* **1**(2) (1995) 73–88.

[2] D. Bates, "Testing option pricing models", in *Statistical Methods in Finance*, Volume 14 of *Handbook of Statistics*, G. Maddala and C. Rao, eds., (North Holland, Amsterdam, 1996) Chapter 20, pp. 567–611.

[3] J. Cvitanić, H. Pham and N. Touzi, "Super-replication in stochastic volatility models under portfolio constraints", *Journal of Applied Probability* **36**(2) (1999) 523–545.

[4] H. Föllmer and P. Leukert, "Quantile hedging", *Finance and Stochastics* **3**(3) (1999).

[5] J.-P. Fouque, G. Papanicolaou and K. R. Sircar, "Asymptotics of a two-scale stochastic volatility model", in *Equations aux derivees partielles et applications, in honour of Jacques-Louis Lions* (Gauthier-Villars, May 1998), pp. 517–525.

[6] J.-P. Fouque, G. Papanicolaou and K. R. Sircar, "Mean-reverting stochastic volatility", *International Journal of Theoretical and Applied Finance* **3**(1) (2000) 101–142.

[7] R. Frey, *Derivative asset analysis in models with level-dependent and stochastic volatility, CWI Quarterly* **10**(1) (1996) 1–34.

[8] J. Hull and A. White, "The pricing of options on assets with stochastic volatilities", *J. Finance* **42**(2) (1987) 281–300.

[9] N. El Karoui, M. Jeanblanc-Picqué and S. Shreve, "Robustness of the Black and Scholes formula", *Math. Finance* **8**(2) (1998) 93–126.

[10] G. C. Papanicolaou, "Asymptotic analysis of stochastic equations", in *Studies in Probability Theory, Mathematical Association of America*, Murray Rosenblatt, eds., **18** (1978) 111–179.

[11] E. Renault and N. Touzi, "Option hedging and implied volatilities in a stochastic volatility model", *Mathematical Finance* **6**(3) (1996) 279–302.

[12] M. Rubinstein, "Non-parametric tests of alternative option pricing models", *J. Finance* **40**(2) (1985) 455–480.

[13] K. R. Sircar and G. C. Papanicolaou, "Stochastic volatility, smile and asymptotics", *Applied Mathematical Finance* **6**(2) (1999) 107–145.

[14] A. Skorohod, *Asymptotic Methods in the Thory of Stochastic Differential Equations*, Translations of Mathematical Monographs, American Mathematical Society, 1989.

DETERMINING VOLATILITY SURFACES AND OPTION VALUES FROM AN IMPLIED VOLATILITY SMILE

PETER CARR

Banc of America Securities, 9 West 57th Street,
40th Floor, New York, NY 10019
E-mail: pcarr@bofasecurities.com

DILIP MADAN

Robert H. Smith School of Business, University of Maryland,
College Park, MD 20742, (301) 405-2127
E-mail: dbm@mbs.umd.edu

Received 2 October 1998

Using only the implied volatility smile of a single maturity T and an assumption of path-independence, we analytically determine the risk-neutral stock price process and the local volatility surface up to an arbitrary horizon $T' \geq T$. Our path-independence assumption requires that each positive future stock price S_t is a function of only time t and the level W_t of the driving standard Brownian motion (SBM) for all $t \in (0, T')$. Using the T-maturity option prices, we identify this stock pricing function and thereby analytically determine the risk-neutral process for stock prices. Our path-independence assumption also implies that local volatility is a function of the stock price and time which can be explicitly represented in terms of the known stock pricing function. Finally, we derive analytic valuation formulae for standard and exotic options which are consistent with the observed T-maturity smile.

Keywords: Option pricing, implied volatility arbitrage pricing theory, path-independence.

1. Introduction

The Black–Scholes model can be described as an ambitious attempt to explain the prices of all standard and exotic options using only a single parameter for each underlying. It is well-known that this volatility parameter can either be estimated from historical spot prices or implied from the price of a single option. In general, the latter approach is preferred in practice since it is "forward-looking" and at least partially captures model error, albeit crudely. When an option price is used to imply volatility, all standard and exotic options on the same underlying are valued using a risk-neutral probability measure consistent with the price of this option, as well as with the prices of the primitive assets.

The persistence of a volatility smile in options markets has lead to several proposed alternatives to the standard Black–Scholes model. For example, modeling the price of the underlying as a CEV process,[3] a displaced diffusion,[6] or as a call price itself,[9] have all been proposed as parsimonious generalizations of the Black–Scholes model which retain the model's analytic flavor. All of these approaches introduce an additional parameter and thus require at least two options for calibration.

Continuing the trend of expanding the set of basis options, Rubinstein[17] introduced the implied tree approach which values claims relative to standard options with strikes at every terminal node in a binomial tree. In his framework, standard options of any nearer maturity and exotic options are valued using a risk-neutral measure consistent with the prices of the primitive assets and of all options of the terminal maturity. Further continuing the trend of expanding the set of basis options, Derman and Kani[6] use standard options with strikes and maturities corresponding to every node in their implied tree. Similarly, working in a continuous setting, Dupire[8] requires a continuum of strikes and maturities. Since all standard option prices are used in the specification of the risk-neutral process, these last two approaches can only be used to determine the values of exotic options.

The expansion of the set of assets used to determine the risk-neutral measure has the important theoretical advantage of generating consistency with a wider set of assets and allowing weaker assumptions on the stochastic process governing the underlying spot prices. When all strikes and maturities are used as in Dupire, the diffusion coefficient can be any reasonable function of the contemporaneous spot price and time. Furthermore, this diffusion coefficient, which is termed the local volatility, need not be specified *ex ante*, since it can be analytically determined from the double continuum of European option prices.

One obvious drawback of these more recent developments is the requirement of an option strike and maturity for every price and time at which dynamic trading is possible. For listed stock options, the minimum distance between strikes and maturities is $5 and one month respectively. Consequently, interpolation is required to implement these approaches in the equity case. Furthermore, extrapolation is always required both from the last maturity out to perpetuity and from the nearest maturity in to the initial time. When implemented naïvely, these interpolations or extrapolations can engender option price quotes which are not arbitrage-free.

A second drawback associated with the expansion of the set of basis options is the comparative silence of these models regarding mis-pricings of standard options. Although the models require that the input option prices be arbitrage-free, these models blithely accept option prices inconsistent with any reasonable equilibrium framework. Since options markets are often illiquid and options' prices have wide bid ask spreads, the expansion of the set of basis options to all strikes and maturities can often induce model mis-specification. To illustrate this point, note that all of the above models embed the Black–Scholes model as a special case. Thus, in the unlikely event that the spot price actually does follow geometric Brownian motion, the incorporation of noisy option prices into the determination of the risk-neutral measure will lead to incorrect valuations of any remaining options. In fact, empirical evidence supporting this observation in the case of Dupire's local volatility model has been recently promulgated by Dumas, Fleming, and Whaley.[7]

Even if the spot price follows a more complicated stochastic process than geometric Brownian motion, certain models may not permit independent specification of initial option prices at every strike and maturity. For example, if Rubinstein's implied tree describes the actual underlying spot price, then intermediate maturity option values are determined by the model. If the market prices of these options differ from the implied tree value for any reason, then incorporating these option prices into the determination of the risk-neutral process will result in mis-pricing, both for these options and exotic options.

In general, the quantity and quality of the options data should determine the appropriate tension between ex-ante process specification on the one hand and calibration to market prices on the other. The purpose of this paper is to propose an analytic and arbitrage-free approach to option valuation when only one maturity is liquid. Working in a continuous setting, we assume as in Rubinstein that one can observe option prices of all strikes maturing at some common expiration T.

In order to determine the entire local volatility surface out to an arbitrarily distant date, we further assume that each positive future spot price S_t depends only on the contemporaneous level W_t of the driving standard Brownian motion (SBM) and not on its path. This path-independence property was also assumed by Black and Scholes, who further assumed that S_t is an exponential function of W_t.[a] In contrast to Black–Scholes, we infer this spot pricing function as a consequence of our three primary assumptions, namely path-independence, no arbitrage, and the liquidity of the T-maturity options.

In contrast to the primary assumption of the Black–Scholes model, stock prices sometimes hit zero in reality. To allow for this contingency, we allow for the possibility of an absorbing lower barrier on the standard Brownian motion (SBM) and assume that the stock pricing function evaluates to zero along this boundary.[b] For analytic tractability, we restrict attention to boundaries for the SBM which are constant over time.[c] Given our path-independence assumption, both the stock pricing function and the constant absorbing boundary can be inferred from the option prices. For underlyings such as commodities and currencies, future spot prices of zero are generally considered impossible and in this case, the option prices will imply that the absorbing barrier for the SBM is negative infinity. Whether or not bankruptcy can occur, we characterize the *entire* class of risk-neutral path-independent stock price processes, which vanish when the driving SBM hits a constant lower boundary (possibly negative infinity) and which can be supported over an arbitrarily long time horizon. We also identify the unique member of this class consistent with the option prices observed at maturity T. Once the stock price becomes a known function of the driving SBM, the stock price process inherits the analytical tractability of this SBM.

[a]In a second model, they instead assumed that the spot is a call on the assets of the firm, and thus a different function of the SBM.

[b]Our model can be generalized to allow for the possibility that option prices imply a positive liquidation value in bankruptcy.

[c]It would be straightforward to consider linear boundaries.

For example, the probability density function (PDF) for the first passage time (if any) of the stock price to the origin is just the known PDF for the first passage time of SBM to its absorbing barrier.

Our path-independence assumption also implies that local volatility is a function of stock price and time which can be explicitly represented in terms of the known stock pricing function. Consequently, we can generate the entire local volatility surface from the observation time out to an arbitrary date occuring after the last maturity. We also show that our path-independence assumption implies that local volatility satisfies a new non-linear partial differential equation (PDE). Conversely, we show that if the local volatility function satisfies this PDE, then spot prices are path-independent and we provide the map transforming spot prices into the SBM.

Given that spot prices are path-independent, or equivalently that local volatility has the required form, we derive analytic formulas relating the values of standard and exotic options to the contemporaneous spot price and time. These analytic formulas exist for a wide range of volatility surfaces and as in Rubinstein, the values are fully consistent with the liquid option prices at maturity T. In fact, our results may be characterized as analytically extending Rubinstein's results to continuous-time diffusion processes, and to the valuation of options maturing both after and before T.

The outline for this paper is as follows. The next section shows how the twin assumptions of path-independence (PI) and no arbitrage (NA) lead to a simple linear PDE governing the spot pricing function and how the T-maturity option prices determine the spot pricing function at time T. We present the general solution to this problem and illustrate it with a simple example. The next section shows how PI and NA also determine the volatility function in terms of the known stock pricing function. We show that our volatility function satisfies a fully non-linear PDE. We also present a converse result showing that the non-linear PDE is both a necessary and sufficient condition for prices to be path-independent. Assuming that spot prices are path-independent, or equivalently that local volatility solves the non-linear PDE, the penultimate section shows how to analytically determine the pricing functions for path-independent standard and exotic options of any maturity. A final section summarizes and discusses extensions. Tedious derivations are banished to the appendices.

2. The Stock Pricing Function

2.1. *Assumptions and PDE*

We assume that European options of a fixed maturity $T > 0$ and their underlying asset (termed the stock) trade in a frictionless market with no arbitrage (NA). We assume that the initial stock price is positive and also assume that the limited liability of the stock induces absorption of the stock price at the origin. We let τ denote the first passage time of the stock price to the origin, if any.[d] To determine

---

[d]If the stock price never hits the origin as in the Black–Scholes model, then we set $\tau = \infty$.

volatility surfaces and option values out to some arbitrary fixed horizon $T' \in [T, \infty)$, we assume that the stock price process is *path-independent* (PI). More specifically, we assume that over the horizon $[0, T' \wedge \tau]$, each future spot price S_t is related to the contemporaneous level B_t of the driving standard Brownian motion (SBM) and time t by a non-negative[e] $C^{2,1}$ function $s(x, t), x \in \Re, t \in [0, T']$

$$S_t = s(B_t, t), \quad t \in [0, T' \wedge \tau]. \tag{1}$$

Our assumption of a positive initial stock price implies $s(0, 0) = S_0 0$. We will later show that NA and PI imply that $s(x, t)$ is increasing in $x \in \Re$ for each $t \in [0, T']$. Thus, for each t, there exists at most one root $L_t \in (-\infty, 0)$ of the stock pricing function

$$s(L_t, t) = 0, \quad t \in [0, T'].$$

For analytic tractability, we restrict attention to the class of stock pricing functions such that the lower liquidation boundary[f] L_t is constant at L over time.

By Itô's lemma, the spot price dynamics can be written as:

$$dS_t = \left[\frac{\partial s}{\partial t}(B_t, t) + \frac{1}{2}\frac{\partial^2 s}{\partial x^2}(B_t, t) \right] dt + \frac{\partial s}{\partial x}(B_t, t)dB_t, \quad t \in [0, \tau \wedge T'). \tag{2}$$

Since we can invert (1) for B_t, (2) implies that both the drift and diffusion coefficients are functions $b(S, t)$ and $\sigma(S, t)$ of only the spot price S and the calendar time t. Thus, over the time interval $(0, \tau \wedge T')$, the stock price process is Markov in the pair (B_t, t) or in the pair (S_t, t).

For simplicity, we assume a constant riskless rate $r \geq 0$ and a constant dividend yield $q \geq 0$ from the stock. Under these assumptions, it is well-known that there exists a unique "risk-neutral" probability measure Q, which is defined by the following spot price dynamics

$$dS_t = (r - q)S_t dt + \sigma(S_t, t)S_t dW_t, \quad t \in [0, \tau \wedge T'], \tag{3}$$

where W is a Q-SBM. Letting W_t^a denote SBM absorbing at a fixed $L < 0$, (1) and (2) may be re-written as

$$S_t = s(W_t^a, t), \quad t \in [0, T'], \tag{4}$$

and

$$dS_t = \left[\frac{\partial s}{\partial t}(W_t^a, t) + \frac{1}{2}\frac{\partial^2 s}{\partial x^2}(W_t^a, t) \right] dt + \frac{\partial s}{\partial x}(W_t^a, t)dW_t^a, \quad t \in [0, T']. \tag{5}$$

Equating coefficients on dt in (3) and (5) and using (4) gives a partial differential equation (PDE) for $s(x, t)$

$$\frac{\partial s}{\partial t}(x, t) + \frac{1}{2}\frac{\partial^2 s}{\partial x^2}(x, t) = (r - q)s(x, t), \quad t \in [0, T'], \ x > L, \tag{6}$$

---

[e]For options on underlyings which can go negative (e.g., spread options), this assumption can be dropped.

[f]If there is no root, we set the liquidation boundary $L = -\infty$.

subject to the boundary conditions

$$\lim_{x \downarrow L} s(x,t) = 0, \quad t \in [0, T'], \tag{7}$$

and

$$\lim_{x \uparrow \infty} s(x,t) = o(e^x), \quad t \in [0, T']. \tag{8}$$

The last boundary condition ensures that the payoff is integrable.

2.2. *Implied stock payoff function*

This subsection show how our assumption of path-independence can be combined with the implied volatility smile at maturity T to determine the stock pricing function at T, $s(x,T) \equiv f(x)$. Using the Black–Scholes formula, the implied volatility smile can be converted to the strike structure of T-maturity option prices. Ross[15] and Breeden and Litzenberger[2] have pointed out that differentiating this strike structure twice and future-valuing yields the "risk-neutral" probability density for the future stock price. Interpreting this density as arising from a change of variables of the known density describing future levels of an absorbing SBM will yield the desired stock payoff function $f(x)$.

To operationalize these observations in our continuous context, we assume that a continuum of T-maturity put prices $\{P_0(K), K > 0\}$, are observable. We further assume that these prices are given by a C^2 function, which by the absence of arbitrage must be positive, upward sloping, and convex. Let $P_k(K) \equiv \lim_{\triangle K \downarrow 0} \frac{P_0(K+\triangle K)-P_0(K)}{\triangle K}$ denote the observed slope in strike at K, which is also the value of an infinitesimal vertical spread struck at K. Given our path-independence assumption, the following relationship exists between this vertical spread and the inverse of the implied spot payoff, $f^{-1}(K)$

$$e^{rT} P_k(K) = \int_L^{f^{-1}(K)} \frac{1}{\sqrt{2\pi T}} \left\{ \exp\left[-\frac{1}{2}\left(\frac{z}{\sqrt{T}}\right)^2 \right] \right.$$
$$\left. - \exp\left[-\frac{1}{2}\left(\frac{z-2L}{\sqrt{T}}\right)^2 \right] \right\} dz, \quad K > 0, \tag{9}$$

since the integrand is the risk-neutral probability density for SBM absorbing at L. Solving the LHS for K gives the implied payoff function

$$f(x) = P_k^{-1} \left\{ e^{-rT} \int_L^x \frac{1}{\sqrt{2\pi T}} \left\{ \exp\left[-\frac{1}{2}\left(\frac{z}{\sqrt{T}}\right)^2 \right] - \exp\left[-\frac{1}{2}\left(\frac{z-2L}{\sqrt{T}}\right)^2 \right] \right\} dz \right\}$$
$$= P_k^{-1} \left\{ e^{-rT} \left[N\left(\frac{x}{\sqrt{T}}\right) - N\left(\frac{L}{\sqrt{T}}\right) \right. \right.$$
$$\left. \left. - N\left(\frac{x-2L}{\sqrt{T}}\right) + N\left(\frac{-L}{\sqrt{T}}\right) \right] \right\}, \quad x > L, \tag{10}$$

where $N(d) \equiv \int_{-\infty}^d \frac{e^{-z^2/2}}{\sqrt{2\pi}} dz$ is the standard normal distribution function.

Since P_k is a positive increasing function of K, P_k^{-1} is a positive increasing function of its argument. Since the argument of P_k^{-1} is increasing in x for $x > L$, it follows that $f(x)$ is a positive increasing function on this domain. If $f(x) = 0$ for x less than some point, then we identify this point as the lower liquidation level L.

2.3. *Solving for the stock pricing function*

If one is interested in obtaining the stock pricing function on the time interval $[0, T)$, then the Feynman–Kac theorem can be used to find the continuous solution to the BVP consisting of the PDE (6) with $t \in [0, T)$, the boundary conditions (7) and (8), and the terminal condition $f(\cdot)$ determined in (10)

$$s(x,t) = e^{-(r-q)(T-t)} E_{x,t}^Q [f(W_T)1(\tau > T)], \quad x > L, \ t \in [0, T \wedge \tau], \tag{11}$$

where $\{W_u; u \in [t,T]\}$ is a Q-SBM starting at x. Since $f(\cdot)$ is a positive increasing function of x, so is $s(x,t)$ for each t.

We now consider the determination of the stock pricing function on the time interval $[0, T')$, where $T' > T$ can be arbitrarily large. If we can obtain the horizon payoff at T' from $f(\cdot)$, then we can again use the Feynman–Kac theorem to represent the solution on $[0, T')$. It is tempting to write

$$f(x) = e^{-(r-q)(T'-T)} \int_L^\infty \frac{1}{\sqrt{2\pi(T'-T)}} \left\{ \exp\left[-\frac{1}{2}\left(\frac{z-x}{\sqrt{T'-T}} \right)^2 \right] \right.$$

$$\left. - \exp\left[-\frac{1}{2}\left(\frac{z-x-2L}{\sqrt{T'-T}} \right)^2 \right] \right\} g(z)dz, x > L \tag{12}$$

as an implicit definition for the horizon payoff $g(\cdot)$ arising at T'. Since $f(x)$ $e^{(r-q)(T'-T)}$ is a convolution of the transition density of absorbing SBM with $g(z)$ $1(z > L)$, Fourier transforms can be used to explicitly express $g(\cdot)$ in terms of $f(\cdot)$. However, this formulation may be ill-posed in that there is no result on the existence of $g(\cdot)$. To appreciate that this is not merely a technical concern, suppose that $f(x)$ described the value at T of a call on the SBM paying $(W_M - K)^+$ at some date M fixed strictly between T and T'. Then it will be impossible to obtain the stock pricing function for any time after M. On the other hand, in the Black–Scholes model, an implied payoff $f(x) = S_0 e^{(r-q-\frac{\sigma^2}{2})T+\sigma W_T}$ would generate a horizon payoff $g(z) = S_0 e^{(r-q-\frac{\sigma^2}{2})T'+\sigma W_{T'}}$ for any $T' \geq T$ without difficulty.

Thus, the question arises as to whether there exist other processes besides geometric Brownian motion which are sustainable over an arbitrarily long horizon. Fortunately, an extension of a result of Robbins and Siegmund[14] (see Karatzas and Shreve,[12] p. 262) proves that the *entire* set of non-negative functions $\phi(z,t)$ which satisfy the backward diffusion equation $\frac{\partial \phi}{\partial t}(z,t) + \frac{1}{2}\frac{\partial^2 \phi}{\partial z}(z,t) = 0$ on the half plane

$(0, \infty) \times \Re$ subject to:

$$\lim_{t \downarrow 0} \phi(0, t) = 1, \quad \lim_{t \uparrow \infty} \phi(0, t) = 0, \quad \text{and} \quad \lim_{z \downarrow -\infty} \phi(z, t) = 0,$$

$$\lim_{z \uparrow \infty} \phi(z, t) = \infty, \quad 0 < t < \infty,$$

are given by

$$\phi(z, t) = \int_0^\infty e^{\theta z - \frac{\theta^2}{2} t} dG(\theta), \tag{13}$$

where $G(\theta)$ is a distribution function with $G(\infty) = 1$ and $G(0+) = 0$. Surprisingly, the process $\phi(W_t, t), t > 0$ is just the familiar geometric Brownian motion for the futures price relative $\frac{F_t}{F_0}$, where the volatility parameter has been randomized[g] by a probability measure G defined on $\Re^+$.

Adapting these results to a stock price process with absorption at zero, Appendix 1 proves that the entire set of stock pricing functions $s(x, t)$ which vanish at $x = L$, are positive, increasing, and unbounded in x on $x > L$, and which satisfy the PDE (6) for $x > L$ and $t \in (0, \infty)$ are given by

$$s(x, t) = S_0 e^{(r-q)t} \int_0^\infty \frac{\sinh[\theta(x - L)]}{\sinh[-\theta L]} e^{-\frac{\theta^2}{2} t} dG(\theta), \quad x > L, \ t \in (0, T'), \tag{14}$$

where $\sinh(x) \equiv \frac{e^x - e^{-x}}{2}$. Assuming that $G(\cdot)$ is the distribution function of a continuous random variable with a probability density fuction $G'(\theta)$, evaluation of (14) at $t = T$ relates the implied stock payoff $f(x)$ to the density function $G'(\theta)$:

$$s(x, T) \equiv f(x) = S_0 e^{(r-q)T} \int_0^\infty \frac{\sinh[\theta(x - L)]}{\sinh[-\theta L]} e^{-\frac{\theta^2}{2} T} G'(\theta) d\theta, \quad x > L. \tag{15}$$

Appendix 1 proves that for any function $f(x)$ such that $f(x)e^{-x^2}$ is L^1, a unique density function $G'(\theta)$ exists which satisfies (15). Appendix 1 also shows how to invert (15) to obtain this density function $G'(\theta)$ from the known implied payoff $f(x)$

$$G'(\theta) = \begin{cases} \dfrac{\sqrt{T}}{S_0 e^{(r-q)T}} \dfrac{\sinh(-\theta L)}{\pi \sqrt{2\pi}} \displaystyle\int_{-\infty}^\infty e^{-i\omega\theta\sqrt{T} + \frac{\omega^2}{2}} \int_{-\infty}^\infty e^{i\omega z - \frac{z^2}{2}} \\ \quad \times f(L + z\sqrt{T}) dz d\omega & \text{if } \theta \geq 0; \\[2ex] 0 & \text{if } \theta < 0. \end{cases} \tag{16}$$

Thus, the stock pricing function $s(x, t)$ is given by substituting (16) in (14):

[g]This randomized process would describe the futures price relative in a filtering context if the futures price process were assumed to be geometric Brownian motion with an unknown positive volatility.

$$s(x,t) = \frac{e^{-(r-q)(T-t)}}{\pi\sqrt{2\pi/T}} \int_0^\infty \sinh[\theta(x-L)]e^{\frac{-\theta^2}{2}t} \int_{-\infty}^\infty e^{-i\omega\theta\sqrt{T}+\frac{\omega^2}{2}}$$

$$\times \int_{-\infty}^\infty e^{i\omega z - \frac{z^2}{2}} f(L+z\sqrt{T})dz\,d\omega\,d\theta$$

$$= \frac{e^{-(r-q)(T-t)}\sqrt{T}}{2\pi} \int_{-\infty}^\infty e^{\frac{\omega^2+tv_t^2}{2}} \int_{-\infty}^\infty$$

$$\times e^{i\omega z - \frac{z^2}{2}} f(L+z\sqrt{T})[N(y_t) - N(-y_t)]dz\,d\omega\,, \tag{17}$$

for $x > L, t \in (0,T')$, where $y_t \equiv \frac{x-L-i\omega\sqrt{T}}{\sqrt{t}}$ and $T' \geq T$ is finite but can be arbitrarily large. For certain specifications of the implied payoff function $f(\cdot)$, both integrals in (17) can be done analytically, as the next subsection demonstrates. It can be shown that $s(x,t)$ is a positive increasing function of x on $x > L$ and $t \in (0,T')$. The horizon payoff is obtained by evaluating $s(x,t)$ at $t = T'$

$$g(x) \equiv s(x,T') = \frac{e^{(r-q)(T'-T)}\sqrt{T}}{2\pi} \int_{-\infty}^\infty e^{\frac{\omega^2+T'v_{T'}^2}{2}} \int_{-\infty}^\infty e^{i\omega z - \frac{z^2}{2}}$$

$$\times f(L+z\sqrt{T})[N(y_{T'}) - N(-y_{T'})]dz\,d\omega\,. \tag{18}$$

Furthermore, the risk-neutral process is given by $S_t = s(W_t^a, t), t > (0,T')$, where recall that W_t^a is SBM absorbing at a fixed $L < 0$.

2.4. *Example*

We now illustrate our result with a simple example. Suppose that the initial T-maturity option prices are such that vertical spreads have the form

$$e^{rT}P_k(K) = \int_L^{\frac{\sinh^{-1}}{\alpha}(\frac{K}{\beta})+L} \frac{1}{\sqrt{2\pi T}} \left\{ \exp\left[-\frac{1}{2}\left(\frac{z}{\sqrt{T}}\right)^2\right] \right.$$

$$\left. - \exp\left[-\frac{1}{2}\left(\frac{z-2L}{\sqrt{T}}\right)^2\right] \right\} dz\,, \quad K > 0\,. \tag{19}$$

Then from (9)

$$f^{-1}(S) = \begin{cases} \dfrac{\sinh^{-1}}{\alpha}\left(\dfrac{S}{\beta}\right) + L & \text{if } S > 0\,; \\ 0 & \text{if } S < 0\,. \end{cases}$$

Inverting this function

$$f(x) = \begin{cases} \beta\sinh[\alpha(x-L)] & \text{if } x > L\,; \\ 0 & \text{if } x < L\,. \end{cases} \tag{20}$$

Figure 1 graphs the implied payoff function and its inverse.

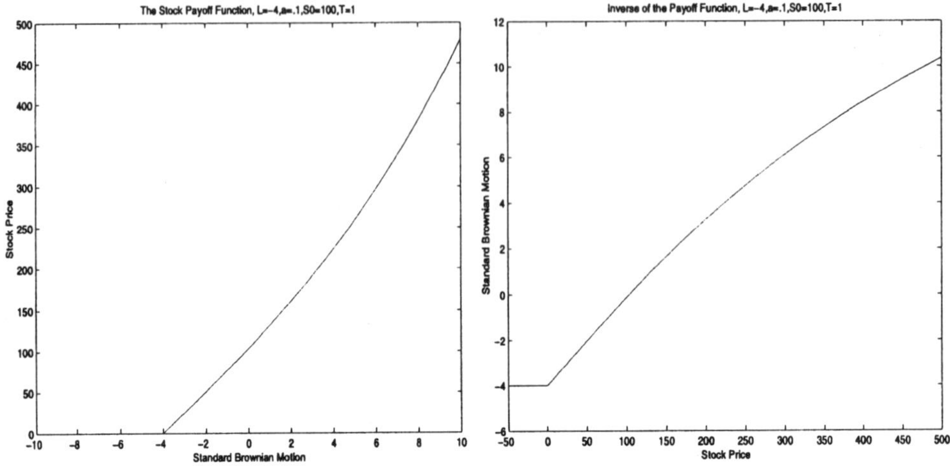

Fig. 1. The implied stock payoff function and its inverse.

Using (16), we obtain that

$$G'(\theta) = \delta(\theta - \alpha)\,, \tag{21}$$

where $\delta(\cdot)$ is Dirac's delta function. Substituting (21) into (18) implies that

$$s(x, T') = g(x) = \beta' \sinh[\alpha(x - L)]\,. \tag{22}$$

The solution of (6) subject to (7), (8), and (22) is

$$s(x, t) = \beta' e^{-\mu(T'-t)} \sinh[\alpha(x - L)]\,, \quad t \in [0, T']\,, \ x > L\,, \tag{23}$$

where $\mu \equiv r - q - \alpha^2/2$. Setting $s(0,0) = S_0$ implies

$$\beta' = S_0 e^{\mu T'} \operatorname{csch}(-\alpha L)\,, \tag{24}$$

where $\operatorname{csch}(x) \equiv \frac{1}{\sinh(x)}$. Hence from (23)

$$s(x, t) = S_0 e^{\mu t} \operatorname{csch}(-\alpha L) \sinh[\alpha(x - L)]\,, \quad t \in [0, T']\,. \tag{25}$$

Figure 2 graphs this stock pricing function against the driving SBM and time. For each time, the stock price is an increasing convex function of the SBM, although the slope and convexity are *larger* than in the Black–Scholes model. The volatility surface corresponding to the risk-neutral process $S_t = S_0 e^{\mu t} \operatorname{csch}(-\alpha L) \sinh[\alpha(W_t^a - L)], t \in [0, T']$ is determined in the next section.

3. Local Volatility

The next subsection shows how the local volatility can be represented in terms of the known stock pricing function derived in the last section from the given implied volatility smile. The following subsection illustrates our results with the same

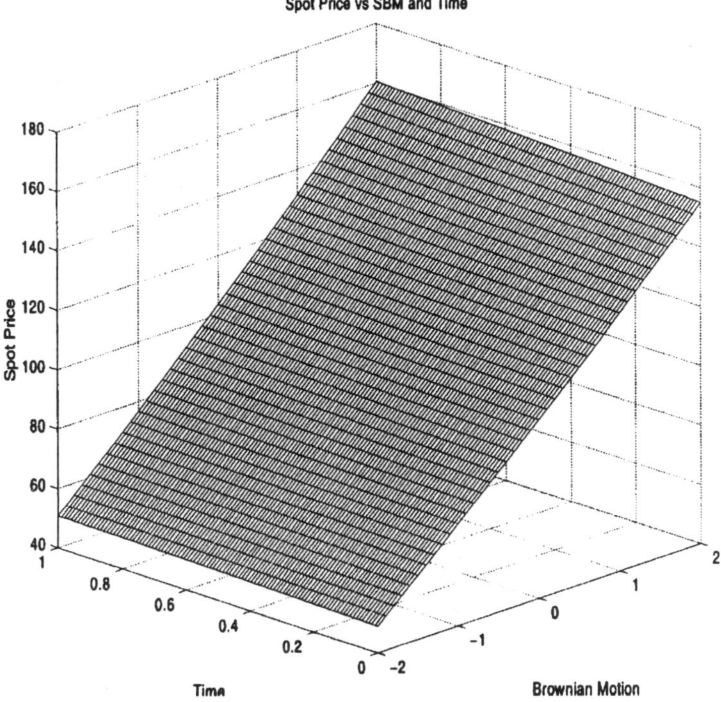

Fig. 2. The stock pricing function.

example developed in the last section. The last subsection shows that our volatility function $\sigma(S, t)$ satisfies a certain non-linear PDE. Conversely, it shows that if the local volatility function satisfies the non-linear PDE, then the stock price process is path-independent and the stock pricing function can be determined.

3.1. *The local volatility surface*

Since the stock pricing function $s(x, t)$ is increasing in x on $x > L$, it can be inverted for $x = s^{-1}(S, t)$. Equating coefficients on dW_t^a in (2) and (3) determines local volatility $\sigma(S, t)$ in terms of this inverse and $\frac{\partial s}{\partial x}(x, t)$:

$$\sigma(S, t) = \frac{1}{S}\frac{\partial s}{\partial x}(s^{-1}(S, t), t), \quad t \in [0, T'], x \in \Re.$$ (26)

Since $s(\cdot, t)$ is increasing in x for each t, $\sigma(S, t)$ is positive. Local volatility is an explicit function of S and t if $s_x(x, t)$, and $s^{-1}(S, t)$ can both be written explicitly in terms of their arguments. We have thus shown that the path-independence assumption combined with the T-maturity implied smile fully determines the local volatility surface. By evaluating the function $\sigma(S, t)$ at historical spot prices and times, the theory can be tested using historical local volatilities.

3.2. *Example*

To illustrate our results, again suppose that T-maturity option prices satisfy (19), so that the stock pricing function is given in (23). Differentiating (23) w.r.t. x implies that

$$s_x(x,t) = \alpha\beta' e^{\mu(T'-t)} \cosh[\alpha(x-L)] \,, \tag{27}$$

where $\cosh(x) \equiv \frac{e^x + e^{-x}}{2}$. Multiplying and dividing by $\sinh[\alpha(x-L)]$ gives

$$s_x(x,t) = \alpha S \coth[\alpha(x-L)] \,, \quad t \in (0,T') \,, \tag{28}$$

from (23), where $\coth(x) \equiv \frac{\cosh(x)}{\sinh(x)}$. Inverting (23) w.r.t. x implies that

$$s^{-1}(S,t) = L + \frac{\sinh^{-1}}{\alpha}\left(\frac{S}{\beta'}e^{\mu(T'-t)}\right) \,, \quad t \in (0,T') \,. \tag{29}$$

Substituting (28) and (29) in (26) determines the local volatility surface

$$\sigma(S,t) = \alpha\coth\left[\sinh^{-1}\left(\frac{S}{\beta'}e^{\mu(T'-t)}\right)\right] \,, \quad t \in (0,T') \,, \tag{30}$$

where β' is given in (24). Figure 3 graphs this local volatility surface.

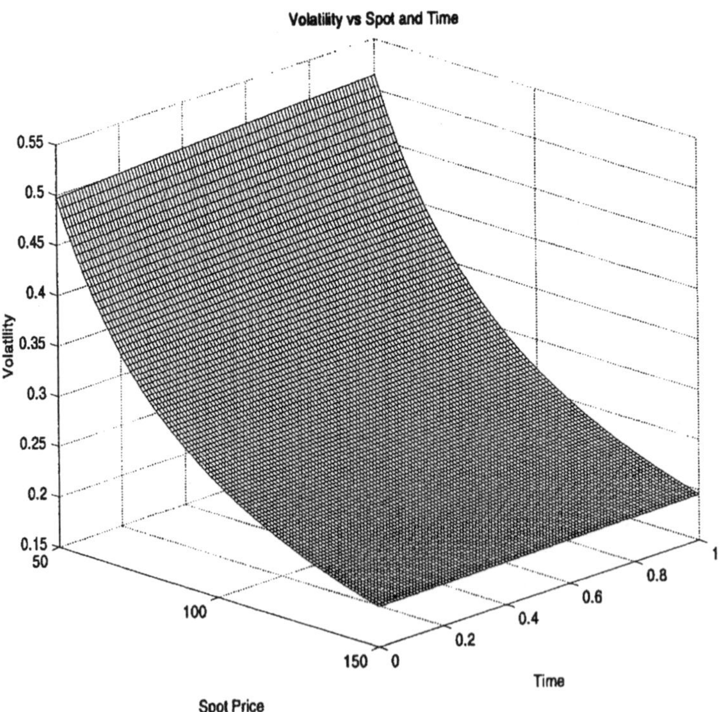

Fig. 3. The local volatility surface.

To understand the behavior of this local volatility as a function of the stock price, note that the stock pricing function is proportional to $\sinh[\alpha(x - L)]$, which behaves linearly in x for x near L and exponentially in αx for x large. Thus at each future date, the volatility smile is approximately hyperbolic in S ("normal volatility") for S near zero, while it is asymptoting to the constant α ("lognormal volatility") for S high. Mathematically, as $S \downarrow 0$ in (30), $\sinh^{-1}(\cdot) \downarrow 0, \coth(\cdot) \uparrow \infty$, as does volatility. Conversely, as $S \uparrow \infty$ in (30), $\sinh^{-1}(\cdot) \uparrow \infty, \coth(\cdot) \downarrow 1$, and volatility asymptotes to α. As S increases from 0 to ∞, the volatility smile slopes downward in a convex fashion.

To understand the behavior of this local volatility as a function of the future time t, substitute β' given in (24) into (30)

$$\sigma(S, t) = \alpha \coth \left[\sinh^{-1} \left(\frac{S}{S_0 e^{\mu(T'-t)}} \sinh(-\alpha L) \right) \right], \quad t \in [0, T'). \tag{31}$$

Thus, along the path $S = S_0 e^{\mu(T'-t)}$, volatility is constant at $\alpha \coth(-\alpha L)$. Thus, just as α controls the asymptotic height of the volatility smile, the parameter L controls this "at-the-money" volatility. For fixed S_0, as $L \downarrow -\infty$, $\coth(\cdot) \uparrow 1$ and at-the-money volatility approaches α. Conversely, as $L \uparrow 0$, $\coth(\cdot) \uparrow \infty$ as does volatility. Thus, in our model, the more likely is bankruptcy, the greater is the relative volatility perhaps due to shareholders' greater willingness to gamble.

As the terminal time T' approaches infinity, then for any fixed level of S, the terminal volatility smile $\sigma(S, T')$ asymptotes down to α if $\mu < 0$ and becomes unbounded above if $\mu > 0$. Since $\mu = r - q - \alpha^2/2$ is increasing in the risk-neutral drift $r - q$, it appears counter-intuitive that raising μ from a negative value to a positive one would raise asymptotic volatility from α to infinity. The explanation is that along a fixed stock price path, shifting μ from a negative value to a positive value lowers the corresponding path of the driving SBM, making bankruptcy more likely.

3.3. *PDE and converse*

Using (26) as a change of variables in (6), Appendix 2 proves that $\sigma(S, t)$ obeys the following non-linear PDE[h]

$$\frac{\sigma^2(S, t) S^2}{2} \frac{\partial^2 \sigma(S, t)}{\partial S^2} + [r - q + \sigma^2(S, t)]$$

$$\times S \frac{\partial \sigma}{\partial S}(S, t) + \frac{\partial \sigma}{\partial t}(S, t) = 0, \quad S > 0, \ t \in (0, T'). \tag{32}$$

Setting $t = T'$ in (26), the T'-maturity (local) volatility smile is

$$\sigma(S, T') = \frac{g'(g^{-1}(S))}{S} \equiv v(S), \quad S > 0, \tag{33}$$

---

[h]If the local volatility of the spot price S is modeled as a function of the $T'-$maturity forward price F instead of S, then PDE (32) is altered by replacing F with S and setting $r = q = 0$.

where $g'(x)$ and $g^{-1}(S)$ are determined by respectively differentiating and inverting $g(x)$, determined in (18). Since $g(x)$ is an increasing function for $x > L$, $g^{-1}(S)$ is well-defined for $S > 0$ and $\sigma(S, T')$ is positive. Equation (26) represents a class of solutions of the non-linear PDE (32) subject to (33).

We have seen that the assumptions of path-independence and no arbitrage imply that the volatility function satisfies a fully non-linear PDE with a known solution class. Conversely, Appendix 3 proves that if a given local volatility function $\sigma(S, t)$ satisfies the non-linear PDE (33), then the stock price is path-independent. The stock pricing function $s(x, t)$ is given by (12), where the horizon payoff $g(S)$ is still given by (18). Alternatively, Appendix 4 proves that the implied payoff can be determined from the T-maturity local volatility smile $v(S) \equiv \sigma(S, T)$

$$f(x) = I_v^{-1}(x + c_0), \quad x \in \Re,$$

where

$$I_v(f) \equiv \int_{c_1}^{f} \frac{1}{v(S)S} dS, \quad f \geq 0,$$

and c_0 and c_1 are constants. Finally, Appendix 3 shows that the inverse of the stock pricing function is given by

$$s^{-1}(S, t) = \int_{S_0}^{S} \frac{1}{\sigma(Z, t)Z} dZ + \int_{0}^{t} \left[\frac{1}{2} \frac{\partial}{\partial S}[\sigma(S, u)S] \Big|_{S=S_0} - \frac{(r - q)}{\sigma(S_0, u)} \right] du,$$

$$S > 0, \ t \in [0, T'].$$

4. Pricing Path-Independent Options

The next subsection derives explicit valuation formulas for standard options and risk-neutral densities of any maturity under a wide class of volatility functions consistent with the given smile and path-independence. The following subsection illustrates our results with our running example. The final section shows how the driving Brownian motion and the local volatility can each be interpreted as a path-independent exotic option written on the stock price.

4.1. *Standard options*

To value European calls of some intermediate maturity $M \in [0, T']$ in terms of the contemporaneous spot price, we first determine the function $\gamma(x, t)$ relating the call value to the SBM W_t and time t. By Itô's lemma, this pricing function solves

$$\frac{1}{2} \frac{\partial^2 \gamma}{\partial x^2}(x, t) + \frac{\partial \gamma}{\partial t}(x, t) = r\gamma(x, t), \quad x > L, \ t \in (0, M), \tag{34}$$

subject to the boundary conditions

$$\lim_{x \downarrow L} \gamma(x, t) = 0, \quad \lim_{x \uparrow \infty} \gamma(x, t) = s(x, t)e^{-q(M-t)} - Ke^{-r(M-t)}, \quad t \in [0, T],$$

$$\tag{35}$$

where $K > 0$ is the call strike, and subject to the terminal condition

$$\gamma(x, M) = [s(x, M) - K]^+, \quad x > L. \tag{36}$$

By the Feynman–Kac theorem, the continuous solution to this BVP is

$$\gamma(x, t) = e^{-r(M-t)} E^Q_{x,t}[s(W_M, M) - K]^+, \quad x > L, \ t \in [0, M \wedge \tau), \tag{37}$$

where $\{W_u, u \in [t, M]\}$ is a Q-SBM starting at x at time t. Equation (37) relates the call value to the SBM W_t and time t. To instead relate the call value to the stock price and time, let $c(S, t) = \gamma(x, t)$ in (37), where $x = s^{-1}(S, t)$

$$c(S, t) = e^{-r(M-t)} \int_L^\infty \frac{[s(z, M) - K]^+}{\sqrt{2\pi(M - t)}} \left\{ \exp\left[-\frac{1}{2}\left(\frac{z - s^{-1}(S, t)}{\sqrt{M - t}} \right)^2 \right] \right.$$

$$\left. - \exp\left[-\frac{1}{2}\left(\frac{z + s^{-1}(S, t) - 2L}{\sqrt{M - t}} \right)^2 \right] \right\} dz, \tag{38}$$

for $S \geq 0, t \in [0, M \wedge \tau)$. This solution will be an explicit function of S and t if $s^{-1}(S, t)$ can be written explicitly in terms of its arguments.

Differentiating (38) twice w.r.t. strike gives the implied risk neutral probability that the stock price is at K at time M, given that the stock is at S at time t. Alternatively, the change of variables $S = s(x, t)$ for the absorbing SBM transition density expresses this probability as

$$q(Z, M; S, t) = \frac{1}{\sqrt{2\pi(M - t)}} \left\{ \exp\left\{ -\frac{1}{2}\left[\frac{s^{-1}(Z, M) - s^{-1}(S, t)}{\sqrt{M - t}} \right]^2 \right\} \right.$$

$$\left. - \exp\left\{ -\frac{1}{2}\left[\frac{s^{-1}(Z, M) + s^{-1}(S, t)}{\sqrt{M - t}} \right]^2 \right\} \right\} \frac{\partial s^{-1}}{\partial S}(Z, M),$$

$$S > 0, \ t \in [0, M). \tag{39}$$

The density will be positive only if $s^{-1}(S, M)$ is increasing in S and it will be explicit in Z only if $s^{-1}(Z, M)$ is explicit in Z.

An alternative derivation of the call pricing function $C(S, t), S \geq 0, t \in [0, M \wedge \tau)$ is obtained by integrating its payoff $(S - K)^+$ against this density, and discounting at the riskfree rate

$$C(S, t; K, M)$$

$$= e^{-r(M-t)} \int_K^\infty \frac{Z - K}{\sqrt{2\pi(M - t)}} \left\{ \exp\left\{ -\frac{1}{2}\left[\frac{s^{-1}(Z, M) - s^{-1}(S, t)}{\sqrt{M - t}} \right]^2 \right\} \right.$$

$$\left. - \exp\left\{ -\frac{1}{2}\left[\frac{s^{-1}(Z, M) + s^{-1}(S, t)}{\sqrt{M - t}} \right]^2 \right\} \right\} \frac{\partial s^{-1}}{\partial S}(Z, M) dZ. \tag{40}$$

To the extent that markets for intermediate maturity calls exist, our theory can be tested in the usual manner by comparing market prices with model values, obtained from either (38) or (40). To obtain an initial *implied* volatility surface, one must numerically solve $c_{BS}(S_0, 0, \sigma_i(K, M); K, M) = c(S_0, 0; K, M)$ for $\sigma_i(K, M)$ at each $K > 0$ and $M \in (0, T')$, where $c_{BS}(S, t, \sigma_i; K, M)$ is the Black–Scholes call pricing formula.

4.2. *Example*

To illustrate our results, suppose again that T-maturity option prices satisfy (19), so that the stock pricing function is given by (23) and the inverse of this function is given in (29). Evaluating this inverse at $(S, t) = (Z, M)$ yields

$$s^{-1}(Z, M) = L + \frac{\sinh^{-1}}{\alpha}\left(\frac{Z}{\beta'}e^{\mu(T' - M)}\right).\tag{41}$$

Differentiating w.r.t the first argument yields

$$\frac{\partial s^{-1}}{\partial S}(Z, M) = \frac{1}{\alpha\sqrt{Z^2 + (\beta')^2 e^{-2\mu(T' - M)}}}.\tag{42}$$

Substituting (41), (29), and (42) in (39) implies that the risk-neutral stock pricing density is

$$q(Z, M; S, t) = \frac{1}{\sqrt{2\pi(M - t)}}$$

$$\times \left\{\exp\left\{-\frac{1}{2}\left[\frac{\sinh^{-1}\left(\frac{Z}{\beta'}e^{\mu(T' - M)}\right) - \sinh^{-1}\left(\frac{S}{\beta'}e^{\mu(T' - t)}\right)}{\alpha\sqrt{M - t}}\right]^2\right\}\right.$$

$$\left. - \exp\left\{-\frac{1}{2}\left[\frac{\sinh^{-1}\left(\frac{Z}{\beta'}e^{\mu(T' - M)}\right) + \sinh^{-1}\left(\frac{S}{\beta'}e^{\mu(T' - t)}\right)}{\alpha\sqrt{M - t}}\right]^2\right\}\right\}$$

$$\times \frac{1}{\alpha\sqrt{Z^2 + (\beta')^2 e^{-2\mu(T' - M)}}},\tag{43}$$

for $S > 0, t \in [0, M \wedge \tau)$. Figure 4 graphs this density (termed the arcsinhnormal) against the future spot price and time. The downward sloping volatility surface graphed in Fig. 3 cancels much of the positive skewness of the lognormal density leading to a close approximation of a Gaussian density.

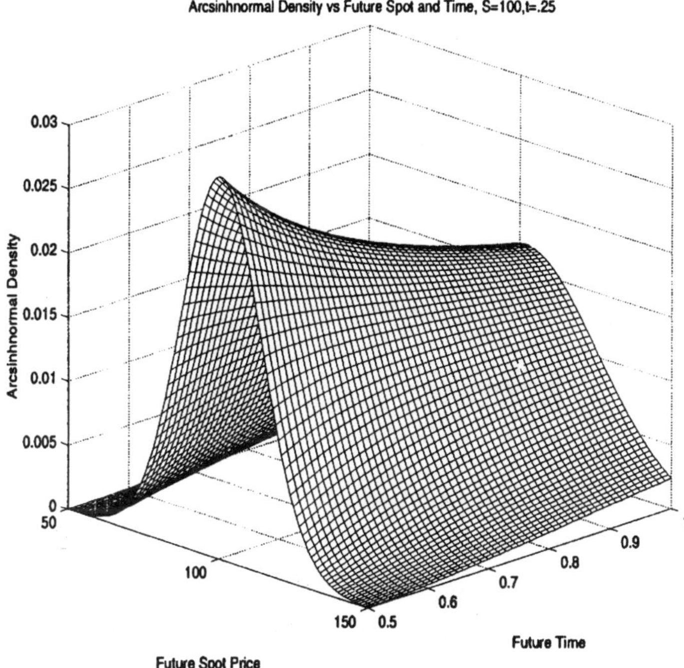

Fig. 4. The arcsinhnormal probability density function.

Integrating the call's payoff against this density yields the following pricing formula

$$C(S,t) = \frac{e^{-q(M-t)}}{2} \left(S + \sqrt{S^2 + \beta^2 e^{-2\mu(T-t)}} \right)$$

$$\times [N(d_+ + \alpha\sqrt{M-t}) + N(d_- - \alpha\sqrt{M-t})]$$

$$- \frac{\beta^2 e^{-q(M-t)}}{2e^{2\mu(T-t)}} \frac{1}{S + \sqrt{S^2 + \beta^2 e^{-2\mu(T-t)}}}$$

$$\times [N(d_+ - \alpha\sqrt{M-t}) + N(d_- + \alpha\sqrt{M-t})]$$

$$- Ke^{-r(M-t)}[N(d_+) - N(d_-)],$$

where

$$d_\pm \equiv \frac{\pm \sinh^{-1}\left(\frac{S}{\beta}e^{\mu(T-t)}\right) - \sinh^{-1}\left(\frac{K}{\beta}e^{\mu(T-M)}\right)}{\alpha\sqrt{M-t}}.$$

Figure 5 graphs the call value and time values of this model against the corresponding values in the Black–Scholes model with the same at-the-money implied volatility. The negative skewness apparent in the volatility surface and arcsinhnormal density function is manifested in higher out-of-the-money put prices and lower out-of-the-money call prices.

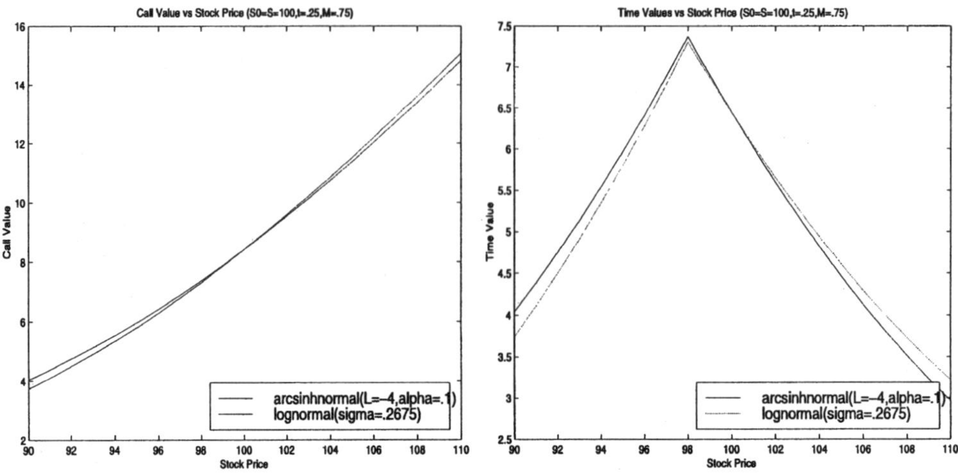

Fig. 5. The arcsinhnormal call value and time value vs. Black–Scholes.

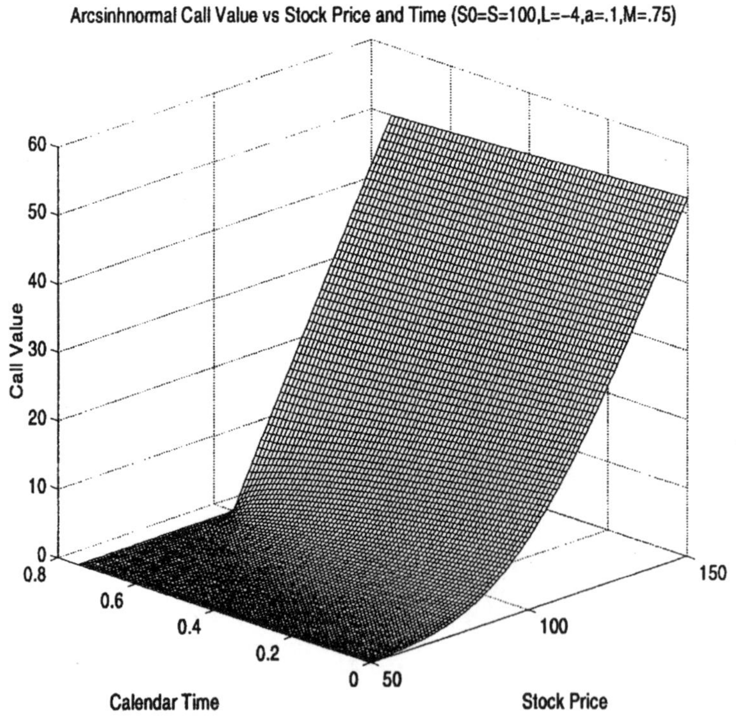

Fig. 6. The arcsinhnormal call valuation function.

Figure 6 plots the arcsinhnormal call value against the current stock price and time.

4.3. *Path-independent exotic options*

The SBM W_t is the forward price of a path-independent exotic option with the payoff $g^{-1}(S_{T'})$ at T', where $g^{-1}(\cdot)$ is the inverse of $g(x)$ determined in Appendix 1 from $f(\cdot)$. This follows from the observation that deferring the payment of this exotic's premium to T' induces zero drift under Q and the payoff induces unit volatility. Thus, $s(x,t)$ is properly called the spot pricing function, since it relates the spot price of the underlying to the (forward) value of an asset. Similarly, $s^{-1}(S_t, t)$ is a standard pricing function relating the time t forward price of the Brownian exotic paying $g^{-1}(S_{T'})$ at T' to the time t spot price and time.

The local volatility $\sigma(S_t, t)$ can also be interpreted as the price process for an exotic equity derivative. To determine the payoff, note from (33) that the T'-maturity *absolute volatility* $a(S, T') \equiv \sigma(S, T')S$ is the following function of the time T' spot price $S_{T'}$

$$a(S_{T'}, T') = g'(W^a_{T'}) = \frac{1}{\frac{\partial g^{-1}}{\partial S}(S_{T'})} = \lim_{h \downarrow 0} \frac{h}{g^{-1}(S_{T'} + h) - g^{-1}(S_{T'})}.$$

The T'-maturity (relative) volatility smile $v(S)$ arises from quantoing this dollar payoff into shares i.e., the payoff in shares is

$$\sigma(S_{T'}, T') = v(S_{T'}) = \frac{g'(W^a_{T'})}{S_{T'}} = \frac{1}{S_{T'}} \lim_{h \downarrow 0} \frac{h}{g^{-1}(S_{T'} + h) - g^{-1}(S_{T'})}.$$

If we also assume that the premium on this exotic is specified in shares and paid at T', then this share-denominated forward price matches the local volatility $\sigma(S_t, t)$ at all times up to T'. The appearance of $\sigma^2(S, t)$ in the middle term of the PDE (33) is now easily interpreted as a quanto correction. The absence of the usual discounting term in (33) is due to the deferral of the premium payment to maturity.

5. Summary and Future Research

By interpreting the stock as essentially a path-independent derivative on the driving Brownian motion (or conversely), we were able to develop a PDE for the stock pricing function. Adding the information from a complete strike structure of option prices and the economic restrictions implied by no arbitrage and limited liability allowed us to uniquely determine this stock pricing function. Similarly, by interpreting local volatility as an exotic derivative on the stock, we were able to develop a non-linear PDE for the volatility function, which we solved analytically using the known stock pricing function. Conversely, if local volatilities are assumed to satisfy the non-linear PDE, then stock prices are path-independent and the function linking the SBM to the stock price can be determined analytically. Finally, we derived closed form formulas for option prices and risk-neutral densities consistent with a wide class of local volatility functions.

In principle, this work can be extended to valuing American and other path-dependent options. For standard American and standard barrier options, the PDE's developed in this paper apply on restricted domains. Similarly, for exotic options with mild path-dependence such as compound, chooser and discretely sampled barrier options, the PDE's developed in this paper would be solved recursively. For options with stronger path-dependence such as Asian, lookback and passport options, additional state variables would be required. Since numerical solutions may be required in any case, the question of the optimal discretization of the PDE's should be considered.

If we relax the assumption of this work that the T-maturity implied volatility smile is given, then a second extension would involve calibrating the smile to a matrix of initial option prices. Evaluating (38) at $(S, t) = (S_0, 0)$ determines initial call values of any maturity M and strike K in terms of the functions $s(S, M)$ and $s^{-1}(S_0, 0)$, which can both be expressed in terms of $f(S)$ via (17). Thus, the inverse problem of finding $f(\cdot)$ from a matrix of initial option prices can be posed once we select an objective function such as minimizing squared distance from a prior or maximizing smoothness. Under the assumptions of this paper, the identification of $f(\cdot)$ is tantamount to the determination of the risk-neutral process and volatility surface for all times. Given that $f(\cdot)$ relates the stock price at T to normally distributed SBM, a natural basis representation of $f(\cdot)$ arises in the form of Hermite polynomials. As these techniques are fairly involved, this extension is best left for future research.

A minor extension of these results which would generate a wider class of analytic volatility surfaces is to assume path-independence after a deterministic time change. To illustrate, suppose that the stock price is an Ornstein–Uhlenbeck process absorbing at the origin as in Cox and Ross.[5] Then the stock price depends on the path of the driving SBM, and as a consequence does not fit the path-independence property described in this paper. However, it is well known (see e.g., Cox and Miller,[4] p. 229) that such a stock price process is proportional to a standard Brownian motion run on a deterministically different clock.

A major extension of these results which would generate a much wider class of analytic volatility surfaces would involve replacing the one-dimensional SBM with an alternative process (e.g., Bessel squared) or even a d-dimensional diffusion process whose solution is known. One can also add jumps in order to generate the casual empirical observation that implied volatilities at low spot prices tend to be higher at short maturities than long ones. One way to generate jumps is to do a stochastic time change[i] on the standard Brownian motion driving all assets. For example, if a gamma time change was used, then the fundamental driver becomes a symmetric VG process (see Madan, Carr and Chang[13]), while the spot price process would be a generalized VG process. As in the continuous case, a generalized VG Levy measure when written as a function of jumps in the spot price would revert to

---

[i]In fact, a stochastic time change on a GBM is an alternative to the scale change performed in this paper. Both approaches can generate a volatility smile.

the standard symmetric VG Levy measure when written in terms of jumps of the driving variable. In the interest of brevity, such extensions are best left for future research.

Acknowledgments

We thank the participants of workshops at the Courant Institute of NYU and the University of Wisconsin, Madison. We also thank Marco Avellaneda, Ilia Bouchev, Stephen Chung, Georges Courtadon, Raphael Douady, Tom Gould, David Hait, Ed Hsu, Guang Jiang, Tom Little, Bill Margrabe, Harry Mendell, and especially Claudio Albanese and Thaleia Zariphopolou for comments. They are not responsible for any errors.

Appendix 1. Characterizing Stock Pricing Functions

In this appendix, we characterize the entire set of stock pricing functions $s(x, t)$ which vanish at $x = L$, are positive, increasing, and unbounded in x for $x > L$, and are consistent with no arbitrage on the time interval $(0, T')$, where T' can be arbitrarily large. Robbins and Siegmund[14] have shown that the entire class of positive martingale functions $m(y, u)$ which satisfy the backward diffusion equation $\frac{\partial m}{\partial u} + \frac{1}{2} \frac{\partial^2 m}{\partial y^2} = 0$ on the region $\Re^+ \times (0, U)$ are given by

$$m(y, u) = \int_0^\infty \frac{e^{-\frac{1}{2}\left(\frac{\theta - y}{\sqrt{U - u}}\right)^2} - e^{-\frac{1}{2}\left(\frac{\theta + y}{\sqrt{U - u}}\right)^2}}{\sqrt{2\pi(U - u)}} dG(\theta) + \int_u^{U^+} \frac{y e^{-\frac{1}{2}\left(\frac{y}{\sqrt{v - u}}\right)^2}}{\sqrt{2\pi(v - u)^3}} dH(v),$$

$$(44)$$

for some measures G on $(0, \infty)$ and H on $(0, U)$. The kernel of the first integral is recognized as the transition density for SBM absorbing at the origin. The kernel of the second integral is recognized as the density for the first passage time of SBM to the origin. If we set $H = 0$, then (44) describes the set of all non-negative martingale functions satisfying the backward diffusion equation on $\Re^+ \times (0, U)$ and vanishing at the origin.

To extend this class of functions to the entire time axis $(0, \infty)$, let

$$v(z, t) \equiv \frac{m(y, u)}{n(y, u)}, \tag{45}$$

where $z \equiv \frac{y}{U - u}$, $t = \frac{1}{U - u}$, and

$$n(y, u) \equiv \frac{e^{-\frac{1}{2}\left(\frac{y}{\sqrt{U - u}}\right)^2}}{\sqrt{2\pi(U - u)}}$$

is a normal density function. Then for any solution m to the backward diffusion equation, it is straightforward to verify that $v(z, t)$ also solves the backward diffusion equation $\frac{\partial v}{\partial t} + \frac{1}{2}\frac{\partial^2 v}{\partial z^2} = 0$ on the region $\Re^+ \times (\frac{1}{U}, \infty)$. Letting U approach infinity in the original problem implies that $v(z,t)$ is defined on the open quarter plane $\Re^+ \times \Re^+$. When m is given by (44) with $H(\cdot) = 0$, then (45) implies that

$$v(z,t) = 2 \int_0^\infty \sinh(\theta z) e^{-\frac{1}{2}\theta^2 t} dG(\theta),$$

which vanishes at $z = 0$.

To obtain (46), we let $x = z + L$ so that the vanishing arises at $x = L < 0$ and we also account for the stock's initial value S_0 and risk-neutral drift $r - q$

$$s(x,t) = S_0 e^{(r-q)t} \int_0^\infty \frac{\sinh[\theta(x-L)]}{\sinh[-\theta L]} e^{-\frac{\theta^2}{2}t} dG(\theta), \quad x > L, \ t \in (0,T'). \tag{46}$$

We next show how to determine the distribution function $G(\theta)$ from the known implied payoff $f(x)$. We assume that the distribution function is that of a continuous random variable, so that $dG(\theta) = G'(\theta)d\theta$. Evaluating (46) at $t = T$ and rearranging implies

$$\frac{f(x)}{F_0} = \int_0^\infty \sinh[\theta(x-L)] e^{-\frac{\theta^2}{2}T} \frac{G'(\theta)}{\sinh(-\theta L)} d\theta,$$

where $F_0 = S_0 e^{(r-q)T}$ is the T-maturity forward price. Letting $z = \frac{x-L}{\sqrt{T}}$ and $y = \theta\sqrt{T}$ yields

$$\sqrt{T}\frac{f(L+z\sqrt{T})}{F_0} = \int_0^\infty \sinh(yz) e^{-\frac{y^2}{2}} \frac{G'(y/\sqrt{T})}{\sinh(-Ly/\sqrt{T})} dy.$$

Substituting in the definition of $\sinh(\cdot)$

$$2\sqrt{T}\frac{f(L+z\sqrt{T})}{F_0} = \int_0^\infty e^{zy-\frac{y^2}{2}} \frac{G'(y/\sqrt{T})}{\sinh(-Ly/\sqrt{T})} dy$$

$$- \int_0^\infty e^{-zy-\frac{y^2}{2}} \frac{G'(y/\sqrt{T})}{\sinh(-Ly/\sqrt{T})} dy.$$

Multiplying by $e^{-\frac{z^2}{2}}$ and changing variables as $y = -y$ in the second integral yields

$$\hat{f}(z) \equiv 2\sqrt{T}\frac{f(L+z\sqrt{T})}{F_0} e^{-\frac{z^2}{2}} = \int_{-\infty}^\infty e^{-\frac{(z-y)^2}{2}} \phi(y)dy, \tag{47}$$

where

$$\phi(y) = \begin{cases} \dfrac{G'(y/\sqrt{T})}{\sinh(-Ly/\sqrt{T})} & \text{if } y \geq 0; \\[2ex] -\dfrac{G'(-y/\sqrt{T})}{\sinh(Ly/\sqrt{T})} & \text{if } y < 0. \end{cases} \tag{48}$$

Equation (47) expresses $\hat{f}(\cdot)$ as a linear combination of translates of a standard normal density function. Wiener proved that the span of all such translates is L_1 (see Goldberg,[10] p. 33). Thus, so long as $f(x)e^{-x^2}$ is L_1, there exists a $\phi(\cdot)$ generating it. To identify this ϕ, note that the right hand side of (47) is a convolution. Assuming that $e^{-\frac{(z-y)^2}{2}}\phi(y)$ is a square-integrable function of z for all $y \in \Re$, taking Fourier transforms of both sides yields

$$\int_{-\infty}^{\infty} e^{i\omega z}\hat{f}(z)dz = \int_{-\infty}^{\infty} e^{i\omega y}\phi(y)dy\sqrt{2\pi}e^{-\omega^2/2}\,,$$

since $e^{-\omega^2/2}$ is the Fourier transform of $\frac{e^{-\frac{z^2}{2}}}{\sqrt{2\pi}}$. Re-arranging this expression yields

$$\int_{-\infty}^{\infty} e^{i\omega y}\phi(y)dy = \frac{e^{\omega^2/2}}{\sqrt{2\pi}}\int_{-\infty}^{\infty} e^{i\omega z}\hat{f}(z)dz\,.$$

Inverting for $\phi(y)$ yields

$$\phi(y) = \frac{1}{\sqrt{(2\pi)^3}}\int_{-\infty}^{\infty} e^{-i\omega y + \frac{\omega^2}{2}}\int_{-\infty}^{\infty} e^{i\omega z}\hat{f}(z)dzd\omega\,.$$

Solving (48) for $G'(\cdot)$ when $y \geq 0$ yields

$$G'(\theta) = \sinh(-L\theta)\phi(\theta\sqrt{T})$$

$$= \frac{\sqrt{T}}{F_0}\frac{\sinh(-\theta L)}{\pi\sqrt{2\pi}}\int_{-\infty}^{\infty} e^{-i\omega\theta\sqrt{T}+\frac{\omega^2}{2}}\int_{-\infty}^{\infty} e^{i\omega z-\frac{z^2}{2}}f(L+z\sqrt{T})dzd\omega\,,$$

$$\tag{49}$$

for $\theta \geq 0$ and we set $G'(\theta) = 0$ otherwise. We are thus able to identify the risk-neutral stock price process consistent with the T-maturity option prices

$$S_t = s(W_t, t) = S_0 e^{(r-q)t}\int_0^{\infty} \frac{\sinh[\theta(x-L)]}{\sinh[-\theta L]}e^{-\frac{\theta^2}{2}t}G'(\theta)d\theta\,,\quad t \in (0, \tau \wedge T')\,,$$

$$\tag{50}$$

where $G'(\theta)$ is given in (49).

Appendix 2. The PDE for Local Volatility as a Consequence of Path-Independence

Recall the PDE (6) describing the spot pricing function $s(x, t)$

$$\frac{\partial s}{\partial t}(x, t) + \frac{1}{2}\frac{\partial^2 s}{\partial x^2}(x, t) = (r-q)s(x, t)\,,\quad t \in [0, T']\,,\ x \in \Re\,. \tag{51}$$

Let $s_x(x, t) \equiv \frac{\partial s}{\partial x}(x, t)$. Differentiating (51) w.r.t. x implies

$$\frac{\partial s_x}{\partial t}(x, t) + \frac{1}{2}\frac{\partial^2 s_x}{\partial x^2}(x, t) = (r-q)s_x(x, t)\,,\quad t \in [0, T']\,,\ x \in \Re\,. \tag{52}$$

Consider the change of variables derived in (26)

$$\sigma(S,t) = \frac{1}{S}\frac{\partial s}{\partial x}(x,t) = \frac{s_x(x,t)}{s(x,t)}\,, \quad \text{where } S = s(x,t)\,, \quad t \in [0,T']\,, \ x \in \Re\,. \tag{53}$$

Thus

$$s_x(x,t) = \sigma(s(x,t),t)s(x,t)\,. \tag{54}$$

$$\frac{\partial s_x}{\partial x}(x,t) = \frac{\partial \sigma}{\partial S}(S,t)\frac{\partial s}{\partial x}(x,t)s(x,t) + \sigma(S,t)\frac{\partial s}{\partial x}(x,t)\,.$$

$$\frac{\partial^2 s_x}{\partial x^2}(x,t) = \frac{\partial^2 \sigma}{\partial S^2}(S,t)\left(\frac{\partial s}{\partial x}(x,t)\right)^2 s(x,t) + \frac{\partial \sigma}{\partial S}(S,t)\frac{\partial^2 s}{\partial x^2}(x,t)s(x,t)$$

$$+ 2\frac{\partial \sigma}{\partial S}(S,t)\left(\frac{\partial s}{\partial x}(x,t)\right)^2 + \sigma(S,t)\frac{\partial^2 s}{\partial x^2}(x,t)\,.$$

$$\frac{\partial s_x}{\partial t}(x,t) = \frac{\partial \sigma}{\partial S}(S,t)\frac{\partial s}{\partial t}(x,t)s(x,t) + \frac{\partial \sigma}{\partial t}(S,t)s(x,t) + \sigma(S,t)\frac{\partial s}{\partial t}(x,t)\,. \tag{55}$$

Substituting (55) to (55) in (52)

$$\frac{\partial \sigma}{\partial S}(S,t)\frac{\partial s}{\partial t}(x,t)s(x,t) + \frac{\partial \sigma}{\partial t}s(x,t) + \sigma(S,t)\frac{\partial s}{\partial t}(x,t)$$

$$+ \frac{1}{2}\frac{\partial^2 \sigma(S,t)}{\partial S^2}\left(\frac{\partial s}{\partial x}(x,t)\right)^2 s(x,t) + \frac{\partial \sigma}{\partial S}(S,t)\frac{\partial^2 s}{\partial x^2}(x,t)s(x,t)$$

$$+ 2\frac{\partial \sigma}{\partial S}(S,t)\left(\frac{\partial s}{\partial x}(x,t)\right)^2 + \sigma(S,t)\frac{\partial^2 s}{\partial x^2}(x,t)$$

$$= (r-q)\sigma(S,t)s(x,t)\,, \quad t \in [0,T]\,, \ x \in \Re\,. \tag{56}$$

Substituting (51) in (56)

$$\frac{\partial \sigma}{\partial S}(S,t)\frac{\partial s}{\partial t}(x,t)s(x,t) + \frac{\partial \sigma}{\partial t}s(x,t) + \frac{1}{2}\frac{\partial^2 \sigma(S,t)}{\partial S^2}\left(\frac{\partial s}{\partial x}(x,t)\right)^2 s$$

$$+ \frac{\partial \sigma}{\partial S}(S,t)\frac{\partial^2 s}{\partial x^2}(x,t)s(x,t) + 2\frac{\partial \sigma}{\partial S}(S,t)\left(\frac{\partial s}{\partial x}(x,t)\right)^2 = 0\,, \quad t \in [0,T']\,, \ x \in \Re\,. \tag{57}$$

Now, multiply (51) by $\sigma(S,t)$

$$\sigma(S,t)\frac{\partial s}{\partial t}(x,t) + \frac{\sigma(S,t)}{2}\frac{\partial^2 s}{\partial x^2}(x,t) = (r-q)\sigma(S,t)s(x,t)\,, \quad t \in [0,T']\,, \ x \in \Re\,. \tag{58}$$

Also multiply (51) by $\frac{\partial \sigma}{\partial S}(S,t)s(x,t)$

$$\frac{\partial \sigma}{\partial S}(S,t)\frac{\partial s}{\partial t}(x,t)s(x,t) + \frac{1}{2}\frac{\partial \sigma}{\partial S}(S,t)\frac{\partial^2 s}{\partial x^2}(x,t)s(x,t)$$

$$= (r-q)\frac{\partial \sigma}{\partial S}(S,t)s^2(x,t)\,, \quad t \in [0,T'],\ x \in \Re\,. \tag{59}$$

Substituting (58) and (59) in (57)

$$\frac{1}{2}\frac{\partial^2 \sigma(S,t)}{\partial S^2}\left(\frac{\partial s(x,t)}{\partial x}\right)^2 s(x,t) + (r-q)S\frac{\partial \sigma}{\partial S}(S,t)s^2(x,t)$$

$$+ \frac{\partial \sigma(S,t)}{\partial S}\left(\frac{\partial s(x,t)}{\partial x}\right)^2 + \frac{\partial \sigma}{\partial t}(S,t)s(x,t) = 0\,, \tag{60}$$

for $S \geq 0, t \in [0,T')$. Dividing by $s(x,t)$

$$\frac{1}{2}\left(\frac{\partial s(x,t)}{\partial x}\frac{1}{s(x,t)}\right)^2 s^2(x,t)\frac{\partial^2 \sigma(S,t)}{\partial S^2} + \left[r - q + \left(\frac{\partial s(x,t)}{\partial x}\frac{1}{s(x,t)}\right)^2\right]$$

$$\times s(x,t)\frac{\partial \sigma}{\partial S}(S,t) + \frac{\partial \sigma}{\partial t}(S,t) = 0\,, \tag{61}$$

for $S \geq 0, t \in [0,T')$. Substituting in $\sigma(S,t) = \frac{\partial s(x,t)}{\partial x}\frac{1}{s(x,t)}$ gives the following non-linear PDE for $\sigma(S,t)$

$$\frac{\sigma^2(S,t)S^2}{2}\frac{\partial^2 \sigma(S,t)}{\partial S^2}$$

$$+ [r - q + \sigma^2(S,t)]S\frac{\partial \sigma}{\partial S}(S,t) + \frac{\partial \sigma}{\partial t}(S,t) = 0\,, \quad S \geq 0,\ t \in [0,T')\,. \tag{62}$$

Appendix 3. The PDE for Local Volatility as a Sufficient Condition for Path-Independence

This appendix proves that if a positive function $\sigma(S,t)$ satisfies the non-linear PDE (33), then the stock price is path-independent. More specifically, there exists a function $s^{-1}(S,t)$ increasing in S, such that the process

$$B_t \equiv s^{-1}(S_t,t)\,, \quad t \in [0, \tau \wedge T']\,, \tag{63}$$

is the driving SBM, i.e., $B_t = W_t$ for all $t \in [0, \tau \wedge T']$. This function is given by

$$s^{-1}(S,t) = \int_{S_0}^{S}\frac{1}{\sigma(Z,t)Z}dZ$$

$$+ \int_0^t \left[\frac{1}{2}\frac{\partial}{\partial S}[\sigma(S,u)S]\Big|_{S=S_0} - \frac{(r-q)}{\sigma(S_0,u)}\right]du\,, \quad S > 0,\ t \in [0,T']\,. \tag{64}$$

To prove this result, Itô's lemma applied to (63) implies that for $t \in [0, \tau \wedge T']$

$$dB_t = \left[\frac{\sigma^2(S_t, t)S_t^2}{2} \frac{\partial^2 s^{-1}}{\partial S^2}(S_t, t) + (r - q)S_t \frac{\partial s^{-1}}{\partial S}(S_t, t) + \frac{\partial s^{-1}}{\partial t}(S_t, t) \right] dt$$

$$+ \frac{\partial s^{-1}}{\partial S}(S_t, t)\sigma(S_t, t)S_t dW_t \,. \tag{65}$$

For B to be SBM, we need to show that its drift is zero

$$\frac{\sigma^2(S, t)S^2}{2} \frac{\partial^2 s^{-1}}{\partial S^2}(S, t)$$

$$+ (r - q)S \frac{\partial s^{-1}}{\partial S}(S, t) + \frac{\partial s^{-1}}{\partial t}(S, t) = 0 \,, \quad S > 0, \ t \in [0, \tau \wedge T'] \,, \tag{66}$$

that its volatility is unity

$$\frac{\partial s^{-1}}{\partial S}(S, t)\sigma(S, t)S = 1 \,, \quad S > 0, \ t \in [0, \tau \wedge T'] \,, \tag{67}$$

and that its initial level is zero

$$B_0 = s(S_0, 0) = 0 \,. \tag{68}$$

Differentiating (64) w.r.t. S

$$\frac{\partial s^{-1}}{\partial S}(S, t) = \frac{1}{\sigma(S, t)S} \,, \quad S > 0, \ t \in [0, \tau \wedge T'] \,, \tag{69}$$

so (67) holds. Furthermore, since $\sigma(S, t)$ is assumed positive, the function s^{-1} is increasing in S. Evaluating (64) at $(S, t) = (S_0, 0)$ implies (68) also holds. To show that (66) also holds, divide (33) by $\sigma^2(S, t)S$

$$\frac{S}{2} \frac{\partial^2 \sigma(S, t)}{\partial S^2} + \frac{\partial \sigma}{\partial S}(S, t) + \frac{(r - q)S}{\sigma^2(S, t)S} \frac{\partial \sigma}{\partial S}(S, t)$$

$$+ \frac{1}{\sigma^2(S, t)S} \frac{\partial \sigma}{\partial t}(S, t) = 0 \,, \quad S \geq 0, \ t \in [0, \tau \wedge T') \,. \tag{70}$$

Noting that

$$\frac{\partial}{\partial S}[S\sigma(S, t)] = \sigma(S, t) + S\frac{\partial \sigma}{\partial S}(S, t) \,, \quad S > 0, \ t \in [0, \tau \wedge T'] \,,$$

and

$$\frac{\partial^2}{\partial S^2}[S\sigma(S, t)] = 2\frac{\partial \sigma}{\partial S}(S, t) + S\frac{\partial^2 \sigma}{\partial S^2}(S, t) \,, \quad S > 0, \ t \in [0, \tau \wedge T'] \,,$$

(70) can be re-written as

$$\frac{1}{2} \frac{\partial^2}{\partial S^2}[S\sigma(S, t)] + \frac{(r - q)S}{\sigma^2(S, t)S^2} \frac{\partial}{\partial S}[S\sigma(S, t)] - \frac{r - q}{\sigma(S, t)}$$

$$+ \frac{1}{\sigma^2(S, t)S} \frac{\partial \sigma}{\partial t}(S, t) = 0 \,, \quad S \geq 0, \ t \in [0, \tau \wedge T') \,. \tag{71}$$

Integrating both sides w.r.t. S implies

$$\frac{1}{2}\frac{\partial}{\partial Z}[\sigma(Z,t)Z]\Big|_{Z=S_0}^{Z=S} - \frac{(r-q)Z}{\sigma(Z,t)Z}\Big|_{Z=S_0}^{Z=S}$$

$$+ \int_{S_0}^{S}\frac{1}{\sigma^2(Z,t)Z}\frac{\partial\sigma}{\partial t}(Z,t)dZ = 0, \quad S \geq 0,\ t \in [0,\tau \wedge T'), \tag{72}$$

where the constant of integration has been set to zero. Multiplying through by -1 implies that for $S \geq 0, t \in [0, \tau \wedge T')$

$$-\frac{\sigma^2(S,t)S^2}{2}\frac{\frac{\partial}{\partial S}[\sigma(S,t)S]}{\sigma^2(S,t)S^2} + (r-q)S\frac{1}{\sigma(S,t)S} + \frac{1}{2}\frac{\partial}{\partial S}[\sigma(S,t)S]\Big|_{S=S_0}$$

$$-\frac{r-q}{\sigma(S_0,t)} - \int_{S_0}^{S}\frac{1}{\sigma^2(Z,t)Z}\frac{\partial\sigma}{\partial t}(Z,t)dZ = 0. \tag{73}$$

Differentiating (69) w.r.t. S

$$\frac{\partial^2 s^{-1}}{\partial S^2}(S,t) = -\frac{\frac{\partial}{\partial S}[\sigma(S,t)S]}{[\sigma(S,t)S]^2}, \quad S > 0,\ t \in [0,\tau \wedge T']. \tag{74}$$

Differentiating (64) w.r.t. t

$$\frac{\partial s^{-1}}{\partial t}(S,t) = \frac{1}{2}\frac{\partial}{\partial S}[\sigma(S,t)S]\Big|_{S=S_0} - \frac{(r-q)}{\sigma(S_0,t)}$$

$$- \int_{S_0}^{S}\frac{1}{\sigma^2(Z,t)Z}\frac{\partial\sigma}{\partial t}(Z,t)dZ, \quad S > 0, t \in [0,\tau \wedge T']. \tag{75}$$

Substituting (69),(74), and (75) in (73) gives (66).

Appendix 4. Implying the Stock Payoff From the Volatility Smile

When the local volatility surface $\sigma(S,t)$ is given for all $t \in [0,T']$, then the time T' volatility smile $v(S) \equiv \sigma(S,T')$ is also given. Consequently, the implied payoff $g(S)$ must solve the first-order non-linear differential Eq. (33). Under certain conditions on $v(\cdot)$, an implicit solution for $g(\cdot)$ is given by

$$I_v(g) = x + c_0, \quad x \in \Re, \tag{76}$$

where

$$I_v(g) \equiv \int_{c_1}^{g}\frac{1}{v(S)S}dS, \quad g \geq 0, \tag{77}$$

and c_0 and c_1 are constants. Since g can be less than c_1, we define $\int_b^a \frac{1}{v(S)S}dS = -\int_a^b \frac{1}{v(S)S}dS$ for $b > a$.

In order that the implied payoff $g(x)$ be well-defined by the solution (77), the volatility smile $v(S)$ must be restricted such that $I_v(g)$ is invertible. To obtain these restrictions, note that as $x \downarrow -\infty$, the volatility smile should permit the implied stock payoff g to approach zero

$$I_v(0) \equiv - \int_0^{c_1} \frac{1}{v(S)S} dS = -\infty , \quad \forall \, c_1 > 0 . \tag{78}$$

This condition holds so long as $v(S)$ is bounded as $S \downarrow 0$ or at least does not approach infinity so fast that the integral of $\frac{1}{v(S)S}$ is bounded.

Differentiating the definition of I_v in (77) w.r.t. g implies that $I_v'(g) = \frac{1}{v(g)g}$, which is strictly positive for $g > 0$. In order that $g(x)$ be well-defined by (77) for all real x, we further require that there exist a positive constant c_1 such that $v(S), S > 0$ satisfies

$$I_v(\infty) \equiv \int_{c_1}^{\infty} \frac{1}{v(S)S} dS = \infty . \tag{79}$$

This condition holds so long as $v(\cdot)$ is bounded above as $S \uparrow \infty$, or at least does not approach infinity so fast that the integral of $\frac{1}{v(S)S}$ over the positive half line is bounded.

To summarize, if the given positive volatility smile $v(S)$ satisfies (78), and (79), then $I_v(0) = -\infty$, $I_v'(g) > 0$ for $g > 0$, and $I_v(\infty) = \infty$. Consequently, the inverse of I_v exists, and from (77), the stock payoff $g(x)$ is given by

$$s(x, T) \equiv g(x) = I_v^{-1}(x + c_0) , \quad x \in \Re . \tag{80}$$

Since I_v is an increasing function defined on $\Re^+$ with range $\Re$, I_v^{-1} is an increasing function defined on $\Re$ with range $\Re^+$. By (80), g is also an increasing function defined on $\Re$ with range $\Re^+$.

References

[1] F. Black and M. Scholes, "The pricing of options and corporate liabilities", *The Journal of Political Economy* **81** (1973) 637–659.

[2] D. Breeden and R. Litzenberger, "Prices of state contingent claims implicit in option prices", *Journal of Business* **51** (1978) 621–651.

[3] J. Cox, "Notes on Option Pricing I: Constant Elasticity of Variance Diffusions", Stanford University Working Paper, 1975.

[4] D. R. Cox and H. D. Miller, *The Theory of Stochastic Processes*, Chapman and Hall, London, 1965.

[5] J. C. Cox and S. A. Ross, "The valuation of options for alternative stochastic processes", *Journal of Financial Economics* **3**(1–2) (1976) 145–166.

[6] E. Derman and I. Kani, "Riding on a smile", *Risk* **7**(2) (1994) 32–39.

[7] B. Dumas, J. Fleming and R. Whaley, "Implied volatility functions: Empirical tests", *Journal of Finance* **53** (1998) 2059–2106.

[8] B. Dupire, "Pricing with a smile", *Risk* **7**(1) (1994) 18–20.

[9] R. Geske, "The valuation of compound options", *Journal of Financial Economics* **7** (1979) 63–81.

[10] R. Goldberg, *Fourier Transforms*, Cambridge University Press, Cambridge, UK, 1965.

[11] I. Hirschman and D. Widder, *The Convolution Transform*, Princeton University Press, Princeton, NJ, 1955.

[12] I. Karatzas and S. Shreve, *Brownian Motion and Stochastic Calculus*, Springer-Verlag, NY, 1988.

[13] D. Madan, P. Carr and D. Chang, "The variance gamma process and option pricing", *European Finance Review* **2** (1998) 79–105.

[14] H. Robbins and D. Siegmund, "Statistical tests of power one and the integral representation of solutions of certain partial differential equations", *Bulletin of the Institute of Mathematics Academia Sinica* **1**(1) (1973) 93–120.

[15] S. Ross, "Options and efficiency", *Quarterly Journal of Economics* **90** (1976) 75–89.

[16] M. Rubinstein, "Displaced diffusion option pricing", *Journal of Finance* **38**(1) (1983) 213–217.

[17] M. Rubinstein, "Implied binomial trees", *Journal of Finance* **49**(3) (1994) 771–818.

Reprinted from The Journal of Computational Finance
2(3) (1999) 77–100

RECONSTRUCTING THE UNKNOWN LOCAL VOLATILITY FUNCTION

THOMAS F. COLEMAN*, YUYING LI* and ARUN VERMA*

Using market European option prices, a method for computing a *smooth* local volatility function in a 1-factor continuous diffusion model is proposed. Smoothness is introduced to facilitate accurate approximation of the local volatility function from a finite set of observation data. Assuming that the underlying indeed follows a 1-factor model, it is emphasized that accurately approximating the local volatility function prescribing the 1-factor model is crucial in hedging even simple European options, and pricing exotic options. A spline functional approach is used: the local volatility function is represented by a spline whose values at chosen knots are determined by solving a constrained nonlinear optimization problem. The optimization formulation is amenable to various option evaluation methods; a partial differential equation implementation is discussed. Using a synthetic European call option example, we illustrate the capability of the proposed method in reconstructing the unknown local volatility function. Accuracy of pricing and hedging is also illustrated. Moreover, it is demonstrated that, using different implied volatilities for options with different strikes/maturities can produce erroneous hedge factors if the underlying follows a 1-factor model. In addition, real market European call option data on the S&P 500 stock index is used to compute the local volatility function; stability of the approach is demonstrated.

Keywords: Local volatility function, implied volatility, option pricing model, 1-factor continuous diffusion.

1. Introduction

An option pricing model establishes a relationship between the traded derivatives, the underlying asset and the market variables, e.g., volatility of the underlying asset [4, 25]. Option pricing models are used in practice to price derivative securities given knowledge of the volatility and other market variables.

The celebrated constant-volatility Black–Scholes model [4, 25] is the most often used option pricing model in financial practice. This classical model assumes constant volatility; however, much recent evidence suggests that a constant volatility model is not adequate [27, 28]. Indeed, numerically inverting the Black–Scholes

*Computer Science Department and Cornell Theory Center, Cornell University, Ithaca, NY 14850. Research partially supported by the Applied Mathematical Sciences Research Program (KC-04-02) of the Office of Energy Research of the US Department of Energy under grant DE-FG02-90ER25013.A000 and NSF through grant DMS-9505155 and ONR through grant N00014-96-1-0050.
This research was conducted using resources of the Cornell Theory Center, which is supported by Cornell University, New York State, the National Center for Research Resources at the National Institutes of Health, and members of the Corporate Partnership Program.

formula on real data sets supports the notion of asymmetry with stock price (volatility skew), as well as dependence on time to expiration (volatility term structure). Collectively this dependence is often referred to as the volatility smile. The challenge is to accurately (and efficiently) model this volatility smile.

In practice, the constant-volatility Black–Scholes model is often applied by simply using different volatility values for options with different strikes and maturities. In this paper, we refer to this approach as the constant implied volatility approach. Although this method works well for pricing European options, it is unsuitable for more complicated exotic options and options with early exercise features. Moreover, as will be illustrated in Sec. 4, this approach can produce incorrect hedge factors even for simple European options, assuming that the underlying follows a 1-factor model.

A few different approaches have been proposed for modeling the volatility smile. One class of methods (Merton [26]) assumes a Poisson jump diffusion process for the underlying asset. Stochastic volatility models (Hull and White [20]) have also been used. Das and Sundaram [10] indicate that neither of these types of models sufficiently explains the implied volatility structure.

Finally, there is the 1-factor continuous diffusion approach: an underlying asset with the initial value S_{init} is assumed to satisfy:

$$\frac{dS_t}{S_t} = \mu(S_t, t)\, dt + \sigma^*(S_t, t)\, dW_t\,, \quad t \in [0, \tau]\,, \ \tau > 0\,, \tag{1}$$

where W_t is a standard Brownian motion, τ is a fixed trading horizon, and μ, σ^*: $\Re^+ \times [0, \tau] \to \Re$ are deterministic functions. The function $\sigma^*(s, t)$ is called the *local volatility function*. The advantages of the 1-factor continuous diffusion model, compared to the jump or stochastic model, include that no non-traded source of risk such as the jump or stochastic volatility is introduced [17]. Consequently, the completeness of the model, i.e., the ability to hedge options with the underlying asset, is maintained. Completeness is ultimately important since it allows for arbitrage pricing and hedging [17].

There may be dispute regarding whether a 1-factor model (1) is the best way to model an underlying process. Our research will not shed light to this dispute. Instead, we demonstrate the importance of accurately approximating the local volatility function in pricing and hedging derivatives when the underlying follows a 1-factor model (1).

In order to price complex exotic options using a 1-factor diffusion model (1), the volatility function $\sigma^*(s, t)$ needs to be approximated. Volatility is the only variable in this 1-factor model which is not directly observable in the market. Similar to the implied volatility in the constant volatility model, one possible idea is to imply this local volatility function from the market option price data. Indeed, it is established [1, 17] that the local volatility function can be uniquely determined from the European call options of all strikes and maturities, under the no arbitrage assumption of the observable European call option prices. Unfortunately, the market European option prices are typically limited to a relatively few different strikes and maturities. Therefore the problem of determining the local volatility function can be regarded as a function approximation problem from a finite data set with a nonlinear observation functional. Due to insufficient market option price data, this is a well-known ill-posed problem.

Computational methods have been proposed to solve this ill-posed problem [1, 2, 5, 13, 14, 17, 22, 23, 27]. Most of these methods [1, 5, 13, 14, 17, 22, 27] overcome the illposedness of the problem by assuming the existence of a complete spanning set of European call option prices, which, in practice, requires use of extrapolation and interpolation of the available market option prices [5, 13, 22, 27]. This can be problematic because potentially erroneous non-market information are introduced into the data. Rubinstein proposes to compute the implied probability without any exogenous assumption on the model for the local volatility function [22, 27]. In [1] the local volatility is computed at each discretization nodal point with a PDE approach. The methods [2, 23] use a regularization approach to the ill-posed local volatility approximation problem. The closeness of the local volatility to a prior is used in [2] and smoothness is used in [23].

The local volatility function approximation problem is ill-posed: there are typically an infinite number of solutions to the problem. It is not difficult to find a local volatility function $\sigma(s,t)$ that matches the market option price data. However, for accurately pricing exotic options, we are not merely concerned with matching the market option prices but would like to reconstruct as accurately as possible the volatility function $\sigma^*(s,t)$ in the diffusion model (1). Accurately approximating this volatility function is especially important for computing hedge factors, even for simple European call/put options, see Sec. 4.

Smoothness of the function has long been used as a regularization criterion for function approximation with a finite observation data [29, 30, 31]. Splines have known to possess good approximation theoretical properties for a model both when the function is fixed and smooth and when it is a sample function from a stochastic process [31]. However, approximating the local volatility function from a finite set of option prices is more complex, compared to a standard function approximation problem, since the (observation) option price functional is nonlinear. Nevertheless, it is intuitive that smoothness regularization will play a similar role here.

In [23] the lack of sufficient market option price data is overcome by regularizing with smoothness of the local volatility function. The local volatility is computed at *each discretization point* to match the given option prices with an additional objective of minimizing the change of the derivative $\nabla\sigma(s,t)$. Unfortunately, this approach requires the solution of a very large-scale nonlinear optimization problem: the dimension is equal to the total number of discretization points. In addition, it requires determination of a regularization parameter.

In this paper, we propose a spline functional approach: a local volatility function $\sigma(s,t)$ is explicitly represented by a spline with a fixed set of spline knots and end condition. The volatilities at the spline knots uniquely determine a local volatility function. We choose the number of spline knots to be no greater than the number of option prices and they are placed with respect to the given data. The spline is determined by solving a constrained nonlinear optimization problem to match the market option prices as closely as possible. The dimension of the optimization problem is typically small, depending on the number of option prices available. The approximation properties of the spline allow an accurate and smooth approximation of the local volatility function prescribing the 1-factor model in a region within which the volatility values are significant for pricing available options.

We start with the motivation for our proposed inverse spline approximation formulation for the local volatility in Sec. 2. Computational issues for solving the proposed optimization problem are discussed in Sec. 3. Numerical examples illustrating the reconstructed local volatility surfaces from the European call option prices are described in Sec. 4. Using a European call option example with the underlying following the known absolute diffusion process, we illustrate the capability of the proposed method for accurately reconstructing the local volatility function. A S&P 500 European index call option example with the real market data is also used to illustrate the smoothness of the local volatility function and the stability of the proposed approach. In Sec. 5, concluding remarks are given.

2. Local Volatility Function Approximation with Splines

Assume that the underlying asset follows a continuous 1-factor diffusion process with the initial value S_{init}:

$$\frac{dS_t}{S_t} = \mu(S_t, t)\, dt + \sigma^*(S_t, t)\, dW_t\,, \quad t \in [0, \tau]\,,$$

for some fixed time horizon $[0, \tau]$, W_t is a Brownian motion, and $\mu(s,t)$, $\sigma^*(s,t)$: $\Re^+ \times [0, \tau] \to \Re$ are deterministic functions sufficiently well behaved to guarantee that (1) has a unique solution [24]. Note that in this notation $\sigma^*(s,t)$ can be negative as well as positive. (The conventional notion of positive volatility corresponds to $\sqrt{\sigma^*(s,t)^2}$ in our notation.) For simplicity, we assume that the instantaneous interest rate is a constant $r > 0$ and the dividend rate is a constant $q > 0$ (a general stochastic interest derivative pricing can be priced, e.g., [19]). Given S_{init}, r and q, and under the no arbitrage assumption [25], an option with the volatility $\sigma(s,t)$, strike price K, and maturity T has a unique price $v(\sigma(s,t), K, T)$.

Assume that we are given m market option (bid, ask)-pairs, $\{(\text{bid}_j, \text{ask}_j)\}_{j=1}^{m}$, corresponding to strike prices/expiration times $\{(K_j, T_j)\}_{j=1}^{m}$. Let

$$v_j(\sigma(s,t)) \stackrel{\text{def}}{=} v(\sigma(s,t), K_j, T_j)\,, \quad j = 1, \ldots, m\,.$$

We want to approximate, as accurately as possible, the local volatility function $\sigma^*(s,t) : \Re^+ \times [0, \tau] \to \Re$ from the requirement that

$$\text{bid}_j \leq v_j(\sigma(s,t)) \leq \text{ask}_j\,, \quad j = 1, \ldots, m\,. \tag{2}$$

Since the observation data $\{(\text{bid}_j, \text{ask}_j, K_j, T_j)\}_{j=1}^{m}$ is finite and the restriction is on the option values $\{v_j(\sigma(s,t))\}_{j=1}^{m}$, problem (2) can be considered an *inverse function approximation problem* from a finite observation data. Let $\mathcal{H}$ denote the space of measurable functions in the region $[0, +\infty) \times [0, \tau]$. The inverse function approximation problem (2) can be written as an optimization problem:

$$\min_{\sigma(s,t) \in \mathcal{H}} \sum_{j=1}^{m} [\text{bid}_j - v_j(\sigma(s,t)))]^+ + \sum_{j=1}^{m} [v_j(\sigma(s,t)) - \text{ask}_j)]^+ , \tag{3}$$

where $x^+ \overset{\text{def}}{=} \max(x, 0)$. This is a nonlinear piecewise differentiable optimization problem: to overcome nondifferentiability in (3), one can alternatively solve a variational least squares problem:

$$\min_{\sigma(s,t) \in \mathcal{H}} \sum_{j=1}^{m} (v_j(\sigma(s,t)) - \bar{v}_j)^2 , \tag{4}$$

where $\bar{v}_j \overset{\text{def}}{=} \frac{\text{bid}_j + \text{ask}_j}{2}$. Since the observation data is finite, problems (2, 3, 4) are severely underdetermined: there are typically an infinite number of solutions. It is easy to find a function $\sigma(s,t)$ that matches the market option price data [2, 5, 13, 14, 17, 22, 23, 27].

The local volatility reconstruction problem (2, 3, 4) is a complicated nonstandard function approximation problem. The option price functional $v(\sigma(s,t), K, T)$ is nonlinear in the local volatility function $\sigma(s,t)$. It is a nonlinear inverse function approximation problem.

In most of the proposed methods [1, 2, 5, 13, 14, 17, 22, 27] matching the market option price data has been emphasized; it is often the only objective. However, a function $\sigma(s,t)$ which matches the finite set of market option prices can be very different from the local volatility $\sigma^*(s,t)$ which prescribes the 1-factor model for the underlying, see Sec. 4 for an example. Moreover, the price $\bar{v}_j$ generally has error (for example, when a bid-ask spread exists). In addition, the option value $v_j(\sigma(s,t))$ can only be computed numerically using a tree method or a PDE approach (there is no closed form solution for a general 1-factor model (1)). Hence, it may not be desirable to insist that $v_j(\sigma(s,t))$ match exactly the observed market price $\bar{v}_j$ for $j = 1, \ldots, m$. For pricing and hedging of exotic options, it is more important to compute a local volatility function $\sigma(s,t)$ which is as close as possible to the local volatility function $\sigma^*(s,t)$. In other words, in addition to calibrating the market option price data sufficiently accurately, we would like to *reconstruct*, as accurately as possible, the local volatility function $\sigma^*(s,t)$ of the diffusion model (1).

Smoothness has long been used [29, 30, 31] as a regularization condition for a function approximation problem with a limited observation data. In addition, smoothness of the local volatility function can be important in computational option valuation schemes. Convergence of a PDE finite difference method, for example, depends on the smoothness of the function $\sigma(s,t)$.

In [23] it is proposed to use smoothness as a regularization condition to approximate the local volatility function. The regularized optimization problem

$$\min_{\sigma(s,t)\in\mathcal{H}} \sum_{j=1}^{m}(v_j(\sigma(s,t)))-\bar{v}_j)^2 + \lambda\|\nabla\sigma(s,t)\|_2 \tag{5}$$

is used in [23] where λ is a positive constant and $\|\cdot\|_2$ denotes the L^2 norm. The change of the first-order derivative is minimized depending on the regularization parameter λ for which determining a suitable value may not be easy. In addition, computational implementation of this method requires solving a large-scale discretized optimization problem: for a PDE implementation, the dimension is NM where N is the number of discretization points in s and M is the number of discretization points in t. A simple gradient descent algorithm is used in [23]. Since the optimization problem is (5) highly nonlinear, with such a method, the computed solution is typically inaccurate. To use a more sophisticated optimization algorithm, the Jacobian matrix of the vector function $(v_1,\ldots,v_m)$ needs to be evaluated but this becomes extremely costly due to the large dimension of the discretized problem.

Splines have long been used in approximating smooth curves and surfaces (see, e.g., [16]). They have also been used as a tool for regularizing ill-posedness of function approximations from finite observation data [31]. In a typical one-dimensional spline interpolation setting, assuming values $f_i, i = 1,\ldots,m$, of the dependent variable $f(x)$ corresponding to values $x_i, i = 1,\ldots,m$, are given, a spline is chosen to fit the data $(f_i, x_i), i = 1,\ldots,m$. Given the number of knots p and their locations, the freedom of the spline is the coefficient of each spline segment. The cubic spline has long been used by craftsman and engineers as the mechanic spline. It is the smoothest twice continuously differentiable function that matches the observations; the minimizer of

$$\min_{f(x)\in\mathcal{S}} \int_a^b (f''(x))^2\,dx, \quad \text{subject to} \quad f(x_i)=f_i, \quad i=1,\ldots,m,$$

is a natural cubic spline, where $\mathcal{S}$ is the Sobolev space of functions whose first-order derivatives are continuously differentiable and the second-order derivatives are square integrable (assuming $m \geq 2$). For mechanical splines, this corresponds to minimize the elastic strain energy. For two-dimensional surface fitting, the bicubic spline defined on a regular grid is twice continuously differentiable [3, 16]. The bicubic spline has a similar variational minimization property. Advantages of spline interpolation include its fast convergence on many types of meshes, computational efficiency, and insensitivity to roundoff errors [3].

Approximating the local volatility function by a spline is particularly reasonable if the local volatility function is smooth. Is this a reasonable expectation for the local volatility function? Assume that the underlying follows the 1-factor diffusion process (1). Let there be given observable arbitrage-free market European call prices $v(K,T)$ for all strikes $K \in [0,\infty)$ and all maturities $T \in (0,\tau]$. From Proposition 1 in [1], the local volatility function $\sigma^*(s,t)$ of the diffusion process (1) that is consistent with the market is given uniquely by

$$(\sigma^*(K,T))^2 = 2\frac{\dfrac{\partial v}{\partial T} + qv(K,T) + K(r-q)\dfrac{\partial v}{\partial K}}{K^2\dfrac{\partial^2 v}{\partial K^2}}. \tag{6}$$

This formula suggests that, assuming $v(K,T)$ is sufficiently smooth (note that $\frac{\partial^2 v}{\partial K^2}$ and $\frac{\partial v}{\partial T}$ already exist) and $\frac{\partial^2 v}{\partial K^2} \neq 0$, $(\sigma^*(K,T))^2$ is sufficiently smooth in the region $(0,\infty) \times (0,\tau]$ as well.

In this paper, we use a two-dimensional spline functional to directly approximate a local volatility function.[a] Let the number of spline knots $p \leq m$. We choose a set of fixed spline knots $\{(\bar{s}_j,\bar{t}_j)\}_{j=1}^p$ in the region $[0,\infty) \times [0,\tau]$. Given $\{(\bar{s}_i,\bar{t}_i)\}_{i=1}^p$ spline knots with corresponding local volatility values $\bar{\sigma}_i \overset{\text{def}}{=} \sigma(\bar{s}_i,\bar{t}_i)$, an interpolating cubic spline $c(s,t)$ with a fixed end condition (in our computation the natural spline end condition is used) is uniquely defined by setting $c(\bar{s}_i,\bar{t}_i) = \bar{\sigma}_i, i = 1,\ldots,p$. We then determine the local volatility values $\bar{\sigma}_i$ (hence the spline) by calibrating the market observable option prices. The freedom in this problem is represented by the volatility values $\{\bar{\sigma}_i\}$ at the given knots $\{(\bar{s}_i,\bar{t}_i)\}$. If $\bar{\sigma}$ is a p-vector, $\bar{\sigma} = (\bar{\sigma}_1,\ldots,\bar{\sigma}_p)^T$, then we denote the corresponding interpolating spline with the specified end condition as $c(s,t;\bar{\sigma})$.

Let
$$v_j(c(s,t;\bar{\sigma})) \overset{\text{def}}{=} v(c(s,t;\bar{\sigma}),K_j,T_j), \quad j = 1,\ldots,m.$$

To allow the possibility of incorporating additional a priori information, l and u are lower and upper bounds that can be imposed on the local volatilities at the knots. Thus, we define the *inverse spline local volatility approximation problem*: Given p spline knots, $(\bar{s}_1,\bar{t}_1),\ldots,(\bar{s}_p,\bar{t}_p)$, solve for the p-vector $\bar{\sigma}$

$$\min_{\bar{\sigma} \in \mathfrak{R}^p} \ f(\bar{\sigma}) \overset{\text{def}}{=} \frac{1}{2}\sum_{j=1}^m w_j[v_j(c(s,t;\bar{\sigma})) - \bar{v}_j]^2 \quad \text{subject to} \quad l \leq \bar{\sigma} \leq u, \tag{7}$$

where positive constants $\{w_j\}_{j=1}^m$ are weights, allowing account to be taken of different accuracies of $\bar{v}_j$ or computed v_j. The determination of an approximation in the l_1 or l_∞ norm instead may be a valuable alternative although the problem becomes even more difficult to solve computationally. Note also that the formulation (7) is quite general: European call/put or even more complicated option prices can be used to compute the spline approximation to the local volatility function $\sigma^*(s,t)$.

The inverse spline local volatility problem (7) is a minimization problem with respect to the local volatility $\bar{\sigma}$ at the spline knots. The computed volatility function has some dependence on the number of knots p and the location of the knots

---

[a]If it is known that $\sigma(s,t)$ is a function of s or t only, then one can use one-dimensional spline.

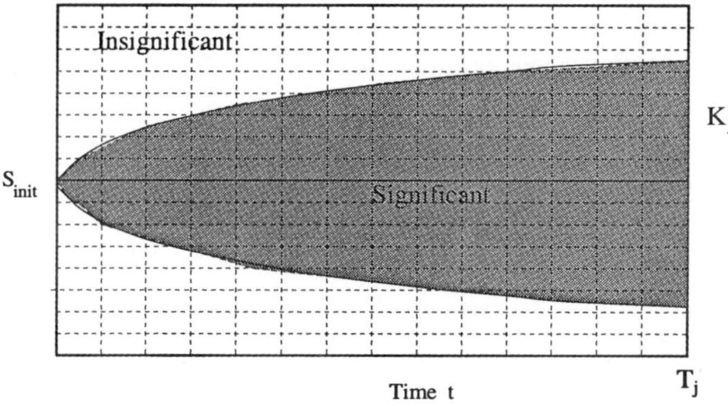

Fig. 1. The local volatility in the shaded **region** $\mathcal{D}_j$ **is** significant in pricing and hedging.

$\{(\bar{s}_i, \bar{t}_i)\}_{i=1}^{p}$. The choice of the number of knots and their placement in spline approximation is generally a complicated issue [16, 31]. The situation here is not typical for spline approximation due to the fact that the dependent option price function is not the function to be approximated. Rather, it depends on the values of the unknown volatility function in the region $R^+ \times [0, \tau]$. Moreover, the dependence on the unknown volatility values is not uniform in the region $R^+ \times [0, \tau]$. The option premium depends little on the volatility values with small t and s far from S_{init}. It is convenient to view this as follows [1]: there exists a region centered around S_{init} within which the volatility values are significant in pricing and hedging: we denote this region as $\mathcal{D}_j$ for the option v_j, see Fig. 1 for illustration of its typical shape. We can at most expect to approximate well the local volatilities in the region $\mathcal{D} \stackrel{\text{def}}{=} \cup_{j=1}^{m} \mathcal{D}_j$ from the market option data. In our experiments, we often choose the number of knots equal to the number of observations. In order to construct and evaluate a spline efficiently, the spline knots can be placed in a rectangular mesh covering the region $\mathcal{D}$ and bicubic spline interpolation [3] can be used.

If the number of spline knots are chosen to be no more than the number of observation data points, the degrees of freedom, compared to that of a (discretized) formulation of (3), is significantly decreased (several orders of magnitude). In addition to gaining smoothness of the local volatility function, formulation (7) significantly decreases the computational cost compared to that of the (discretized) formulations (3, 5) due to reduction of the dimension of the optimization problem.

It is not appropriate to choose p much larger than m since (7) may become underdetermined. If one decides to use more spline knots, additional regularization, e.g.,

$$\min_{\bar{\sigma} \in \Re^p} f(\bar{\sigma}) \stackrel{\text{def}}{=} \frac{1}{2} \sum_{j=1}^{m} w_j [v_j(c(s,t;\bar{\sigma})) - \bar{v}_j]^2 + \lambda \eta(c(s,t;\bar{\sigma}))$$

$$\text{subject to} \quad l \leq \bar{\sigma} \leq u, \tag{8}$$

is more appropriate: here $\lambda > 0$ is a regularization parameter and $\eta(\sigma(s,t;\sigma))$ is a smoothing norm for the tensor product splines [15]. (A referee has pointed out that this has recently been considered in [21].)

In this paper, we focus on the formulation (7) and assume p is not greater than m. In order to solve the inverse local volatility problem (7), an optimization method will be needed to evaluate the values of options $v_j(c(s,t;\bar\sigma))$ for any spline $c(s,t;\bar\sigma)$; the derivatives may also be computed. We discuss this next.

3. The Computational Procedure

Our proposal is to approximate the local volatility surface, $\sigma^*(s,t)$, with a cubic spline $c(s,t;\bar\sigma)$ by solving (7) for the vector $\bar\sigma = (\bar\sigma_1,\dots,\bar\sigma_p)^T$. Problem (7), when $p \le m$, is defined once the p knots $(\bar s_1,\bar t_1),\dots,(\bar s_p,\bar t_p)$ have been chosen appropriately. To express (7) more succinctly, define a vector-valued function $F : \Re^p \to \Re^m$ where component j of F is given by $w_j^{\frac{1}{2}}[v_j(c(s,t;\bar\sigma)) - \bar v_j]$, for $j = 1,\dots,m$. Therefore (7) can be rewritten:

$$\min_{\bar\sigma \in \Re^p} f(\bar\sigma) \stackrel{\text{def}}{=} \frac{1}{2}\|F(\bar\sigma)\|_2^2 \quad \text{subject to} \quad l \le \bar\sigma \le u. \tag{9}$$

Problem (9) is a box-constrained nonlinear least-squares problem in $\bar\sigma$; there are a variety of optimization methods available to enable its solution. In our implementation we use a trust region/interior point method [6, 7], in which a sequence of strictly feasible points are generated: $\{\bar\sigma^{(k)}\} \in int\{\mathcal{F}\}$, where $\mathcal{F} = \{\bar\sigma \in \Re^p : l \le \bar\sigma \le u\}$. Moreover, the sequence corresponds to a monotonically decreasing sequence of function values, i.e., $f^{(k+1)} < f^{(k)}, k = 1,\dots,\infty$, where $f^{(k)} = f(\bar\sigma^{(k)})$. Under mild assumptions this approach guarantees convergence, i.e., $\bar\sigma^{(k)} \to \bar\sigma^*$, where $\bar\sigma^*$ is a local minimizer for problem (9).

The Jacobian of F with respect to $\bar\sigma$ is required: $J(\bar\sigma) \stackrel{\text{def}}{=} \nabla F(\bar\sigma)$. Note that J is an $m \times p$ matrix. In the square case when $p = m$, it is possible to use a standard secant update to approximate J, e.g., [12], which can significantly reduce the cost. Under reasonable assumptions a superlinear rate of convergence can be achieved. We note that there are optimization approaches that do not require the calculation (or approximation) of the Jacobian matrix J; however, they typically converge very slowly — we have not investigated those methods in this work.

In this paper we explore two possibilities in the framework of our optimization approach:

1. Use of automatic differentiation [8] and/or finite-differencing to compute $J^{(k)} \stackrel{\text{def}}{=} J(\bar\sigma^{(k)})$;

2. Use of a secant update to approximate $J^{(k)}$ when $p = m$.

3.1. *The problem structure*

The evaluation of $f(\bar{\sigma})$ requires the evaluation of each component of F, i.e., $w_j^{\frac{1}{2}}[v_j(c(s,t);\bar{\sigma}) - \bar{v}_j]$, for $j = 1,\ldots,m$. These are generalized Black–Scholes computations. There are several ways to approach this — we choose, as an example, to use a standard PDE-discretization technique.

Given S_{init}, r, q, and $\sigma(s,t)$, let $V(s,t)$ denote the option value of an underlying asset with strike price K and expiry date T at (s,t), $t \in [0,T]$. Under the no arbitrage assumption, the option value satisfies the following generalized Black–Scholes equation [25]

$$\frac{\partial V}{\partial t} + (r-q)s\frac{\partial V}{\partial s} + \frac{1}{2}\sigma(s,t)^2 s^2 \frac{\partial^2 V}{\partial s^2} = rV. \tag{10}$$

The boundary conditions for the European call option are:

$$\lim_{s \to +\infty} \frac{\partial V(s,t)}{\partial s} = e^{-q(T-t)}, \quad t \in [0,T],$$

$$V(0,t) = 0, \quad t \in [0,T],$$

$$V(s,T) = \max(s - K, 0).$$

We use a Crank–Nicholson finite difference solution strategy for solving (10), based on discretization on a uniform grid. Given a two-dimensional grid the numerical solution of (10) is standard and discussed in several texts. Zvan *et al.* [32] have a good discussion of complexity issues. It is possible to increase efficiency by employing a number of computing techniques such as vectorization and pipelining — description of these implementation aspects goes beyond the scope of this paper.

3.2. *Computing the Jacobian and the gradient*

The Jacobian matrix $J(\bar{\sigma})$ satisfies

$$J(\bar{\sigma}) = \frac{\partial v}{\partial c} \times \frac{\partial c}{\partial \bar{\sigma}},$$

where $\frac{\partial v}{\partial c}$ is an m-by-MN matrix, $\frac{\partial c}{\partial \bar{\sigma}}$ is an MN-by-p matrix.

It is useful to note that matrix $C \stackrel{\text{def}}{=} \frac{\partial c}{\partial \bar{\sigma}}$ is constant and therefore needs to be computed just once for the entire problem (given a fixed discretization and spline knot placement). The product $\frac{\partial v}{\partial c} \times C$ can be computed directly using automatic differentiation (forward mode) or approximately using finite differences (differencing v along the columns of C). In either case the work involved is $O(p \cdot \omega(F))$ where $\omega(F)$ is the work (flops) required to evaluate F. (In the finite-differencing case this is a tight bound whereas this bound can be undercut considerably if automatic differentiation is used [18].)

The gradient of f, with respect to $\bar{\sigma}$, is simply $J^T F$. Therefore if the function F and its Jacobian J have been computed as described above the gradient is given by a matrix-vector multiplication.

If the secant method is used in the square case, i.e., $p = m$, then the gradient is approximated by $A \times F$ where A is the secant approximation to the Jacobian. The Jacobian is not computed (except, possibly, for secant method restarts) with this approach.

4. Computational Examples

We now describe some computational experience with our proposed method for reconstructing the local volatility function $\sigma^*(s,t)$ from limited observation data. We illustrate how European call options can be used to approximate the local volatility function.

We have implemented the proposed method in Matlab using a trust region optimization algorithm with a PDE approach for function and Jacobian evaluation. Without precise knowledge of accuracy of the market data, the weights in the inverse spline local volatility approximation problem (7) are simply set to unity: $w_j \overset{\text{def}}{=} 1, j = 1, \ldots, m$. The generalized Black–Scholes PDE (10) is solved with a Crank–Nicolson finite difference method. Given any $\bar{\sigma}$, the bicubic spline $c(s,t;\bar{\sigma})$ with the variational end condition (the second-order derivative at the end is zero) is computed and evaluated using the functions in the Matlab spline toolbox [11]. We use a simple discretization scheme: a uniformly spaced mesh with $N \times M$ grid points in the region $[0, 2S_{\text{init}}] \times [0, \tau]$ where τ is the maximum maturity in the market option data:

$$s_i = i \frac{2S_{\text{init}}}{N - 1}, \quad i = 0, \ldots, N - 1,$$

$$t_j = j \frac{\tau}{M - 1}, \quad j = 0, \ldots, M - 1. \tag{11}$$

For simplicity, we have chosen the spline knots to be on a uniform rectangular mesh covering the region $\mathcal{D}$ in which the volatility values are significant in pricing the market options. Given a European option, we do not have an explicit knowledge of the region $\mathcal{D}$. In our experiments, we have used $[\gamma_1 S_{\text{init}}, \gamma_2 S_{\text{init}}] \times [0, \tau]$ as an estimate of $\mathcal{D}$ with $\gamma_1 \in [.6, .8]$ and $\gamma_2 \in [1.4, 1.6]$ depending on the magnitude of S_{init}. The number of spline knots p typically equals the number of observation m. In the event that the option prices are calibrated to high precision, we have experimented with $p < m$.

4.1. *Reconstructing local volatility, pricing and hedging*

In order to demonstrate the effectiveness of the proposed method in reconstructing the local volatility surface and its accuracy in pricing and hedging, we consider a synthetic European call option example used in [23]. In this example, the underlying is assumed to follow an absolute diffusion process:

$$\frac{dS_t}{S_t} = \mu(S_t, t)\, dt + \sigma^*(S_t, t)\, dW_t, \tag{12}$$

where the local volatility function $\sigma^*(s,t)$ is a function of the underlying only,

$$\sigma^*(s,t) = \frac{\alpha}{s},$$

with $\alpha = 15$, and W_t representing a standard Brownian motion. We use the same parameter setting as in [1]: Let the initial stock index be $S_{\text{init}} = 100$, the risk free interest rate $r = 0.05$ and the dividend rate $q = 0.02$.

We consider, as market option data, 22 European call options on the underlying following the absolute diffusion process (12). Eleven options have half year maturity with strike prices $[90:2:110]$ and another eleven options have one year maturity with the same strikes. Thus the option strike and maturity vectors are given below

$$K = [90; 92; \ldots; 110; 90; 92; \ldots; 110] \in \Re^{22},$$

$$T = [0.5; 0.5; \ldots; 0.5; 1; 1; \ldots; 1] \in \Re^{22}.$$

For the absolute diffusion process (12), the analytic formula for pricing European options exists [9] and we set the market European option call price $\bar{v}_j$ equal to this analytic value. The discretization parameters in (11) are set as $M = 101$ and $N = 51$.

For this example the lower and upper bounds for the local volatility at the knot $\bar{\sigma}_i$ are $l_i = -1$ and $u_i = 1$ respectively (though no variable is at the bound at the computed solution in this case). First, we let $p = m$ and place the spline knots on the grid $[0:20:200] \times [0,1]$. The initial volatility values at knots are specified as $\bar{\sigma}_i^{(0)} = 0.15, i = 1, \ldots, 22$. The resulting optimization problem is relatively easy to solve. The optimization method requires seven iterations (six Jacobian evaluations) and the computed optimal objective function value $f(\bar{\sigma}^*)$ is 10^{-6}.

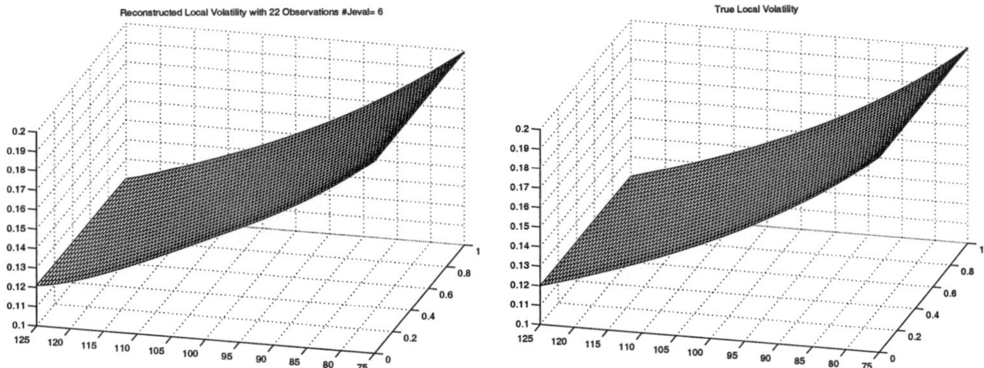

Fig. 2. The reconstructed and true local volatility.

Figure 2 demonstrates the accuracy of this local volatility reconstruction: the reconstructed spline surface $c(s,t;\bar{\sigma}^*)$ very accurately approximates the actual volatility surface $\sigma^*(s,t)$ in the neighborhood of the region $[75,125] \times [0,1]$. To better observe accuracy of reconstruction, the three plots on the left in Fig. 3

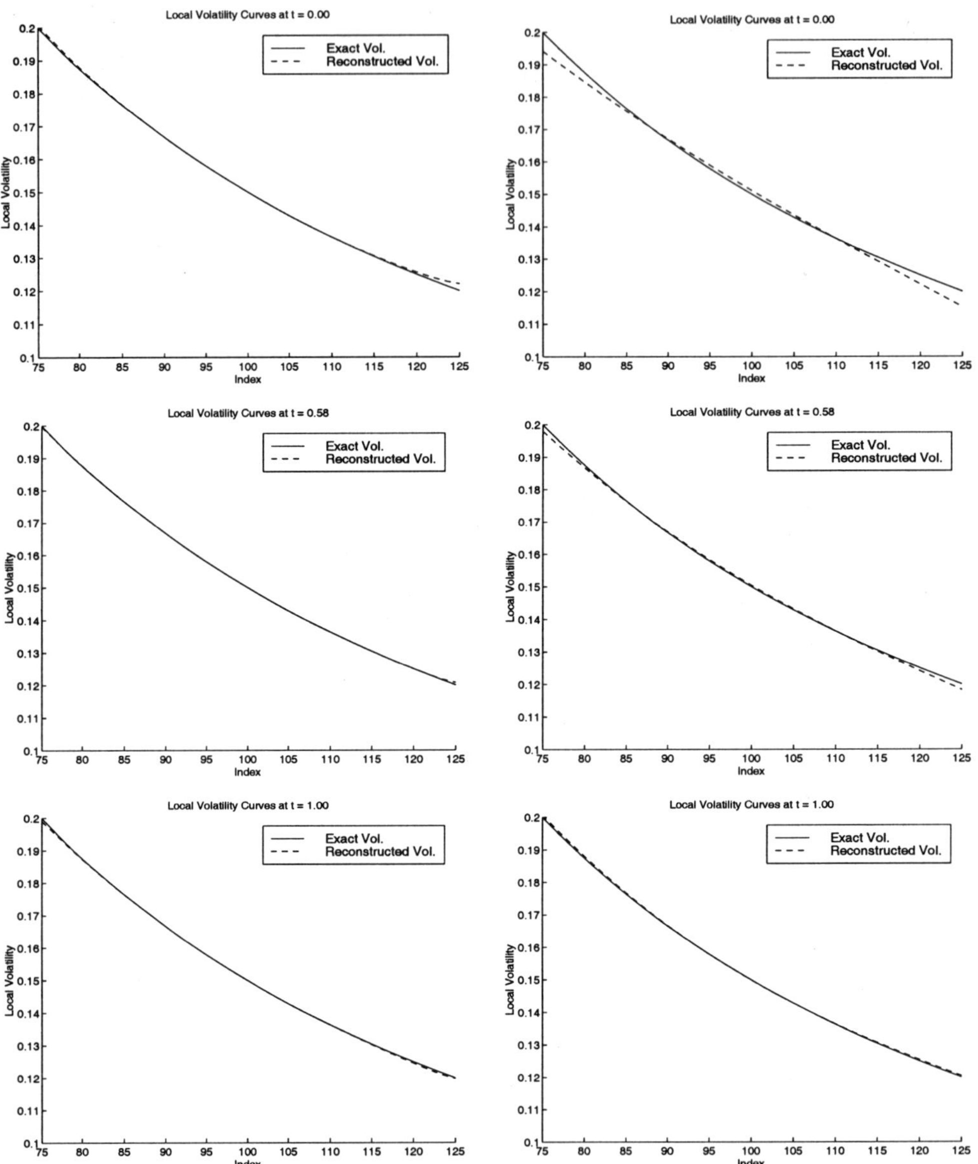

Fig. 3. Left: knots $= 0 : .2S_{\text{init}} : 2S_{\text{int}}] \times [0, 1]$, right: knots $= [.4S_{\text{init}} : .4S_{\text{init}} : 1.6S_{\text{init}}] \times [0, 1]$.

display the local volatility curves for $t = 0$, 0.58 and $t = 1$ respectively. Since the calibration error is very small and the reconstructed volatility surface is nearly linear, we experimented with choosing the number of spline knots less than m. The three plots on the right of Fig. 3 display the local volatility curves reconstructed with eight spline knots placed on the mesh $[.4S_{\text{init}} : .4S_{\text{init}} : 1.6S_{\text{init}}] \times [0, 1]$. We observe that the local volatility reconstruction remains excellent, with a slightly larger deviation when t is small and s is far from S_{init}.

To illustrate the accuracy of pricing using the reconstructed local volatility $c(s,t;\bar{\sigma}^*)$ rather than the true local volatility $\sigma^*(s,t)$, we compare prices and hedge factors of a number of European call options using both the true local volatility and the reconstructed volatility surfaces. The hedge factors vega (sensitivity to the change in the volatility), delta (sensitivity to the change in the underlying), gamma (sensitivity of delta to the change in the underlying), rho (sensitivity to change in the interest rate) and theta (sensitivity to change in the maturity) are computed using a finite difference approximation. A constant shift in both volatility surfaces is used to calculate the vega hedge factor. For European call options with strikes and maturities over the grid $[85:5:110] \times [.4:.1:.7]$, the results are shown in Table 1. These results indicate that fairly accurate prices as well as hedge factors are obtained using the reconstructed volatility surface $c(s,t;\bar{\sigma}^*)$. Note that the PDE option evaluation with the chosen discretization can generate errors of at least these magnitudes.

Table 1. Accuracy of pricing and hedging.

	Max. relative error	Average relative error
Price	$7.8e^{-3}$	$2.1e^{-3}$
Vega	$9.8e^{-3}$	$6.1e^{-3}$
Delta	$4.8e^{-2}$	$1.3e^{-2}$
Gamma	$9.5e^{-2}$	$5.9e^{-2}$
Rho	$4.5e^{-3}$	$2.0e^{-3}$
Theta	$6.9e^{-3}$	$2.2e^{-3}$

We emphasize that the formulation (7) is appropriate when the number of spline knots p is not greater than the number of observations m. If p is much larger than m, then formulation (7) can become severely underdetermined. To illustrate the potential pitfalls of allowing too much freedom in approximating $\sigma^*(s,t)$, we simulate the more realistic market situation when there is a bid-ask spread in the given option prices by setting

$$\bar{v}_j = \text{exact price of option } j + .02\,\textbf{rand}$$

where **rand** is a Matlab generated random number. We compare the local volatility reconstructions using the spline knots on the rectangular meshes $[.4S_{\text{init}} : .4S_{\text{init}} : 1.6S_{\text{init}}] \times [0,1]$ $(p = 8)$ and $[0:.01S_{\text{init}}:2S_{\text{init}}] \times [0,1]$ $(p = 202 < MN)$. The plots on the left in Fig. 4 illustrate the reconstructed local volatility curves using the rectangular mesh $[.4S_{\text{init}} : .4S_{\text{init}} : 1.6S_{\text{init}}] \times [0,1]$ for knots. The plots on the right in Fig. 4 illustrate the reconstructed local volatility curves using the rectangular mesh $[0:2:2S_{\text{init}}] \times [0,1]$ for knots. Although the available option prices are matched with very high accuracy (error about 10^{-6}) using $p = 202$, the computed local volatility surface does not resemble the true local volatility surface $\sigma^*(s,t)$. Using eight knots

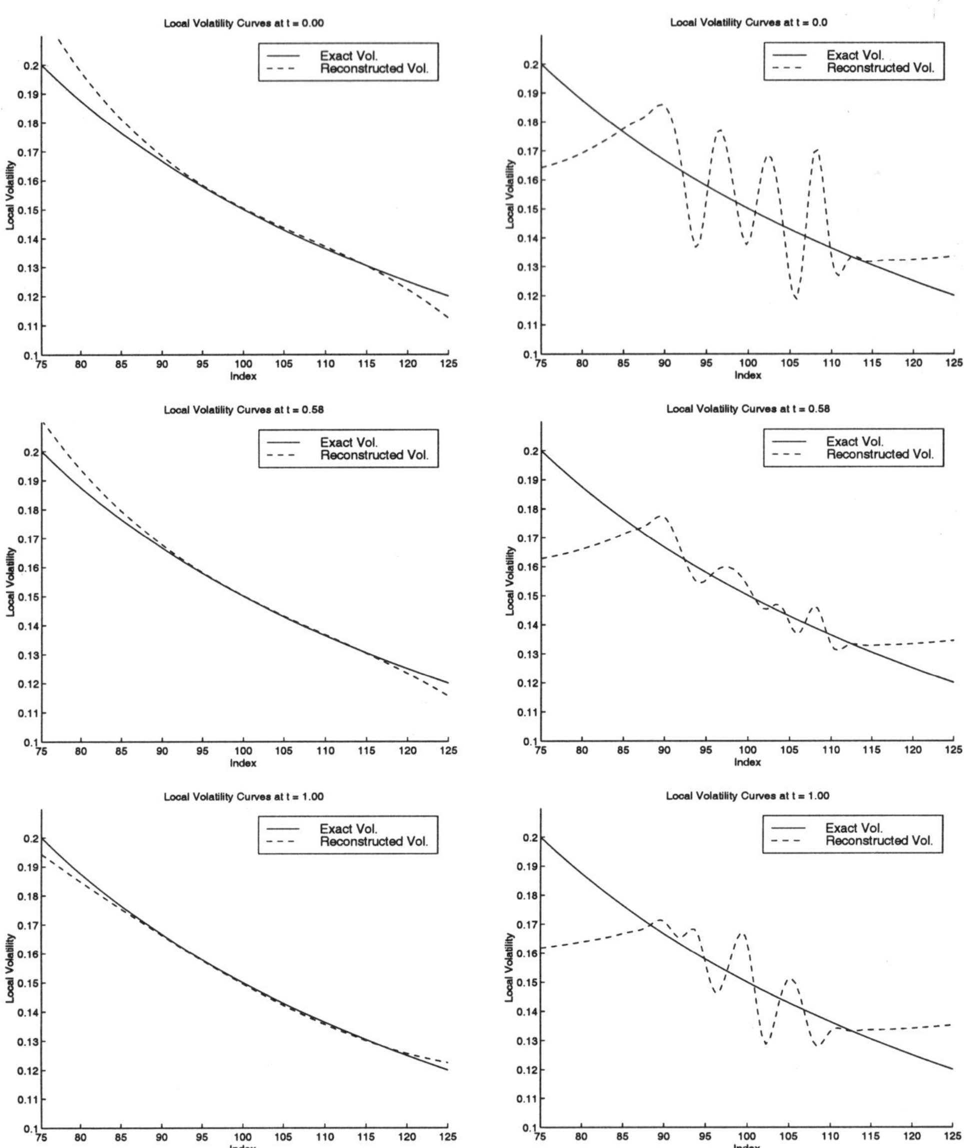

Fig. 4. Left: knots $= .4S_{\text{init}} : .4S_{\text{init}} : 1.6S_{\text{init}}] \times [0, 1]$, right: knots $= [0 : .01S_{\text{init}} : 2S_{\text{init}}] \times [0, 1]$.

on the rectangular mesh $[.4S_{\text{init}} : .4S_{\text{init}} : 1.6S_{\text{init}}] \times [0, 1]$, on the other hand, yields a much more accurate volatility surface, even though the calibration error of the available options is larger (about 10^{-4}).

Next we illustrate that, assuming the underlying follows a continuous 1-factor model (1), a constant implied volatility approach can produce erroneous hedge factors even though the option prices may be computed accurately. We use the same absolute diffusion model (12) but with greater volatility: the constant $\alpha = 75$

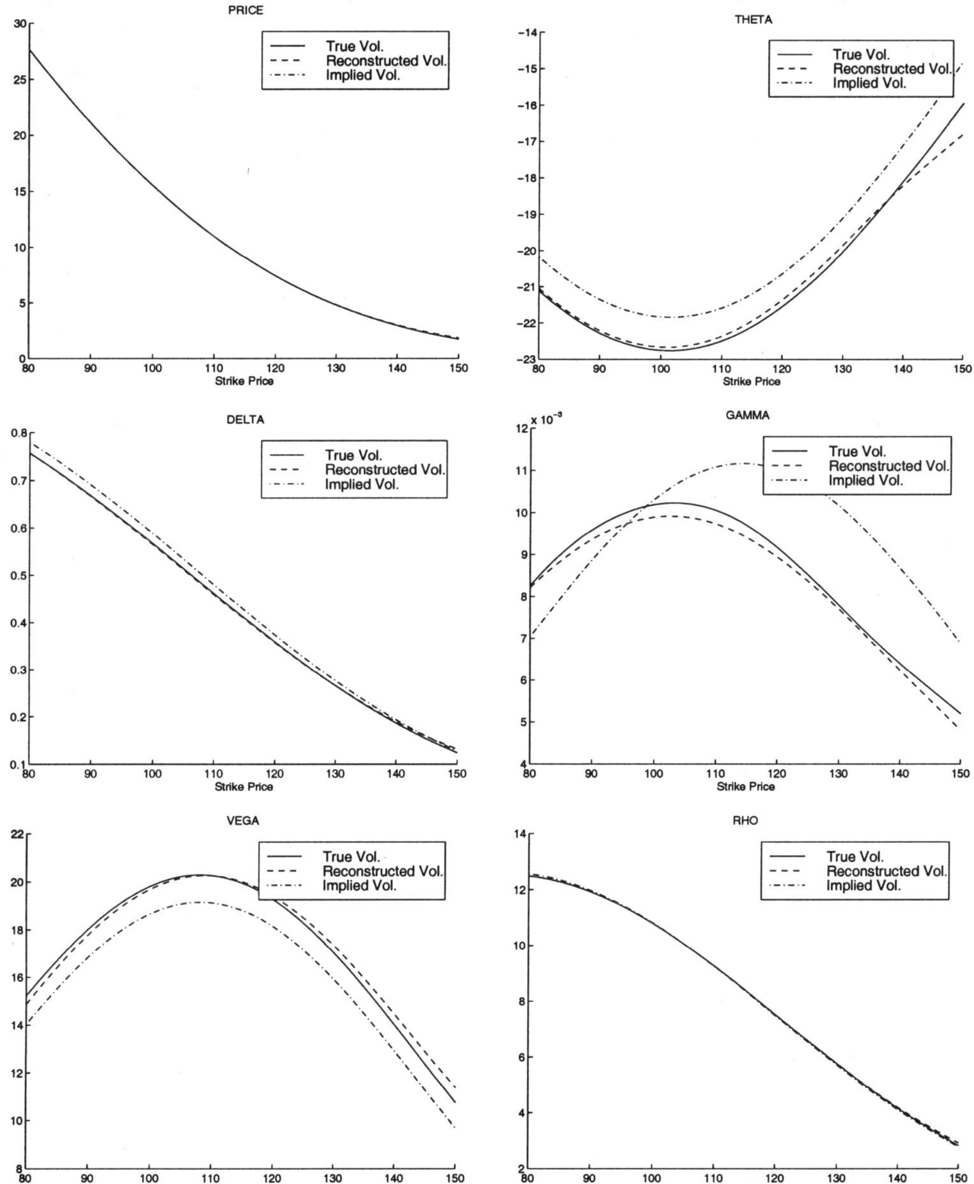

Fig. 5. Comparison between using the true, reconstructed and constant implied volatility functions.

is used instead of $\alpha = 15$. The same initial underlying $S_{\mathrm{init}} = 100$ and the risk free interest rate $r = 0.05$ are used but the dividend rate q is set to zero. We consider European call options with strikes and maturities at the grid $[80\!:\!4\!:\!120] \times [.25, .5, 1]$. The spline knots are at the grid $[0\!:\!20\!:\!2S_{\mathrm{init}}] \times [0, .5, 1]$. Figure 5 displays the price and hedge factors of options with maturity .25 year using the true local volatility, reconstructed volatility, and constant implied volatility. From these plots, we see that the price and all the hedge factors computed using the reconstructed local volatility function are fairly accurate approximation to the true values. Using the constant implied volatility method, however, large errors exist in hedge factors (mostly noticeably in theta, delta, gamma and vega).

In addition to choosing the number of spline knots p, the placement of the knots requires some care as well. The spline knots should be placed to cover the region $\mathcal{D}$ within which the values of the local volatility are significant in the option values. We have used the uniform spacing in the interval $[0, 2S_{\text{init}}]$ and $[.4S_{\text{init}}, 1.6S_{\text{init}}]$ in this synthetic example but an alternative is to place them nonuniformly with a more refined placement around $s = S_{\text{init}}$. Moreover, one need to avoid placing spline knots too closely together since this can lead to ill conditioning of the Jacobian matrix ∇F.

Table 2. Implied volatilities for S&P 500 index options.

Maturity (in years)	Strike (% of spot)									
	85%	90%	95%	100%	105%	110%	115%	120%	130%	140%
.175	.190	.168	.133	.113	.102	.097	.120	.142	.169	.200
.425	.177	.155	.138	.125	.109	.103	.100	.114	.130	.150
.695	.172	.157	.144	.133	.118	.104	.100	.101	.108	.124
.94	.171	.159	.149	.137	.127	.113	.106	.103	.100	.110
1	.171	.159	.150	.138	.128	.115	.107	.103	.099	.108
1.5	.169	.160	.151	.142	.133	.124	.119	.113	.107	.102
2	.169	.161	.153	.145	.137	.130	.126	.119	.115	.111
3	.168	.161	.155	.149	.143	.137	.133	.128	.124	.123
4	.168	.162	.157	.152	.148	.143	.139	.135	.130	.128
5	.168	.164	.159	.154	.151	.148	.144	.140	.136	.132

4.2. *A S&P 500 example illustrating smoothness and stability*

We consider now a more realistic example of approximating the local volatility function $\sigma^*(s, t)$ from the European S&P 500 index European call options. We use the same European option data of October 1995 given in [1]. The market option price data (in the implied Black–Scholes constant volatility) is given in Table 2. Similar to [1], we use only the options with no more than two years maturity in our computation. The initial index, interest rate and dividend rate are set as in [1],

$$S_{\text{init}} = \$590, \quad r = 0.06, \quad \text{and} \quad q = 0.0262.$$

The discretization parameters in (11) are set as,

$$N = 101, \quad \text{and} \quad M = 101.$$

In order to solve the proposed inverse spline volatility problem (7), we compute the market European call option prices with given strikes and maturities using the constant volatility Black–Scholes formula with the corresponding implied volatility. The Matlab function **blsprice** is used.

For this example, the number of spline knots p equals the number of observations m and the spline knots are placed on a rectangular mesh $[.8S_{\mathrm{init}} : .066S_{\mathrm{init}} : 1.4S_{\mathrm{init}}] \times [0 : .33 : 2]$. Using all the call option prices with maturity $T \leq 2$ in Table 2, the reconstructed local volatility surface is given in Fig. 6. This optimization problem seems to be more nonlinear and difficult to solve. After 28 iterations, the average error of $v_j(c(s, t; \bar{\sigma})) - \bar{v}_j$ using the reconstructed local volatility is 0.0076. The average error using the constant implied volatility via the PDE implementation with this discretization, compared to the Black–Scholes analytic formula, is 0.0510.

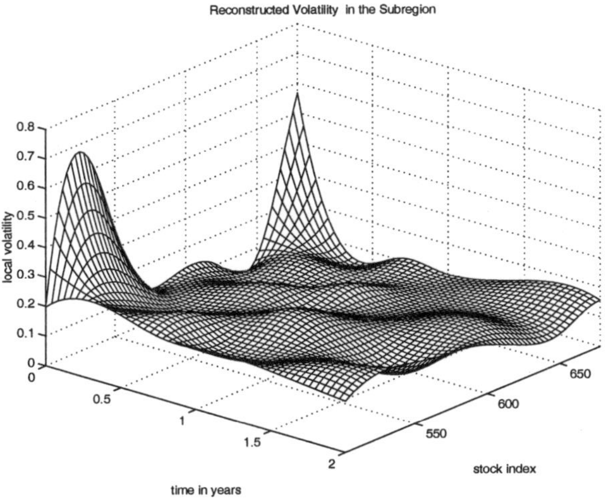

Fig. 6. Spline knot placement is the rectangular mesh $[.8S_{\mathrm{init}} : .066S_{\mathrm{init}} : 1.4S_{\mathrm{init}}] \times [0 : 0.33 : 2]$.

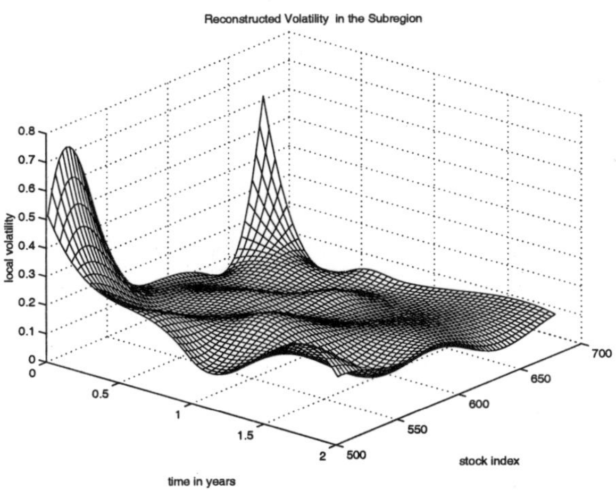

Fig. 7. Spline knot placement is the rectangular mesh $K \times [0 : 0.33 : 2]$.

The reconstructed local volatility surfaces can be slightly different if different spline knots are chosen. In order to show that the local volatility surface reconstruction, pricing and hedging are relatively robust, we consider the second spline knots placement using the rectangular mesh $K \times [0 : .33 : 2]$. The average price calibrating error for the market call options in this case is .0027. The reconstructed volatility surface using this knot placement is shown in Fig. 7. Comparing Fig. 6 with Fig. 7, the reconstructed volatility surfaces are quite similar in the region $\mathcal{D}$, noting the shape of $\mathcal{D}$. For options with strikes and maturities over the grid $[.85 : .1 : 1.15]S_{\text{init}} \times [.85 : .1 : 1.15]$, the relative difference of pricing and hedging factor with the two spline knot placements are shown in Table 3. We observe that indeed they are acceptably close.

Table 3. Differences between using two rectangular meshes for knots.

	Max. relative difference	Average relative difference
Price	$6.8e^{-3}$	$1.4e^{-3}$
Vega	$1.3e^{-2}$	$2.7e^{-3}$
Delta	$4.3e^{-2}$	$1.6e^{-2}$
Gamma	$8.8e^{-2}$	$4.1e^{-2}$
Rho	$5.3e^{-3}$	$2.0e^{-3}$
Theta	$4.9e^{-2}$	$9.2e^{-3}$

For pricing simple European call/put options, different implied volatilities are often used in practice to price options of different strikes/maturities in order to accommodate volatility smile. For pricing an exotic option such as a knock-out option, a constant volatility model is inappropriate since the price of this option depends on volatilities of different strikes and maturities. In order to illustrate the potential error in using a constant volatility in pricing exotic options, we examine here the price and hedge factors differences between using a constant volatility model and the 1-factor model with the reconstructed volatility function. We use the same S&P 500 index option example and choose the the arithmetic average (which is 0.1319) of the implied volatilities with $T \le 2$ as the constant volatility.

Table 4. Relative difference in pricing and hedging using constant volatility.

	Max. relative difference	Average relative difference
Price	11%	6%
Vega	15%	9%
Delta	19%	11%
Gamma	27%	17%
Rho	12%	7%
Theta	29%	16%

We compare the prices at a grid $[.85:.1:1.15]S_{\text{init}} \times [.85:.1:1.15]$ of strike prices and maturity dates (different from given market data). The results are in Table 4. These two methods give significantly different prices: we notice as much as 11% relative difference. Similarly all the hedge factors computed using the constant volatility have a large relative difference, we document the results in Table 4. To visualize the difference in detail, we plot the price and hedge factor curves for options with 1-year maturity in Fig. 8.

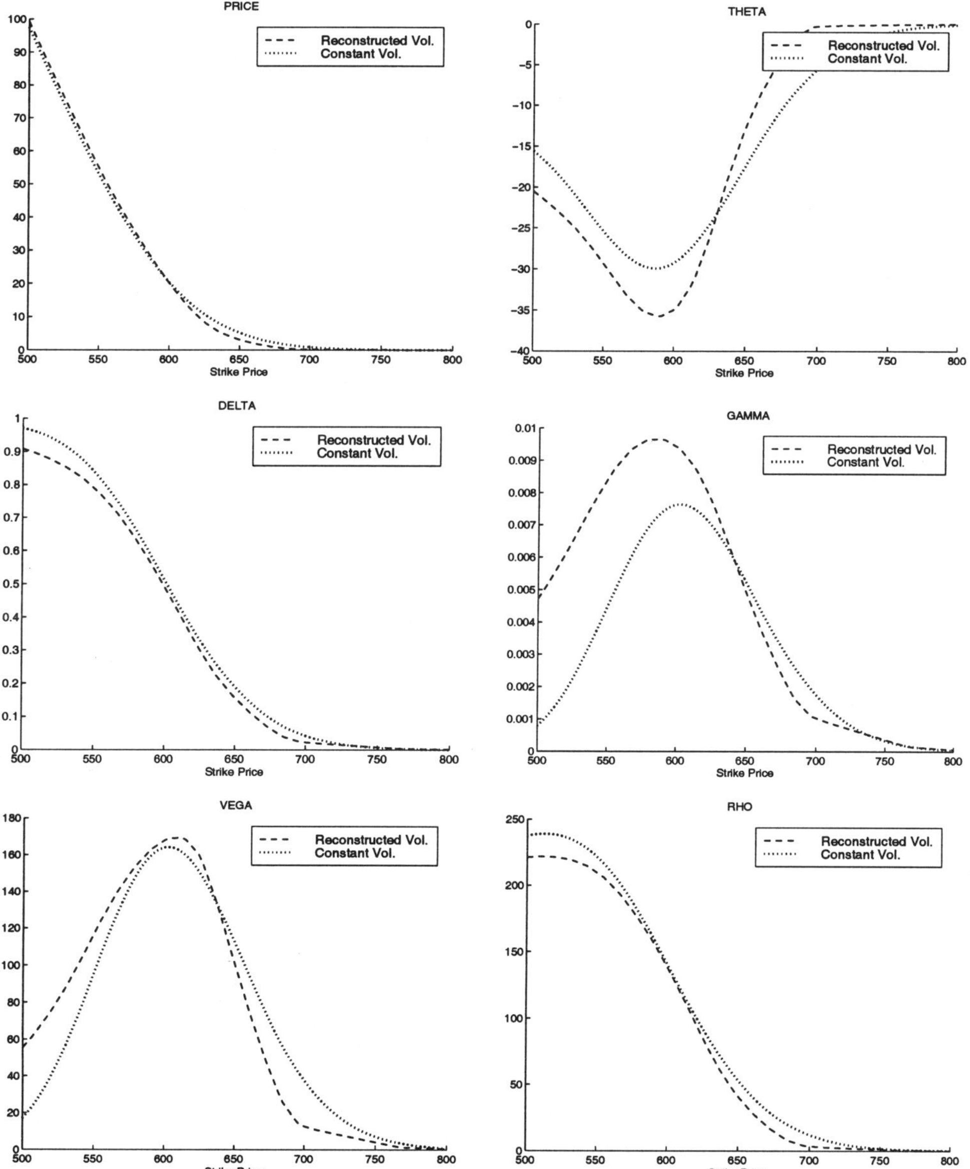

Fig. 8. Comparison using the reconstructed versus constant (average) volatility.

4.3. *Incorporate additional information*

Using market option data to imply the local volatility function in a diffusion model is a look-ahead technique. Frequently, historical data has been used to estimate a constant volatility. The latter is a look-back technique. An interesting question is whether it is possible to combine both techniques to generate better approximation to the local volatility function.

In the proposed spline volatility formulation (7), there are two potential ways that *a priori* information can be incorporated. The first is to use the simple bounds to limit the range of the local volatilities at knots. The second possibility is to specify fixed local volatilities at some chosen knots.

We have experimented with setting tighter bounds on the volatility $\bar{\sigma}$ for the S&P 500 index European call option example. We observe that, as long as the bounds are not too small ($l \leq -.3, u \geq .3$ in the S&P 500 example), they can influence volatility values of small t and s far from S_{init} but do not have much impact in the region $\mathcal{D}$ within which the volatility function is significant in market option prices. However, setting bounds too tight can impede calibrating the market option prices. Therefore, unless one has reliable knowledge on the bounds of the volatilities, they should be sufficiently large to ensure that the calibrating error is sufficiently small. Similar remarks can be made if one wishes to set the volatilities at certain knots to some fixed values.

Finally, we would like to illustrate the potential computational saving by using the quasi-Newton updates. In Table 5, we present Matlab computational results using the finite difference and quasi-Newton update for Jacibian for the S&P 500 index option example with different termination tolerances for optimization. We observe significant total speedup using a quasi-Newton approach. The quasi-Newton approach takes more iterations to converge but requires fewer Jacobian evaluations.

Table 5. Quasi-Newton results.

Tolerance	Finite Difference		quasi-Newton Update		
	Iterations	Time	Iterations	Time	J evaluations
$1e^{-2}$	7	321.29	11	131.85	3
$1e^{-3}$	10	401.43	18	193.91	4
$1e^{-4}$	19	651.89	28	286.78	5
$1e^{-5}$	25	795.45	36	332.13	6

5. Concluding Remarks

Assuming that the underlying asset of options follows a continuous 1-factor diffusion model, we propose a method of accurately approximating the local volatility function $\sigma^*(s, t)$ using a finite set of option prices. We emphasize that accurate approximation of the local volatility function in the 1-factor model is crucial in hedging all options (including simple European options) and pricing exotic options. Moreover, since the market option data typically has bid-ask spreads, exact calibration of the option data (the average of the bid-ask spreads) is not necessary and can be harmful.

Based on the formula (6) established in [1, 17], the local volatility function $\sigma^*(K,T)^2$ is smooth if the European call option value function $v(K,T)$ is sufficiently smooth. We use a spline functional approach to reconstruct this local volatility function. After choosing the number of spline knots and their placement, we represent a local volatility function $\sigma(s,t)$ by an interpolating spline with a fixed end condition. The volatility values at knots are determined by solving a small nonlinear optimization problem subject to simple bounds. The number of variables in the optimization (7) is no greater than the number of option observations.

We solve the proposed inverse spline approximation optimization problem using a trust region method, with the function and Jacobian evaluated using a PDE approach. Computational efficiency through structure exploitation within the framework of finite difference and automatic differentiation is discussed.

We consider two European call options examples illustrating the capability of the proposed method. In the first example, we consider synthetic European call options for which the underlying follows a known absolute diffusion model. Option observation data is simulated by evaluating a set of European call options using the analytic formula. The reconstructed local volatility is compared to the true local volatility, indicating a fairly accurate reconstruction in the region within which the local volatility values are significant for option evaluations.

With the same example, we illustrate that the constant implied volatility approach can produce erroneous hedge factors, compared to that from the 1-factor model, even for simple European options. Moreover, when the observable option prices have bid-ask spreads, calibrating market data exactly by using too many spline knots can lead to poor reconstruction of the true local volatility function. In the second example, S&P 500 index European call options with market option data of October 1995 are considered. We illustrate the smoothness of the reconstructed local volatility and stability of the proposed method in pricing and hedging.

We have demonstrated the potential of the proposed spline volatility approach in discovering, from a finite set of option prices, the local volatility function in the 1-factor process followed by the underlying. We plan to further investigate automatic techniques for the optimal selection of the number of knots $p \leq m$ and their placement. The importance of the proposed local volatility function reconstruction in pricing exotic options or American options will also be explored.

Acknowledgements

We would like to thank Yohan Kim for carefully reading the manuscript and his useful comments. We also thank Bob Jarrow for his encouragement.

References

[1] L. Andersen and R. Brotherton-Ratcliffe, "The equity option volatility smile: An implicit finite-difference approach", *The Journal of Computational Finance* **1** (1998) 5–32.

[2] M. Avellaneda, C. Friedman, R. Holemes and D. Samperi, "Calibrating volatility surfaces via relative entropy minimization", *Applied Mathematical Finance* **4** (1997) 37–64.

[3] G. Birkhoff and C. R. De Boor, "Piecewise polynomial interpolation and approximation", in *Approximation of Functions*, H. L. Garabedian, ed., (Elsevier, New York, 1965), pp. 164–190.

[4] F. Black and M. Scholes, "The pricing of options and corporate liabilities", *Journal of Political Economy* **81** (1973) 637–659.

[5] I. Bouchouev and V. Isakov, "The inverse problem of option pricing", *Inverse Problems* **13** (1997) L11–L17.

[6] M. A. Branch, T. F. Coleman and Y. Li, *A subspace, interior and conjugate gradient method for large-scale bound-constrained minimization*, Tech. Report TR95-1525, Computer Science Department, Cornell University, 1995.

[7] T. F. Coleman and Y. Li, "An interior, trust region approach for nonlinear minimization subject to bounds", *SIAM Journal on Optimization* **6** (1996) 418–445.

[8] T. F. Coleman and A. Verma, "Structure and efficient Jacobian calculation", in *Computational Differentiation: Techniques, Applications, and Tools*, M. Berz, C. Bischof, G. Corliss and A. Griewank, eds. (SIAM, Philadelphia, Penn., 1996), pp. 149–159.

[9] J. C. Cox and S. A. Ross, "The valuation of options for alternative stochastic processes", *Journal of Financial Economics* **3** (1976) 145–166.

[10] S. R. Das and R. K. Sundaram, *Of smiles and smirks: A term-structure perspective*, Tech. Report NBER working paper no. 5976, Department of Finace, Harvard Business School, Harvard University, Boston, MA 02163, 1997.

[11] C. de Boor, *Spline Toolbox for use with MATLAB*, The Math Works Inc., 1997.

[12] J. E. Dennis and R. B. Schnabel, *Numerical Methods for Unconstrained Optimization*, Prentice-Hall, 1983.

[13] E. Derman and I. Kani, "Riding on a smile", *Risk* **7** (1994) 32–39.

[14] ——, "The local volatility surface: Unlocking the information in index option prices", *Financial Analysts Journal* **52** (1996) 25–36.

[15] P. Dierckx, "An algorithm forsurface fitting with spline functions", *IMA Journal of Numerical Analysis* **1** (1981) 267–283.

[16] ——, *Curve and Surface Fitting with Splines*, Oxford Science Publications, 1993.

[17] B. Dupire, "Pricing with a smile", *Risk* **7** (1994) 18–20.

[18] A. Griewank, "Some bounds on the complexity of gradients, Jacobians, and Hessians", in *Complexity in Nonlinear Optimization*, P. Pardalos, ed., (World Scientific Publishers, 1993), pp. 128–161.

[19] D. Heath, R. Jarrow and A. Morton, "Bond pricing and the term structure of interest rates: A new methodology for contingent claims valuation", *Econometrica* **60** (1992) 77–105.

[20] J. Hull and A. White, "The pricing of options on assets with stochastic volatilities", *Journal of Finance* **3** (1987) 281–300.

[21] N. Jackson, E. Süli and S. Howison, *Computation of deterministic volatility surfaces*, Tech. Report No. 98/01, Oxford University Computing Laboratory, 1998.

[22] J. Jackwerth and M. Rubinstein, "Recovering probability distributions from option prices", *The Journal of Finance* **51** (1996) 1611–1631.

[23] R. Lagnado and S. Osher, "Reconciling differences", *Risk* **10** (1997) 79–83.

[24] D. Lamberton and B. Lapeyre, *Introduction to Stochastic Calculus Applied to Finance*, Chapman & Hall, 1996.

[25] R. Merton, "The theory of rational option pricing", *Bell Journal of Economics and Management Science* **4** (1973) 141–183.

[26] ——, "Option pricing when underlying stock returns are discontinuous", *Journal of Financial Economics* **3** (1976) 124–144.

[27] M. Rubinstein, "Implied binomial trees", *The Journal of Finance* **49** (1994) 771–818.

[28] D. Shimko, "Bounds of probability", *Risk* **6**(4) (1993) 33–37.

[29] A. N. Tikhonov and V. Y. Arsenin, *Solutions of Ill-posed Problems*, W. H. Winston, Washington, DC, 1977.

[30] V. N. Vapnik, *Estimation of Dependences Based on Empirical Data*, Springer-Verlag, Berlin, 1982.

[31] G. Wahba, *Splines Models for Observational Data*, Series in Applied Mathematics, Vol. 56, SIAM, Philadelphia, 1990.

[32] R. Zvan, K. Vetzal and P. Forsyth, "Swing high swing low", *Risk Magazine* (1998) 71–74.

BUILDING A CONSISTENT PRICING MODEL FROM OBSERVED OPTION PRICES*

JEAN-PAUL LAURENT

*Center for Research in Economics and Statistics, Finance Department,
15 Boulevard Gabriel Péri, 92245 Malakoff Cedex, France
E-mail: jpl@ensae.fr*

DIETMAR P. J. LEISEN

*Stanford University, Hoover Institution, Stanford, CA 94305, USA
E-mail: leisen@hoover.stanford.edu*

Received 3 September 1999

This paper constructs a model for the evolution of a risky security that is consistent with a set of observed call option prices. It explicitly treats the fact that only a *discrete* data set can be observed in practice. The framework is general and allows for state dependent volatility and jumps. The theoretical properties are studied. An easy procedure to check for arbitrage opportunities in market data is proven and then used to ensure the feasibility of our approach. The implementation is discussed: testing on market data reveals a U-shaped form for the "local volatility" depending on the state and surprisingly, a large probability for strong price movements.

Keywords: Markov chain, no-arbitrage, cross-entropy, model risk.

JEL classification codes: C51, G12, G13.

1. Introduction

The connection between the absence of arbitrage and equivalent martingale measures, first noted by Harrison and Kreps (1979), has led to a deep understanding of the pricing of derivatives *depending* on the proposed model. Models extending

Date: January 1998. This version: 3 September 1999. We are grateful to Marco Avellaneda, Phelim Boyle, Peter Carr, Emmanuel Derman, Christian Henning, Kenneth Judd and Dieter Sondermann for helpful discussions; also, participants in seminars at the University of Bonn, the Humboldt University in Berlin, the London School of Economics, in the Bonn–Aarhus workshop at the University of Bonn, Goldman Sachs, New York University, and at the University of Waterloo offered useful comments. The second author acknowledges financial support from the Deutsche Forschungsgemeinschaft, Sonderforschungsbereich 303 at Rheinische Friedrich–Wilhelms Universität Bonn and a Deutscher Akademischer Austauschdienst scholarship through HSP III (a joint program by the Federal and State Governments in Germany).

the now classical Black–Scholes setup become more and more sophisticated, and allow for time/state dependent and stochastic volatility or even for stock price jumps. These address "model risk" in the presence of "crashes", "smiles", "volatility surfaces" and other empirical shortcomings in the data (Bakshi, Cao and Chen, 1997).

In this paper, we are interested in choosing the correct model from a second-level perspective. Such an approach is known by practitioners as "reverse engineering" or as "inverse problems" (Chriss, 1997). Today's actively traded plain vanilla options can be seen as "correctly" priced. We take them as inputs to infer how the market prices risk. Here, we are interested in a model which explains the *observed* plain vanilla option prices in the market *and* is immediately *implementable* in practice to price other derivatives. Instead of estimating a continuous time model that will be further discretized to price complex derivatives, we think of a direct estimation of discrete models and restrict ourselves to those where time and asset space are discrete.

Here, we suppose that discounted asset prices can be modeled by a discrete Markov chain. This approach first appeared in a paper by Zipkin (1993) for the valuation of mortgage-backed securities. It has recently been used in a series of papers related to credit rating instruments (Jarrrow, Lando and Turnbull, 1997 and Duffie and Singleton, 1998). A Markov chain approach is sufficiently flexible to handle a wide class of different models; it is a natural approximation of models with time and state dependent volatility and jumps (Kushner and Dupuis, 1992). It is also a generalization of binomial and trinomial trees.

Breeden and Litzenberger (1978) were the first to infer information from option prices: the risk-neutral marginal density in terms of the second derivative of option prices with respect to the strike. Jackwerth and Rubinstein (1996), Dumas, Fleming and Whaley, (1998), Buchen and Kelly (1996), and Melick and Thomas (1997), among others, were interested in the practical implementation of this approach. More recently, the problem to infer the dynamics of the underlying has been addressed: Dupire (1996) and Derman and Kani (1995) relate the local volatility of the risky asset price to partial derivatives of the option prices; Rubinstein (1994) constructed a binomial process consistent with observed prices at *one* date; Carr and Madan (1998) addressed this problem in a model with time and state dependent volatility.

Derman and Kani (1995) consider consistent binomial trees and determine both nodes and probabilities implicitly. Unfortunately, their algorithm leads to negative probabilities. They claim this deficiency is due to arbitrage opportunities present in the data and override them by some artificial value (see Barle and Cakici, 1995 for a discussion). In Derman, Kani and Chriss (1996) a grid for a trinomial grid is ad hoc specified. However, even in this case, it may be impossible to fit all market prices, no matter what choice of state space is made, (see Derman, Kani and Chriss, 1996, p. 18). This is due to the fact, that binomial and trinomial models are approximations to diffusion processes and will therefore have difficulties to handle option prices in the presence of jumps or a large volatility.

The assumption of all these approaches is that call option prices are available, whatever strike and exercise date. In practice, a continuous data set is usually obtained by numerical interpolation from a discrete set of observed option prices. As Shimko (1993) pointed out, special care must be taken in order to get proper probability densities. Also, these approaches rely heavily on the choice of the smoothing technique.

In line with our view, Avellaneda, Friedman, Holmes and Samperi (1997) assume only a discrete data set. However, their assumption of the underlying model in the form of a continuous one-dimensional diffusion seems to be too restrictive (Dumas, Fleming and Whaley, 1998). Also, in the diffusion setup of Avellaneda, Friedman, Holmes and Samperi (1997) jumps are not included. Moreover, their empirical works show that the local volatility often hits minimal and maximal levels which are set ad hoc. The local discrepencies of the volatility show the difficulties of a continuous sample path setting. Our major criticism is that all previous approaches assume that the risky security price can be modeled through a scalar diffusion process, *excluding* jumps.

We assume, as did the previous authors, that markets are information efficient in the sense that there are no riskless profits for free, and use this "no-arbitrage" principle to price derivatives. However, we have to guarantee that this assumption actually holds in the data. It is a well-known and necessary condition for absence of arbitrage that the option prices are decreasing and convex in the exercise price and are non-decreasing with the exercise date. A contribution of this paper is to prove the converse implication. This is a simple check for the presence of arbitrage opportunities in data sets.

We extend the notion of Arrow–Debreu securities to our dynamic case and show that the dynamic market can be transformed to a static one by introducing extra assets that correspond to the martingale restriction. This allows us to relate the "no-arbitrage" principle to the feasibility of our approach. Together with our characterization for the absence of arbitrage opportunities, this can be easily used to check that our approach is valid. Our analysis extensively uses superreplication ideas. Taking into account the consistency constraints with observed option prices narrows the bid-ask spread in our case compared to Cvitanic and Karatzas (1993), El Karoui and Quenez (1995), and Cvitanic, Pham and Touzi (1998). Moreover, our direct examination of the underlying probability structure allows for a thorough analysis of arbitrage violations and explicitly prevents the arbitrage problems, resulting from a simple implementation of the approach of Dupire (1996).

Finally, we address in detail the implementation of our approach. Once the feasibility of our approach has been ensured, we apply Bayesian estimation to infer a unique Markov Chain. We prove that standard results in the static case (see, e.g., Buchen and Kelly, 1996) still apply in our discrete framework. This leads to a simple expression of the risk-neutral probabilities and makes the implementation straightforward.

In FX markets, using a parametric approach, Bates (1996a), and Bates (1996b) find a jump-component whose intensity is small and does not depend on the exchange rate level. McCauley and Melick (1996) and Campa, Chang and Reider (1997) document that in these markets, skewness is strongly correlated with the spot rate. Looking at the data set of Avellaneda, Friedman, Holmes and Samperi (1997), we find a U-shaped form for the "local volatility" and evidence that jumps are necessary to explain the data appropriately.

The paper is organized as follows. Section 2 introduces our setup and basic notation: it discusses joint, conditional and marginal probabilities, risk-neutral Markov chains, and their consistency with the set of observed option prices. Section 3 studies tests for arbitrage opportunities, risk-neutral measure, and consistent superreplication prices in the static market case. Section 4 shows how the dynamic market can be transformed to a static one and discusses the set of attainable assets, dynamic arbitrage opportunities and their relation to superreplication. Section 5 presents a tractable way to test for the absence of arbitrage opportunities in the observed set of option prices. Section 6 considers the choice of a martingale measure consistent with option prices in a dynamic market and its implementation using the data set of Avellaneda, Friedman, Holmes and Samperi (1997). The paper concludes with Sec. 7. Proofs are postponed to the appendix.

2. The Market Model

The basis of the paper is today's observation of the future contract and a set $\mathcal{D} = \{(\tau_l, K_l, C_l) | l = 1, \ldots, L\}$ of observed option contracts written on asset S. Here C_l is the price of the call option with maturity τ_l; i.e., the contract paying $(S_{\tau_l} - K_l)^+$ at time τ_l if S is the the price of the asset at that time.[a] We assume that the law of one price holds: for each K there is at most one $(t, K, C) \in \mathcal{D}$. We will also denote by $\mathcal{K}_t \stackrel{\text{def}}{=} \{K | (t, K, C) \in \mathcal{D}\}$ the set of strikes at date t; by $\mathcal{D}_t \stackrel{\text{def}}{=} \{(\tilde{K}, \tilde{C}) | (\tilde{\tau}, \tilde{K}, \tilde{C}) \in \mathcal{D}, \tilde{\tau} = t\}$ the set of contracts with maturity t; and by $\mathcal{D}_K \stackrel{\text{def}}{=} \{(\tilde{t}, \tilde{C}) \in \mathcal{D} | (\tilde{\tau}, \tilde{K}, \tilde{C}) \in \mathcal{D}, \tilde{K} = K\}$ the set of contracts with strike K. We assume that all observed prices are discounted, taking a bond as numeraire. This sets apart interest rates in the analysis.

Definition 1. *Assume there is a finite alphabet $\Omega = \{s_1, \ldots, s_N\}$, a discrete set of dates $\mathcal{T} = \{0, 1, \ldots, T\}$, as well as sequences $\Pi = (\Pi_t)_{t \in \mathcal{T} \backslash T}, \Sigma = (\Sigma_t)_{t \in \mathcal{T}}.$ We further assume for date t, that $\Sigma_t \subset \Omega$ denotes the set of nodes at t and that Π_t is a stochastic matrix on $\Sigma_t \times \Sigma_{t+1}$, i.e. a $|\Sigma_t| \times |\Sigma_{t+1}|$ matrix where elements are non-negative and rows sum to one. We call such a triplet $(\mathcal{T}, \Sigma, \Pi)$ a Markov Chain Market Model (MCMM). The asset grid will be denoted $\mathcal{A} \stackrel{\text{def}}{=} \bigotimes_{t \in \mathcal{T}} \Sigma_t$. A derivative asset $\mathcal{X}$ is a random variable on $\mathcal{A}$.*

In practice $\Sigma_0 = \{S_0\}$, since today's asset price is observed, and Π_0 is a degenerated stochastic matrix. Nevertheless we introduce it for simplicity. Any time

---

[a]We could allow for put options. However by put-call parity all put options in the data set could be replaced by their corresponding call prices. So we adopt this assumption further without loss of generality.

$t \in \mathcal{T}$, nodes are ordered, and we index them either by their corresponding number or directly by their corresponding node.

The set of all derivative assets is a linear state space that can be identified with $\mathbb{R}^{|\mathcal{A}|}$, where its dimension is the number of linearly independent paths. The Euclidean basis of the state space can be seen as the set of Arrow–Debreu securities, i.e., the securities paying one unit of numeraire conditionally on the realization of some path.

An MCMM describes the dynamics of the risky asset process $S = (S_t)_{t \in \mathcal{T}}$ on $\mathcal{A}$ through the probability measure P^{Π} defined by

$$P^{\Pi}[S_{t+1} = s | S_t = s'] \overset{\text{def}}{=} \Pi_t(s, s') \,.$$

Pricing measures which are risk-neutral give rise to viable pricing models (in the Harrison–Kreps sense). They are characterized by the fact that the discounted asset price $(S_t)_t$ is a martingale under P^{Π}. Therefore, we adopt the following:

Definition 2. *A Markov Chain Pricing Model (or MCPM) $(\mathcal{T}, \Sigma, \Pi)$ is an MCMM with*

$$E^{\Pi}[S_{t'} | S_t = s] = s \quad \text{for all} \quad s \in \Sigma_t \quad \text{and} \quad t, t' \in \mathcal{T}, t > t' . \tag{1}$$

For any MCPM with $\Pi_t = \{\Pi_t(s, s')\}_{s \in \Sigma_t, s' \in \Sigma_{t+1}}$, we can introduce the probability P^{Π} over sample paths by

$$P^{\Pi}(S_1 = \bar{s}_1, \ldots, S_T = \bar{s}_T) = \prod_{t=0}^{T-1} \Pi_t(\bar{s}_t, \bar{s}_{t+1})$$

for any $(\bar{s}_1, \ldots, \bar{s}_T) \in \mathcal{A}$, with $\bar{s}_0 \equiv S_0$. For $t, t' \in \mathcal{T}$, a stochastic matrix $\Pi_{t,t'}$ that describes the transition between dates t and t' can be defined by

$$\Pi_{t,t'} = \prod_{u=t}^{t'-1} \Pi_u \,.$$

These probabilities have a simple financial interpretation, which will be useful in the sequel: $P^{\Pi}(S_1 = \bar{s}_1, \ldots, S_t = \bar{s}_T)$ can be viewed as the price of an Arrow–Debreu security paying one unit of numeraire conditionally on the realization of the path $(\bar{s}_1, \ldots, \bar{s}_T)$. Similarly, entry $\Pi_{t,t'}(i, j)$ can be viewed as the price at time t in state i of an asset paying one unit of numeraire at time t' in state j or as the transition matrix between t and t'. In particular, we will further denote by $\lambda_t \overset{\text{def}}{=} \Pi_{0,t}(S_0, \cdot)'$ the marginal distribution at date t. The ith component of λ_t can be interpreted as the price of an asset paying one unit of numeraire if, and only if, the discounted price S is equal to s_i at time t (i.e., a path-independent Arrow–Debreu security).

Please note that any $\mathcal{D}$-MCPM can be embedded into a continuous framework by setting $S_t \overset{\text{def}}{=} S_{\lfloor t \rfloor}$. The jump times of S_t are known and occur at the exercise dates $\tau \in \mathcal{T}$. Meanwhile, S_t remains constant. The transition matrices are such that $\Pi(\tau, t) = Id \quad \forall t \in [\tau, \tau + 1[.$ S_t is a continuous time Markov chain, and a martingale and generates option prices consistent with observed option prices.

Definition 3. *For a given set $\mathcal{D}$ of call options, let us first define $\mathcal{T} \stackrel{\text{def}}{=} \{t \mid (t, K, C) \in \mathcal{D}\}$. A $\mathcal{D}$-MCPM is a tuple (Σ, Π) where $(\mathcal{T}, \Sigma, \Pi)$ is an MCPM with $\mathcal{K}_t \subset \Sigma_t$ and which fulfills for all $(t, K, C) \in \mathcal{D}$:*

$$E^\Pi[(S_t - K)^+] = C.$$

A $\mathcal{D}$-MCPM summarizes our approach in that we take the observed set $\mathcal{D}$ as an input. Condition $\mathcal{K}_t \subset \Sigma_t$ reflects our choice for the states of the Markov chain. Here, we require that for any observed strike at t there is a node which is identical to it. There are certainly many different ways to choose the nodes in relation to strikes. This does, however, not affect the results in this paper. Our choice is just a very simple one to keep track of them in the exposition. Unless otherwise explicitly noted, $\mathcal{K}_t = \Sigma_t$ holds. In general, we omit Σ_t when referring to a $\mathcal{D}$-MCPM, in order to focus on the problem of inferring Π.

The following extension of the set of nodes is necessary. A call (put) option with strike equal to the highest (lowest) node $s_t^{\max}$ ($s_t^{\min}$) has zero payoff. Thus in the absence of arbitrage, their price must be equal to zero. This cannot be the consistent outcome of a $\mathcal{D}$-MCPM. Therefore we introduce two dummy nodes, $s_t^{\min}$ and $s_t^{\max}$, with corresponding prices, $C(t, s_t^{\max}) = 0$ and $C(t, s_t^{\max}) = s_t^{\max} - S_0$.[b] Later in Sec. 6, we explain in detail how to choose the additional strikes $s_t^{\max}, s_t^{\min}$ in relation those already existing one's.

3. The Static Market

This section studies both the transition between today and a *fixed* date $t \in \mathcal{T}$ and then the market with $\mathcal{D} = \mathcal{D}_t$ and $\mathcal{T} = \{0, t\}$. We call this the *static market*, as there is no conditional dynamics in the future. Many of the results here are well-known. However, we recall them here to introduce the main concepts useful in the dynamic case in an easy framework.

A $\mathcal{D}_t$-MCPM in this market is completely defined by a probability measure λ_t on Σ_t (i.e., $\lambda_t(s) \geq 0$ and $\sum_{s \in \Sigma_t} \lambda_t(s) = 1$) that is consistent with the prices of traded assets.

Property 4. *The observed option prices are decreasing and convex, i.e., for any $(C, K), (C', K'), (C'', K'') \in \mathcal{D}_t$ with $K'' > K' > K$:*

$$\frac{C - C'}{K - K'} < \frac{C - C'}{K' - K''}. \tag{2}$$

[b]These prices are such that put-call parity holds, since the interest rate was supposed to equal zero. For a time-homogeneous grid, the two extreme states $s_t^{\min}$ and $s_t^{\max}$ need to be absorbing for the process to be risk-neutral, i.e., $P^\Pi[S_{t'} = s_t^{\min}|S_t = s_t^{\min}] = P^\Pi[S_{t'} = s_t^{\max}|S_t = s_t^{\max}] = 1$ (see also Sondermann, 1987 and Sondermann, 1988). In practical applications, however, the grid will spread out and the extreme states are not absorbing.

Please note that the property of decreasing call prices ensures positive option prices, since the call price at the highest nodes $s_t^{\max}$ was set equal to zero. For an inner node $K \in \Sigma_t \setminus \{s_t^{\min}, s_t^{\max}\}$, let us consider a so-called *butterfly spread* with payoff $B(\cdot)$ based on the three adjacent traded strikes $K_- < K < K_+$ ($K_- \overset{\text{def}}{=} \max_{(k,C) \in \mathcal{D}_t, k<K} k$, $K_+ \overset{\text{def}}{=} \min_{(k,C) \in \mathcal{D}_t, k>K} k$) and a payoff equal to one at K, i.e., $B(K) = 1$ and 0, otherwise. For $K = s_t^{\min}$ and $K = s_t^{\max}$, we consider call spreads. The positive payoff of the butterflies at inner nodes has the price

$$\frac{1}{K - K_-} C_- - \left(\frac{1}{K - K_-} - \frac{1}{K_+ - K} \right) C + \frac{1}{K_+ - K} C_+ \,.$$

This translates directly into Property 4. If there is a complete set of observed option prices, i.e., $\Sigma_t = \mathcal{K}_t$, these butterflies form a basis of the payoff space: it is then straightforward to prove the following Lemma 5. Otherwise, there will be nodes for which no option contract can be observed. Then we can introduce dummy call options for the missing strikes, with a price equal to the linear interpolation of the adjacent ones.

Lemma 5.[c] *There exists a risk-neutral marginal density λ_t at date t if, and only if, Property 4 is fulfilled. Moreover, in a complete market with Property 4, the prices $\lambda_t(s)$, $s \in \Sigma_t$ of path-independent Arrow–Debreu securities with payment date t are uniquely determined. For any $s \in \Sigma_t \setminus \{s_t^{\min}, s_t^{\max}\}$, we have*

$$\lambda_t(s) = \frac{C_+ - C}{K_+ - K} - \frac{C - C_-}{K - K_-} \,.$$

$\lambda_t^{\min}$ and $\lambda_t^{\max}$ have a more complicated form due to boundary effects which we do not exhibit here. So Property 4 allows us to check for arbitrage opportunities and for the existence of a risk-neutral measure λ_t in this static market.

When there are fewer traded strikes than nodes, there is not a unique price at which a call option with strike $K \in \Sigma_t \setminus \mathcal{K}_t$ can be consistently priced without introducing arbitrage opportunities. Similarly λ_t is not uniquely determined. However, it is straightforward to prove the following linear interpolation bounds

Lemma 6. *In the static case, the superreplication price of a call option with strike $K \in \Sigma$ and maturity $t \in \mathcal{T}$ is given by the linear interpolation*

$$\frac{K_+ - K}{K_+ - K_-} C_- + \frac{K - K_-}{K_+ - K_-} C_+ \,.$$

The associated probability puts non-zero probability weights only on traded strikes and zero elsewhere.

[c]This is a discrete analog of the results of Breeden and Litzenberger (1978), since the difference of the fractions is a natural approximation to the second derivative of call options with respect to the strike.

This superreplication price is very different from that obtained in a standard stochastic volatility model, where it is equal to the trivial price S_0. The departure is due to the use of traded options; consistency with these option prices restricts the set of risk-neutral measures and, thus, the no–arbitrage bounds on calls are narrowed.

Remark 7. *In the case of a call option payoff, we have an explicit characterization of the superreplicating portfolio and price. This can be extended to the case of any concave payoff $\mathcal{X}$. The superreplicating portfolio is obtained as the linear interpolation of points $\{(K,C)|(C,t,K) \in \mathcal{D}\}$.*

4. The Dynamic Market and Arbitrage Opportunities

This section transforms the dynamic market into a static one. The martingale condition corresponds to constraints on assets we introduce. We also describe the linear subspace of all path dependent payoffs made of attainable claims by static or dynamic strategies. We restrict ourselves here to the case of a market with $\mathcal{T} = \{0,1,2\}$.

4.1. *Transforming to a fictitious static market*

Let us now introduce our *path-dependent static market*. It is a fictitious market in which the states are $\Sigma_1 \times \Sigma_2$. The payoff in state $(s, s') \in \Sigma_1 \times \Sigma_2$ depends on the joint occurrence of state s at date 1 and state s' at date 2. The payoff structure of all assets is then a matrix. The basic securities are the following: for a pair $(s, s') \in \Sigma_1 \times \Sigma_2$, let us consider the *path-dependent Arrow–Debreu security* $\delta^{\text{dep}}_{s,s'}$, paying 1 unit of numeraire at date 2 if the asset is in state i at date 1 *and* in state j at date 2, and 0 otherwise. Their price will be denoted by $P_{s,s'}$. This can be interpreted as the probability of occurrence for that specific path.

In the fictitious market, the (standard) *path-dependent* Arrow–Debreu security δ^1_s at date 1, paying 1 unit in state s is described by the payoff matrix

$$\delta^1_s(\bar{s}_1, \bar{s}_2) = \begin{cases} 1 & \text{if } \bar{s}_1 = s \\ 0 & \text{otherwise} \end{cases},$$

i.e., has 1's in row s and 0's in all other rows. Similarly the (standard) *path-dependent* Arrow–Debreu security $\delta^2_{s'}$ at date 2, paying 1 unit in state s' is described by the payoff-matrix 1's in column s' and 0's in all other columns.

Let us further introduce for any $s \in \Sigma_1$ the contract σ_s between two parties: if the asset price is in state s at date 1, both exchange a bond with face value s and an asset, where it is specified that the "holder" receives the bond. At date 2 the option seller buys the asset back from the holder at the then prevailing market price. This contract can be interpreted as a trading strategy where the bond is used to transfer the payments such that there are neither initial nor intermediate payments. So in the absence of arbitrage, the price of the contract must be 0. This introduces $|\Sigma_1|$ new assets.

These conditions on the price of assets σ_s are actually sufficient for a probability measure to be a martingale measure. This can be seen as follows: the payoff at date 2 of contract σ_s is $(S_2 - s)1_{S_1=s}$. Its price can be rewritten as

$$\sum_{s \in \Sigma_2} P_{s,s'} \cdot (s' - s) = 0 \,, \tag{3}$$

which is precisely the martingale condition (1).

Definition 8. *The set of investment strategies is* $\mathcal{I} = \mathbb{R}^{|\Sigma_2|} \times \mathbb{R}^{|\Sigma_1|} \times \mathbb{R}^{|\Sigma_1|}$. *For any strategy* $(\alpha, \beta, \gamma) \in \mathcal{I}$, *the claim paying* $\alpha_{s'} + \beta_s + \gamma_s s'$ *in state* (s, s') *will be denoted by* $\mathcal{X}(\alpha, \beta, \gamma)$ *and its price at date 0 by* $\mathcal{P}(\alpha, \beta, \gamma)$.

Proposition 9. *Let us assume that we observe a set of options with* $\Sigma_1 = \mathcal{K}_1$ *and* $\Sigma_2 = \mathcal{K}_2$. *The set of attainable claims through investment in calls, puts, and the asset and numeraire is the linear subspace*

$$\{\mathcal{X}(\alpha, \beta, \gamma) | (\alpha, \beta, \gamma) \in \mathcal{I}\} \subset \mathbb{R}^{|\Sigma_1 \times \Sigma_2|} \,,$$

i.e., equivalent to investments in $\delta_{s,s'}^{\mathrm{dep}}$, δ_s^1 *and* $\delta_{s'}^2$ *(*$(s, s') \in \Sigma_1 \times \Sigma_2$*). In the absence of arbitrage opportunities, the price of the claim* $\mathcal{X}(\alpha, \beta, \gamma)$ *is uniquely determined by*

$$\mathcal{P}(\alpha, \beta, \gamma) = \sum_{s \in \Sigma_1} (\beta_s + \gamma_s s)\lambda_1(s) + \sum_{s \in \Sigma_2} \alpha_s \lambda_2(s) \,.$$

This characterizes the claims attainable through static investments in traded options in more detail: any attainable claim can be replicated by buying $\alpha_{s'}$ units of the path-independent Arrow–Debreu security s' at date 2, β_s units of the path-independent Arrow–Debreu security s at date 1, and γ_s units of stocks conditional on being in state i at date 1. The set of attainable claims in the dynamic market is thus equal to the set of attainable claims in the static market where the extra assets have been introduced. In the sequel, we will always adopt the static viewpoint to the dynamic problems.

Above we explained that the martingale condition translates into a consistency condition on the prices of these extra assets. It is well-known that the existence of a state price density in the fictitious market consistent with $\mathcal{D}$ and additional assets is equivalent to the absence of arbitrage opportunities in the static market. This is a standard result in financial theory which can be found in any textbook (see, e.g., Duffie, 1992, Chapter 1). We have used this approach in the case of static markets. In the two-period case studied in this section, all probabilities in the fictitious market can be associated with a Markov chain: dependence on a path is exactly the dependence on the state at date 1. This implies the following:

Proposition 10. *Existence of a* $\mathcal{D}$-MCPM *is equivalent to no-arbitrage in the fictitious static market.*

4.2. α-*arbitrage opportunities*

The absence of arbitrage opportunities is equivalent to the positivity of the linear operator $\mathcal{P}$, i.e., the optimization problem

$$\min_{\alpha,\beta,\gamma \in \mathcal{I}} \quad \mathcal{P}(\alpha,\beta,\gamma)$$

$$\text{s.t.} \qquad \mathcal{X}(\alpha,\beta,\gamma) \geq 0$$

has a non-negative minimum. Since we have excluded straightforward static arbitrage opportunities *a priori*, the only way an arbitrage opportunity can occur comes from a dynamic trade between 1 and 2. We split up this problem. First, we fix an arbitrary payoff $\alpha \in \mathbb{R}^{|\Sigma_2|}$ at date 2, and then we study the superreplication of α (i.e., we consider only strategies of the form $(0,\beta,\gamma)$ which dominate α). $(0,\beta,\gamma)$ is a self-financed dynamic strategy which corresponds to investing at date 1 through path-independent Arrow–Debreu securities and reinvest the proceeds between date 1 and date 2, conditional on the realized state at time 1 in the risky asset and the bond. This leads to the following:

Definition 11. *The* set of α-investment strategies *is*

$$\mathcal{I}^{\alpha} \overset{\text{def}}{=} \left\{ (\beta,\gamma) \in \mathbb{R}^{|\Sigma_1|} \times \mathbb{R}^{|\Sigma_1|} \,\middle|\, \mathcal{X}(0,\beta,\gamma) \geq \mathcal{X}(\alpha,0,0) \right\}.$$

An α-arbitrage opportunity *is an investment* $(\beta,\gamma) \in \mathcal{I}^{\alpha}$ *such that* $\mathcal{P}(0,\beta,\gamma) < \mathcal{P}(\alpha,0,0)$.

The "no α-arbitrage" condition can then be interpreted in the sense that any dynamic strategy $(0,\beta,\gamma)$ synthesizes the $(\alpha,0,0)$ payoff at a higher price than the static strategy in Arrow–Debreu securities of date 2. The relation to the no-arbitrage condition is then expressed by

Proposition 12. *There are no arbitrage opportunities in the fictitious static market if, and only if, there are no α-arbitrage opportunities for any payoff* $\alpha \in \mathbb{R}^{|\Sigma_2|}$.

5. Testing for the Existence of a $\mathcal{D}$-MCPM

This section studies a condition on data set $\mathcal{D}$ to find an easy way to check for dynamic arbitrage opportunities. Clearly this can only hold if there are no static ones. Therefore, we always assume that Property 4 holds. We take as a starting point a homogeneous grid and study the general case only later in this section.

Assumption 13. *The option grid is time-homogeneous, i.e.,* $\forall t \in \mathcal{T}, \mathcal{K}_1 = \mathcal{K}_t$.

Assumption 14. *The node grid is time-homogeneous, i.e.,* $\forall t \in \mathcal{T}, \Sigma_1 = \Sigma_t$.

Similar to Property 4, the following characterizes the absence of arbitrage opportunities.

Property 15. *Option prices are non-decreasing with the exercise date; i.e., for any time $t \in \mathcal{T}$, $K \in \mathcal{K}_t$ and $(t, C_t), (t+1, C_{t+1}) \in \mathcal{D}_K : C_t \le C_{t+1}$.*

Merton (1973) states that in the absence of arbitrage opportunities, Property 15 holds. It is straightforward to construct the arbitrage strategy if $C_t > C_{t+1}$: we sell the call with maturity t and buy the call with maturity $t+1$, a strategy known as selling a calendar spread. This generates a strictly positive inflow at date 0. At date t if the observed asset price is less than K, we do nothing. Otherwise, the short call will be exercised; in order to self-finance that transaction, we hold the asset short and put the received amount K into the bank account. At date $t+1$, this generates the payoff $(S_{t+1} - K)^+ - (S_{t+1} - K)1_{S_t > K} \ge 0$.

The striking fact is that Property 15 is also sufficient for the absence of arbitrage opportunities, and ensuring existence of a $\mathcal{D}$-MCPM. We will prove this in several steps. First, we prove it between two dates $t, t+1$ for $t \in \mathcal{T}$. Our aim will be to prove that increasing call option prices imply that there are no α-arbitrage opportunities; i.e.,

$$\forall \alpha \in \mathbb{R}^{|\Sigma_{t+1}|} : \inf_{(\beta,\gamma) \in \mathcal{I}^\alpha} \mathcal{P}(0, \beta, \gamma) \ge \mathcal{P}(\alpha, 0, 0).$$

The proof is given in the appendix and uses the following lemma. It proves that there are no α-arbitrage opportunities, when α represents the payoff of a short call with strike K and maturity $t+1$.

Lemma 16. *For any date, $t \in \mathcal{T}$ if $\mathcal{K}_t = \Sigma_t$, $\mathcal{K}_t = \Sigma_{t+1}$, and under Assumptions 13 and 14, and Properties 4 and 15, there are no "α-arbitrage" opportunities in the dynamic market consisting of trading dates $\{t, t+1\}$ for $\alpha = (-(s - K)^+)_{s \in \Sigma_{t+1}, K \in \mathcal{K}_{t+1}}$.*

Any concave payoff is a linear combination with non-negative coefficients of short calls for all positive strikes. Using Proposition 12, it is then a small step to prove:

Theorem 17. *For any date, $t \in \mathcal{T}$, if $\mathcal{K}_t = \Sigma_t$, $\mathcal{K}_t = \Sigma_{t+1}$, under Assumptions 13 and 14, the existence of a $\mathcal{D}$-MCPM in the dynamic market consisting of trading dates $\{t, t+1\}$ is equivalent to the following two conditions:*

(1) *Observed call option prices for a given arbitrary exercise date are positive, decreasing and convex in strike (Property 4).*
(2) *For a given arbitrary exercise price, call option prices are increasing in exercise date (Property 15).*

This also gives an equivalent characterization of the presence of arbitrage opportunities in the data set. We will now generalize the previous theorem to the case $\mathcal{K}_t \subset \Sigma_t, \mathcal{K}_{t+1} \subset \Sigma_{t+1}$. This will allow us to cover the case of the non-homogeneous node grid later, too. Furthermore, we now study the original market consisting of all dates in $\mathcal{T}$ at the same time. We first show the following:

Proposition 18. *Under Assumption 14 there exists a $\mathcal{D}$-MCPM if, and only if, there exists a set of option prices $\tilde{\mathcal{D}}$ which is complete (i.e., $\mathcal{K}_t = \Sigma_t$ for any $t \in \mathcal{T}$), fulfills Properties 4 and 15 such that $\mathcal{D} \subset \tilde{\mathcal{D}}$.*

Proposition 18 states that there exists a $\mathcal{D}$-MCPM if, and only if, we are able to "complete" the set of observed option prices while keeping the conditions that prices are decreasing and convex in the exercise price and monotone in the exercise date. Our remaining problem is then, to find an easy procedure to check whether such an extension exists. We will address this in the remainder of this section and consider all observed options for all maturity dates in $\mathcal{T}$.

Definition 19. *For any strike $K \geq 0$ and any date $t \in \mathcal{T}$, the convex envelope $\mathcal{E}_t(K)$ of the discrete set of points in $\mathcal{D}$ is defined by*

$$\mathcal{S}_t = \left\{ (\lambda, \mu) \in \mathbb{R}^2 \,|\, \forall \tau \geq t, \tau \in \mathcal{T}, (K', C') \in \mathcal{D}_\tau : \lambda + \mu K' \leq C' \right\},$$

$$\mathcal{E}_t(K) = \sup\{\lambda + \mu K \,|\, (\lambda, \mu) \in \mathcal{S}_t\}.$$

Please note that $\mathcal{E}_t(K)$ is convex and fulfills Property 15, the latter since $\mathcal{S}_t \subset \mathcal{S}_{t+1}$ for any t. It is therefore a "first choice" for an extension of $\mathcal{D}$.

Proposition 20. *In the absence of arbitrage opportunities, the superreplication price $\bar{C}_t(K)$ of an option is less or equal than the convex envelope; i.e.,*

$$\forall t \in \mathcal{T} \, \forall K \in [0, \infty[\, : \, \bar{C}_t(K) \leq \mathcal{E}_t(K),$$

$$and \quad \forall (t, K, C) \in \mathcal{D} : \bar{C}_t(K) = C = \mathcal{E}_t(K).$$

In the absence of arbitrage opportunities, according to Proposition 20, the convex envelope $\mathcal{E}_t(K)$ interpolates the observed option prices with maturity t. Moreover, the superreplication price of an option is equal to the convex envelope. The following theorem proposes a simple and tractable way to check for the existence of a $\mathcal{D}$-MCPM in the general case.

Theorem 21. *There exists a $\mathcal{D}$-MCPM if, and only if, the convex envelope $\mathcal{E}_t(K)$ interpolates the observed option prices, i.e., $\forall (t, K, C) \in \mathcal{D} : \mathcal{E}_t(K) = C$.*

This is a tractable way to test for arbitrage opportunities. The convex envelopes of observed option prices are easy to compute. In most standard cases in real applications, they are simply the linear interpolations of observed option prices of a given maturity. Then the simplest construction, where no further nodes have to be introduced is possible.

For example, we can consider the data set of exchange rate option prices used by Avellaneda, Friedman, Holmes and Samperi (1997) (see Table 1 and Fig. 1). Here, straight lines connecting the different option prices do not intersect, so they are the convex envelope; and we deduce that there are no arbitrage opportunities in this specific dataset.

Table 1. USD/DEM OTC market data set of Avellaneda, *et al.*; August 23, 1995.

maturity	type	strike	bid	ask	impl. volat.
	Call	1.5421	0.0064	0.0076	14.9
	Call	1.5310	0.0086	0.0100	14.8
30 days	Call	1.4872	0.0230	0.0238	14.0
	Put	1.4479	0.0085	0.0098	14.2
	Put	1.4371	0.0063	0.0074	14.4
	Call	1.5621	0.0086	0.0102	14.4
	Call	1.5469	0.0116	0.0135	14.5
60 days	Call	1.4866	0.0313	0.0325	13.8
	Put	1.4312	0.0118	0.0137	14.0
	Put	1.4178	0.0087	0.0113	14.2
	Call	1.5764	0.0101	0.0122	14.1
	Call	1.5580	0.0137	0.0160	14.1
90 days	Call	1.4856	0.0370	0.0385	13.5
	Put	1.4197	0.0141	0.0164	13.6
	Put	1.4038	0.0104	0.0124	13.6
	Call	1.6025	0.0129	0.0152	13.1
	Call	1.5779	0.0175	0.0207	13.1
180 days	Call	1.4823	0.0494	0.0515	13.1
	Put	1.3902	0.0200	0.0232	13.7
	Put	1.3682	0.0147	0.0176	13.7
	Call	1.6297	0.0156	0.0190	13.3
	Call	1.5988	0.0211	0.0250	13.2
270 days	Call	1.4793	0.0586	0.0609	13.0
	Put	1.3710	0.0234	0.0273	13.2
	Put	1.3455	0.0173	0.0206	13.2

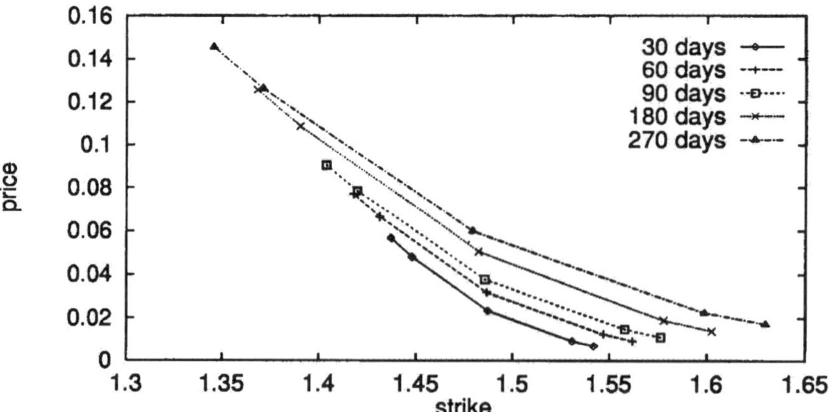

Fig. 1. Prices depending on the strike for different maturities.

Remark 22. *Theorem* 21 *guarantees that once Properties* 4 *and* 15 *are satisfied, it is possible to calibrate option prices with a scalar Markov chain. In particular, stochastic volatility or extra static variables are not required. Of course, one may think that stochastic volatility is a desirable feature and may consider processes consistent with an observed data set based on a larger state space provided that the properties on the call option prices are fulfilled.*

6. Characterization of the $\mathcal{D}$-MCPM

Once the existence of a $\mathcal{D}$-MCPM has been ensured, there is typically a multiplicity of consistent MCPM's.[d] There are various ways to choose one. According to our equivalence between a static and dynamic market, we can treat it as an incomplete two-dimensional market. The superreplication prices studied in Sec. 3 correspond to that $\mathcal{D}$-MCPM that yield the highest price for all non-traded assets.

The Bayesian approach takes a prior, e.g., the Black–Scholes setup, and looks for the minimal departure from this model consistent with the observed option prices. In the finacial literature, L^2-criteria have been used by Rubinstein (1994) and Jackwerth and Rubinstein (1996) among others for pricing options. On the statistical side, these criteria appear in Hansen and Jagannathan (1997) and Luttmer (1997). The cross-entropy criterion has been used by Buchen and Kelly (1996), Jackwerth and Rubinstein (1996), and Avellaneda, Friedman, Holmes and Samperi (1997). Also known as the information optimization criterion, it is well-established in probability theory and the statistical literature for its superiority in filling in missing information. We will therefore adopt it here.

We study the case of all dates by looking at each path over time as a different state. We assume that they are in some linear order and index them by i, $i \in \{1, \ldots, \Pi_{t \in \mathcal{T}} |\Sigma_t|\} \cong \mathcal{A}$. Let us denote by $\mathcal{M}$ the set of all probability measures on the state space $\mathcal{A}$ associated with an MCMM and by $\mathcal{M}_{\mathcal{D}}$, the subset of those probability measures associated with a $\mathcal{D}$-MCPM. We are given a prior probability measure $P \in \mathcal{M}$, allowing explicitly for the fact that P might not be consistent with the observed option price.

The cross-entropy of a probability measure $Q \in \mathcal{M}$ is defined by

$$H(Q) = E^P \left[\frac{dQ}{dP} \log \frac{dQ}{dP} \right] .$$

Here we are interested in the cross-entropy problem

$$\min_{Q \in \mathcal{M}_{\mathcal{D}}} H(Q), \tag{4}$$

which corresponds to the optimization problem under the constraint to reproduce the calls in the observed data set.

[d]Let us consider T dates, N states and observed options prices for all nodes. Apart from the positivity constraints, there are usually $N(2T - 1)$ linear constraints and $N + (T - 1)(N^2 - N)$ unknowns.

Proposition 23. *Let us assume that $\mathcal{M}_{\mathcal{D}} \neq \emptyset$ and that there is $Q_0 \in \mathcal{M}_{\mathcal{D}}$, $Q_0 \sim P$; then the minimal cross-entropy $\mathcal{D}$-MCPM exists, and the associated probability measure Q^* is equivalent to P and is uniquely characterized by*

$$\frac{dQ^*}{dP} = \exp\left\{\mu_0 + \sum_{t \in \mathcal{T}}\left[\lambda_t(S_t)(S_{t+1} - S_t) + \sum_{(t,K,C) \in \mathcal{D}} \mu_{t,K}(S_t - K)^+\right]\right\},$$

where the parameters $\mu_0, \mu_{t,K}$ (for $(t, K, C) \in \mathcal{D}$) and the functions $\lambda_i : \Sigma_t \to \mathbb{R}$ (for $t \in \mathcal{T} \setminus T$) are determined by

$$E^{Q^*}[1] = 1\,,$$

$$E^{Q^*}[S_{t+1} - S_t | S_t] = 0 \quad \forall t \in \mathcal{T} \setminus T\,,$$

$$\text{and} \quad E^{Q^*}[(S_t - K_{i,t})^+] = C \quad \text{for any } (t, K, C) \in \mathcal{D}\,.$$

Existence of such a positive Q_0 follows: e.g., if all marginal distributions are strictly positive and the conditions of the previous section are fulfilled, we can simply redistribute part of the transition probability mass appropriately for and Q to ensure its strict positivity.

We now explain the implementation of our approach using the dataset of Avellaneda, Friedman, Holmes and Samperi (1997) — introduced in Sec. 5 — in detail. We will draw some interesting conclusions from our implementation and compare them with the results obtained by Avellaneda, Friedman, Holmes and Samperi (1997). We deduced from Fig. 1, together with Theorem 21, that there are no arbitrage opportunities and that a $\mathcal{D}$-MCPM exists.

Besides the fact that we chose nodes identical to strikes, as explained in Sec. 2, a full specification of our approach requires introducing "dummy"-nodes at each data above the highest node $s_t^{\max}$ and below the lowest node $s_t^{\min}$ of the grid. To do so, we first choose a constant *factor* μ. Then we study the difference between the highest ones and introduce the new node above the highest one so that the difference is exactly μ times the difference before. If $s(s')$ denotes highest (second highest), then the new one is at $s + \mu(s - s')$. We proceed similarly for the new one below. It then remains to specify μ. The highest (lowest) node has to carry the probability weight for all possible higher (lower) movements. It should therefore not be chosen too small. We tested several factors, but only those with $\mu \geq 5$ worked well, producing adequate results in the marginal distribution. We therefore adopted $\mu = 5$ in this analysis.

The second choice is the prior. We adopt a trinomial prior where for inner nodes $1 < i < \nu_t$

$$P_{i,j} = \begin{cases} \dfrac{1}{3}; & i = j \\[2ex] \dfrac{1}{3} - \dfrac{1}{10(\nu_t - 3)}; & j = i \pm 1\,. \\[2ex] \dfrac{1}{10(\nu_t - 3)}; & \text{otherwise} \end{cases}$$

(For the outer most-nodes adjust by adding the probability of the missing node to the starting node.) Here $\nu_t \overset{\text{def}}{=} |\Sigma_t|$ denotes the number of nodes at date t. This prior will, in general, not be consistent with the observed option prices. However, it supports good the fact that we would like a departure as small as possible from the Black–Scholes setup to make it compatible with the observed option prices. Furthermore, it is similar to a trinomial model which represents the fact that it is only a discrete approximation to some continuous time model. We nevertheless increased the probabilities slightly (to $1/(10(\nu_t - 3))$) to allow for non-zero probability weights also in non-adjoint nodes. A lognormal prior — the Black–Scholes model — yields similar results in the following implementation.

We then proceeded exactly as described in the previous subsection. Tables 2 to 5 contain the transition matrices in the form found throughout the paper. To make them easily accessible, however, we index them directly by the nodes. In parentheses we have put the corresponding probabilities resulting from a Black–Scholes model with constant volatility of 0.15 pa. The last column contains the local volatility of the risk-adjusted process.

Except for the column furthest to the left, we see that the lower left hand part is always different from the Black–Scholes case, where it is zero. The market anticipates a strictly positive probability for large downward movements. The fear

Table 2. Transition between dates 1 and 2.

	1.2673	1.4178	1.4312	1.4866	1.5469	1.5621	1.6381	σ_{loc}
1.3831	0.38 (0.01)	0.43 (0.98)	0.04 (0.01)	0.06 (0.00)	0.03 (0.00)	0.02 (0.00)	0.04 (0.00)	0.36
1.4371	0.10 (0.00)	0.35 (0.30)	0.37 (0.62)	0.05 (0.08)	0.04 (0.00)	0.03 (0.00)	0.07 (0.00)	0.29
1.4479	0.02 (0.00)	0.28 (0.13)	0.34 (0.65)	0.34 (0.21)	0.01 (0.00)	0.00 (0.00)	0.01 (0.00)	0.15
1.4872	0.00 (0.00)	0.01 (0.00)	0.23 (0.09)	0.60 (0.87)	0.12 (0.04)	0.00 (0.00)	0.03 (0.00)	0.16
1.531	0.00 (0.00)	0.01 (0.00)	0.02 (0.00)	0.38 (0.29)	0.33 (0.63)	0.20 (0.08)	0.07 (0.00)	0.16
1.5421	0.06 (0.00)	0.05 (0.00)	0.06 (0.00)	0.05 (0.13)	0.31 (0.67)	0.21 (0.20)	0.26 (0.00)	0.33
1.5976	0.01 (0.00)	0.02 (0.00)	0.02 (0.00)	0.04 (0.00)	0.02 (0.02)	0.33 (0.60)	0.57 (0.38)	0.21

Table 3. Transition between dates 2 and 3.

	1.1833	1.4038	1.4197	1.4856	1.558	1.5764	1.6684	σ_{loc}
1.2673	0.65 (0.97)	0.30 (0.03)	0.02 (0.00)	0.02 (0.00)	0.01 (0.00)	0.01 (0.00)	0.00 (0.00)	0.40
1.4178	0.03 (0.00)	0.55 (0.44)	0.32 (0.55)	0.05 (0.02)	0.02 (0.00)	0.01 (0.00)	0.02 (0.00)	0.22
1.4312	0.01 (0.00)	0.46 (0.18)	0.25 (0.73)	0.27 (0.08)	0.00 (0.00)	0.00 (0.00)	0.00 (0.00)	0.15
1.4866	0.00 (0.00)	0.03 (0.00)	0.15 (0.05)	0.68 (0.93)	0.13 (0.02)	0.00 (0.00)	0.01 (0.00)	0.15
1.5469	0.00 (0.00)	0.01 (0.00)	0.01 (0.00)	0.32 (0.13)	0.37 (0.76)	0.22 (0.11)	0.08 (0.00)	0.19
1.5621	0.02 (0.00)	0.07 (0.00)	0.05 (0.00)	0.06 (0.03)	0.34 (0.64)	0.22 (0.33)	0.25 (0.00)	0.33
1.6381	0.00 (0.00)	0.01 (0.00)	0.01 (0.00)	0.02 (0.00)	0.02 (0.00)	0.27 (0.28)	0.68 (0.72)	0.18

Table 4. Transition between dates 3 and 4.

	0.9697	1.3682	1.3902	1.4823	1.5779	1.6025	1.7255	σ_{loc}
1.1833	0.53 (0.44)	0.36 (0.56)	0.02 (0.00)	0.03 (0.00)	0.02 (0.00)	0.02 (0.00)	0.02 (0.00)	0.46
1.4038	0.03 (0.00)	0.56 (0.39)	0.24 (0.40)	0.06 (0.21)	0.03 (0.01)	0.01 (0.00)	0.06 (0.00)	0.24
1.4197	0.02 (0.00)	0.39 (0.28)	0.22 (0.41)	0.34 (0.30)	0.01 (0.01)	0.01 (0.00)	0.01 (0.00)	0.17
1.4856	0.00 (0.00)	0.04 (0.04)	0.15 (0.20)	0.64 (0.60)	0.15 (0.14)	0.00 (0.02)	0.01 (0.00)	0.14
1.558	0.00 (0.00)	0.02 (0.00)	0.01 (0.02)	0.35 (0.36)	0.36 (0.38)	0.20 (0.21)	0.06 (0.03)	0.15
1.5764	0.05 (0.00)	0.05 (0.00)	0.04 (0.01)	0.04 (0.26)	0.33 (0.39)	0.22 (0.29)	0.27 (0.05)	0.34
1.6684	0.00 (0.00)	0.02 (0.00)	0.01 (0.00)	0.03 (0.02)	0.02 (0.12)	0.35 (0.41)	0.57 (0.45)	0.17

Table 5. Transition between dates 4 and 5.

	0.8335	1.3455	1.371	1.4793	1.5988	1.6297	1.7842	σ_{loc}
0.9697	0.75 (1.00)	0.21 (0.00)	0.01 (0.00)	0.01 (0.00)	0.01 (0.00)	0.00 (0.00)	0.00 (0.00)	0.47
1.3682	0.01 (0.00)	0.63 (0.50)	0.28 (0.40)	0.05 (0.10)	0.02 (0.00)	0.00 (0.00)	0.01 (0.00)	0.16
1.3902	0.02 (0.00)	0.43 (0.33)	0.27 (0.47)	0.28 (0.20)	0.01 (0.00)	0.00 (0.00)	0.00 (0.00)	0.19
1.4823	0.00 (0.00)	0.04 (0.02)	0.16 (0.18)	0.66 (0.69)	0.14 (0.11)	0.00 (0.01)	0.01 (0.00)	0.14
1.5779	0.00 (0.00)	0.01 (0.00)	0.01 (0.01)	0.35 (0.31)	0.43 (0.47)	0.12 (0.20)	0.08 (0.01)	0.18
1.6025	0.02 (0.00)	0.07 (0.00)	0.05 (0.00)	0.06 (0.18)	0.40 (0.46)	0.16 (0.32)	0.24 (0.03)	0.34
1.7255	0.00 (0.00)	0.01 (0.00)	0.01 (0.00)	0.03 (0.00)	0.03 (0.05)	0.25 (0.40)	0.66 (0.54)	0.21

of "crashes" seems to be present in the observed option prices or, differently, an appropriate model should allow for jumps in the underlying prices. We observe a similar pattern in the upper right-hand part, indicating that there is a strictly higher probability for large upward movements. Although this effect seems to be similar to "fat tails," however, here it is much more pronounced since it attributes "high" probability (between 0.01 and 0.07) on zero events, and we see it in the conditional evolution in the future. The observation that the first column is zero might be, attributable to the fact that μ is too large. With the exception of the last row, we find a U-shaped form for the "local volatility".

7. Conclusion

We calibrated a risk-neutral asset process to observe call option prices under the assumption that the discounted asset prices follow a Markov chain. We proved that existence of such a Markov chain is equivalent to the condition that call option prices are decreasing and convex in the strike price and increasing in the exercise date. An important technique we used throughout was to reduce our dynamic market to a static one with extra constraints due to the possibility of dynamically trading in the bond and the risky asset. We characterized the superreplication price of a call in both a static and dynamic framework. The bid-ask spread was shown to be reduced due to the trade-ability of other options. We applied the Bayesian approach to infer the "optimal" measure and revealed surprising results.

A Proofs

Proof of Proposition 9. The set of attainable claims is clearly $\{\gamma(S_1)(S_2 - S_1) + \sum_{(t,K,C\in\mathcal{D})} a_{t,K}(S_t - K)^+ | a \in \mathbb{R}^{|\Sigma_1|} \times \mathbb{R}^{|\Sigma_2|}, \gamma \in \mathbb{R}^{|\Sigma_1|}\}$. For a payoff $V_2 = \gamma(S_1)(S_2 - S_1) + \sum_{s\in\Sigma_1} a_s(S_1 - s)^+ + \sum_{s\in\Sigma_2} b_s(S_2 - s)^+$, the term $-\gamma(S_1)S_1 + \sum_{s\in\Sigma_1} a_s(S_1 - s)^+$ can be written on the basis of path-independent Arrow–Debreu securities as $\sum_{s\in\Sigma_1}\beta_s 1_{S_1=s}, \beta \in \mathbb{R}^{|\Sigma_1|}$; similary, there exist some $\alpha_s \in \mathbb{R}^{|\Sigma_2|}$ such that $\sum_{s\in\Sigma_1} b_S(S_2 - s)^+ = \sum_{s\in\Sigma_2}\beta_s 1_{S_2=s}$. This proves that an investment through calls, puts, and the asset and numberaire can be achieved as an investment in $\delta_{s,s'}^{\mathrm{dep}}, \delta_{s'}^1$ and $\delta_{s'}^2((s, s') \in \Sigma_1 \times \Sigma_2)$. The converse follows in a similar way. Using Eq. (3), it is straightforward to prove the form for the price functional $\mathcal{P}$.

Proof of Proposition 12. If there exists an arbitrage opportunity $\bar{\alpha}, \bar{\beta}, \bar{\gamma}$ such that $\mathcal{X}(\bar{\alpha}, \bar{\beta}, \bar{\gamma}) \geq 0$ and $\mathcal{P}(\bar{\alpha}, \bar{\beta}, \bar{\gamma}) < 0$, then by linearity of $\mathcal{X}$ and $\mathcal{P}$ we get $\mathcal{X}(0, \bar{\beta}, \bar{\gamma}) \geq \mathcal{X}(-\bar{\alpha}, 0, 0)$ and $\mathcal{P}(0, \bar{\beta}, \bar{\gamma}) < \mathcal{P}(-\bar{\alpha}, 0, 0)$, which means that the superreplication price is strictly below the replication price.

Conversely, any α-arbitrage opportunity is obviously an arbitrage opportunity in the strict sense.

Proof of Lemma 16. Let us first fix a call with strike $K \in \mathcal{K}_{t+1}$ and maturity date $t + 1 \in \mathcal{T}$ and denote by C_t, C_{t+1} the two call prices and by i_K the node corresponding to the strike. We address the problem by backward induction and consider in state $\imath$ at data t the minimum amount of numeraire to be held in order to superreplicate the short call payoff at date $t + 1$:

$$V_\imath \overset{\mathrm{def}}{=} \inf_{\beta_\imath, \gamma_\imath} \beta_\imath + \gamma_\imath s_\imath$$

$$\text{s.t.} \quad \forall s \in \Sigma_{t+1} : \beta_\imath + \gamma_\imath s \geq -(s - K)^+.$$

The optimal superreplication strategy splits up into two cases:

(1) For $\imath \leq i_K$: From $\beta_\imath + \gamma_\imath S_\imath \geq -(S_\imath - K)^+ = 0$, we deduce that $V_\imath \geq 0$. Since $\beta_\imath = \gamma_\imath = 0$ satisfies the constraint $\beta_\imath + \gamma_\imath s = 0 \geq -(s - K)^+$ for and $s \in \Sigma_{t+1}$, we deduce that the optimum is attained at $V_\imath = 0$.
(2) For $\imath > i_K$: Similar to the previous case, we deduce from $\beta_\imath + \gamma_\imath s_\imath \geq -(s_\imath - K)^+ = K - s_\imath$, For $\beta_\imath = +K, \gamma_\imath = -1$ the constraint $\beta_\imath + \gamma_\imath s = K - s \geq -(s - K)^+(s \in \Sigma_2)$ is satisfied and so the optimum $V_\imath = K - S_\imath$ is attained.

In both cases, the inflow $V_\imath = -(s_\imath - K)^+$ for the superreplication strategy at time t is simply a short call payoff with strike K. By the dynamic programming principle, $\inf_{(\beta,\gamma)\in\mathcal{I}^\alpha} \mathcal{P}(0, \beta, \gamma)$ corresponds to the superreplication price of

$V_i = -(S_i - K)^+$, and so, by Property 4, it is equal to the replication price of the short call with maturity date one i.e., $-C_t$. So for $\alpha_s = -(s - K)^+$, we have

$$\inf_{(\beta,\gamma)\in\mathcal{I}^\alpha} \mathcal{P}(0,\beta,\gamma) = -C_t\,.$$

By assumption, the price of the short call with strike K and maturity t is larger than $-C_{t+1} = \mathcal{P}(\alpha,0,0)$; this proves the assertion.

Proof of Theorem 17. Merton (1973), cited at the beginning of Sec. 5, proved necessity; therefore we only need to prove sufficiency. First, let us assume that α is concave. Short calls form a basis of the space of (path-independent) payoffs, so there exists $\tilde{\alpha}_l \in \mathbb{R}$ s.t. $\alpha_s = \tilde{\alpha}_0 s - \sum_l \tilde{\alpha}_l (s - K_l)^+$. We quickly check that $\tilde{\alpha}_l = (\alpha_l - \alpha_{l-1}) - (\alpha_{l+1} - \alpha_l) \geq 0$. Thus the superreplication price of $\mathcal{X}(\alpha,0,0)$ can be written as

$$\tilde{\alpha}_0 S_0 - \sum_l \tilde{\alpha}_l C_{t,l} \geq \tilde{\alpha} S_0 - \sum_l \tilde{\alpha}_l C_{t+1,l} = \mathcal{P}(\alpha,0,0)\,.$$

Together with Property 15, this proves that there are no α-arbitrage opportunities for any concave payoff.

In the case of a general payoff α at date $t + 1$, we will denote by $\bar{\alpha}$ the concave envelope of α. Similar to Lemma 16, we test for α-arbitrage and study the optimization problem in every state i at time t:

$$V_i \stackrel{\text{def}}{=} \inf_{\beta_i,\gamma_i} \beta_i + \gamma_i s_i$$

$$\text{s.t.} \quad \forall j : \beta_i s_j \geq \alpha_j\,.$$

The concave envelope $\bar{\alpha}$ is by definition the solution of this optimization problem. Thus, we have $V_i = \bar{\alpha}_i$ and so

$$\inf_{(\beta,\gamma)\in\mathcal{I}^\alpha} \mathcal{P}(0,\beta,\gamma) = \inf_{(\beta,\gamma)\in\mathcal{I}^{\bar\alpha}} \mathcal{P}(0,\beta,\gamma)\,.$$

In other words, the superreplication price of α is equal to the superreplication price of its concave envelope. The absence of arbitrage opportunities in the static market at time $t + 1$ and $\bar{\alpha} \geq \alpha$ imply $\mathcal{P}(\bar{\alpha},0,0) \geq \mathcal{P}(\alpha,0,0)$. By the preceding study of concave payoffs, we get

$$\inf_{(\beta,\gamma)\in\mathcal{I}^\alpha} \mathcal{P}(0,\beta,\gamma) \geq \mathcal{P}(\bar{\alpha},0,0)\,,$$

which guarantees the result. The proof now follows with Proposition 12 and Proposition 10.

Proof of Proposition 18. Necessity is the result of Merton (1973) stated at the beginning of Sec. 5.

In the case of only two exercise dates, sufficiency follows directly from Theorem 17 since a $\tilde{\mathcal{D}}$-MCPM is also a $\mathcal{D}$-MCPM. With several exercise dates, let us first note that the marginal densities of S_t, λ_t are uniquely determined from the option prices $(t, K, \tilde{C}(t, K)) \in \tilde{\mathcal{D}}$. From the above argument, looking at only two dates, follows the existence of a set of joint probabilities $P[S_t = x, S_{t+1} = y], (x, y) \in \Sigma_t \times \Sigma_{t+1}$ such that: $\forall (x, y) \in \Sigma_t \times \Sigma_{t+1} : P[S_t = x] = \lambda_t(x), P[S_{t+1} = y] = \lambda_{t+1}(y)$ and $\forall x \in \Sigma_t : \sum_{y \in \Sigma_t} P[S_t = x, S_{t+1} = y](y - x) = 0$. We can define a stochastic matrix Π_t by

$$\Pi_t(x, y) = \frac{P[S_t = x, S_{t+1} = y]}{\lambda_t(x)} = P[S_{t+1} = y | S_t = x].$$

Now, for $t \in \mathcal{T}$, the MCMM defined by Π_t has marginal densities equal to λ_t and is thus both consistent with the option prices $\tilde{C}(t, K), K \in \Sigma_t$ and with the smaller set of observed option prices $C(t, K), t \in \mathcal{T}, K \in \mathcal{K}_t$. Moreover, it fulfills the martingale restriction and so it is a $\mathcal{D}$-MCPM.

Proof of Proposition 20. Absence of arbitrage opportunities implies there is an arbitrage-free option pricing model which we represent here by P^Π. We denote by $C_\Pi(t, K), K \in [0, \infty[t \in \mathcal{T}$ the corresponding option prices. These arbitrage-free option prices are convex and fulfill:

$$\forall \tau \geq t, \forall K \in \mathcal{K}_\tau : C_\Pi(t, K) \leq C_\Pi(\tau, K) = C(\tau, K).$$

The superreplication price $\bar{C}(t, K)$ is the supremum of C_Π taken over all possible P^Π. Thus,

$$\forall \tau \geq t, \forall K \in \mathcal{K}_\tau : \bar{C}(t, K) \leq C(\tau, K).$$

Moreover, $\bar{C}(t, K)$ inherits the convexity from $C_\Pi(t, K)$. By definition of the convex enveloppe, this implies, $\forall t \in \mathcal{T}, \forall K \in [0, \infty[: \bar{C}(t, K) \leq \mathcal{E}_t(K)$. This proves directly the first part of the assertion.

Whatever P^Π, we have $\forall K \in \mathcal{K}_t : C_\Pi(t, K) = C(t, K)$. Thus $\forall K \in \mathcal{K}_t$, $\bar{C}(t, K) = C(t, K)$. From the first part follows that $C(t, K) = \bar{C}(t, K) \leq \mathcal{E}_t(K)$. On the other hand, $\mathcal{E}_t(K) \leq C(t, K)$; this implies $C = \mathcal{E}_t(K)$.

By Proposition 18, there exists a $\mathcal{D}$-MCPM which then has, by definition, $\bar{C} = C$. This proves the second assertion.

Proof of Theorem 21. To prove the theorem, we apply Proposition 10. First we prove the theorem in the homogeneous case of Assumption 13. By Proposition 20, we need only to study sufficiency: the convex envelope $\{\mathcal{E}_t(K) | t \in \mathcal{T}, K \in \Sigma\}$ provides a set of option prices consistent with observed option prices:

$$\forall t \in \mathcal{T}, K \in \Sigma_t : \mathcal{E}_t(K) = C(t, K).$$

By definition $\mathcal{E}_t(K)$ is convex in K and is also clearly increasing in t. Thus, we can Proposition 18 and prove the absence of arbitrage opportunities.

Now we address the general inhomogeneous case. First, we prove necessity by contradiction: assume the condition would not hold. Then, by definition of the convex envelope, there exists a $(\tilde{C}, \tilde{t}, \tilde{K}) \in \mathcal{D}$ such that $\mathcal{E}_t(K) < \mathcal{L}_t(K)$, where $\mathcal{L}_t(K)$ is the linearly interpolated value corresponding to the next traded strike to the left and right. However this constitutes an arbitrage opportunity as in Merton (1973), cited at the beginning of Sec. 5. This contradiction proves the first part.

To finish we prove sufficiency and study a fictitious model with the thinnest grid corresponding to all nodes in which we set the call option prices corresponding to the additional strikes equal to their price resulting from from interpolating the option prices of the next two options strikes. This will result in the same $\mathcal{E}$. The above argument for the homogeneous case ensures existence. By construction of this fictitious model, the additional nodes have zero probability weight. Therefore, we can translate the fictitious model directly back into our original model.

Proof of Proposition 23. Let us for the moment consider an arbitrary finite set of couples $\tilde{\mathcal{D}} = \{(\mathcal{X}_j, C_j), 1 \leq j \leq J\}$, where $\mathcal{X}_j$ is a derivative asset and C_j, its price. We denote by $\hat{\mathcal{M}}$ the set of probability measures on the state space $\mathcal{A} = \otimes_{t \in \mathcal{T}} \Sigma_t$ consistent with $\tilde{\mathcal{D}}$; they do not necessarily have to be associated to MCMM. We have $\mathcal{M}_\mathcal{D} \subset \mathcal{M} \subset \hat{\mathcal{M}}$. Then we study the following optimization over the set of probability measures absolutely continuous w.r.t. to P and consistent with $\tilde{\mathcal{D}}$:

$$\min_{Q \in \hat{\mathcal{M}}_{\tilde{D}}, Q \ll P} E^P \left[\frac{dQ}{dP} \log \frac{dQ}{dP} \right] .$$

We identify the elements of $\hat{\mathcal{M}} \subset \hat{\mathcal{D}}$ with the elements of the simplex of $\mathbb{R}^{|\mathcal{A}|}$ and, similarly, $\mathcal{X}_j$ with an element in $\mathbb{R}^{|\mathcal{A}|}$. The set of absolutely continuous measure w.r.t P and consistent with $\tilde{\mathcal{D}}$ is a closed bounded set of $\mathbb{R}^{|\mathcal{A}|}$ and the optimization criterion is continuous; since $\mathcal{M}_\mathcal{D} \neq \emptyset$, there exists an optimal probability measure Q^* absolutely continuous w.r.t. P.

Let us assume that Q^* is not equivalent to P. We denote by q_i^* the probability being in state i under Q^*. For any $\epsilon \geq 0$, we define a (signed) measure $Q(\epsilon)$ by $q_i(\epsilon) \stackrel{\text{def}}{=} q_i^* + \epsilon(q_{0,i} - q_i^*)$, where $q_{0,i}$ denotes the state probability i under Q_0. This measure satisfies all equality constraints. Furthermore, with $q_{\min}^* \stackrel{\text{def}}{=} \inf_{q_i^* > 0} q_i^*, \delta \stackrel{\text{def}}{=} \sup_i |q_{0,i} - q_i^*|$, we can check that we have $q_i(\epsilon) \leq 0$ for all $\epsilon \in [0, q_{\min}^*/\delta[$ and $i \in \mathcal{A}$. Now, let us consider for any such $\epsilon \in [0, q_{\min}^*/\delta[$:

$$\frac{H(Q(\epsilon)) - H(Q^*)}{\epsilon} = \sum_{q_i^* \neq 0} \frac{q_i(\epsilon) \log q_i(\epsilon)/p_i - q_i^* \log q_i^*/p_i}{\epsilon} + \sum_{q_i^* = 0} q_{0,i} \log \frac{\epsilon q_{0,i}}{p_i} .$$

The first term admits a limit when $\epsilon \to 0$, while the second term tends to $-\infty$. Thus there exists some ϵ such that $H(Q(\epsilon)) < H(Q^*)$ which contradicts the assumption and proves that Q^* is an equivalent measure to P.

Therefore there exists an interior solution $Q^* > 0$; i.e., the inequality constraints are not binding. Applying Theorem 28.2 and 28.3 of Rockafellar (1970), we find that there exists $\lambda_j \in \mathbb{R}, j = 1, \ldots, J$ such that

$$\frac{dQ}{dP} \exp\left\{ \sum_{j=1}^{J} \lambda_j \mathcal{X}_j \right\},$$

which proves the desired result.

References

M. Avellaneda, C. Friedman, R. Holmes and D. Samperi, "Calibrating volatility surfaces via relative-entropy minimization", *Applied Mathematical Finance* **4** (1997) 37–64.

G. Bakshi, C. Cao and Z. Chen, "Empirical performance of alternative option pricing models", *The Journal of Finance* **50** (1997) 2003–2049.

S. Barle and N. Cakici, "Growing a smiling tree", *RISK* **8** (1995) 76–81.

D. S. Bates, "Dollar jump fears, 1984–1992: Distributional abnormalities implicit in currency futures options", *Journal of International Money and Finance* **15**(1) (1996) 65–93.

———, "Jumps and Stochastic volatility: Exchange rate processes implicit in PHLX Deutschemark options", *The Review of Financial Studies* **9**(1) (1996) 69–107.

D. Breeden and R. Litzenberger, "Prices of state-contingent claims implicit in option prices", *Journal of Business* **51** (1978) 621–651.

P. Buchen and M. Kelly, "The maximum entropy distribution of an asset inferred from options prices", *Journal of Financial and Quantitative Analysis* **31** (1996) 143–159.

J. Campa, K. Chang and L. Reider, "Implied exchange rate distribution: Evidence from OTC option markets", Discussion Paper, New York University, 1997.

P. Carr and D. Madan, "Determining volatility surfaces and option values from an implied volatility smile", Disussion Paper, 1998.

N. A. Chriss, *Black–Scholes and Beyond — Option Pricing Models*, Irwin, 1997.

J. Cvitanic and I. Karatzas, "Hedging contingent claims with constrained portfolios", *The Annals of Applied Probability* **3** (1993) 651–681.

J. Cvitanic, H. Pham and N. Touzi, "Superreplication in stochastic volatility models under portfolio constraints", *Applied Probability Journals* (to appear).

E. Derman and I. Kani, "Riding on the smile", *RISK* **7** (1995) 32–39.

E. Derman, I. Kani and N. Chriss, "Implied trinomial tree of the volatility smile", *The Journal of Derivatives* **3**(4) (1996) 7–22.

D. Duffie, *Dynamic Asset Pricing Theory*, Princeton University Press, 1992.

D. Duffie and K. Singleton, "Rating-based term structure of credit spreads", Discussion Paper, Stanford University, 1998.

B. Dumas, J. Fleming and R. Whaley, "Implied volatility functions: Empirical tests", *The Journal of Finance* **53**(6) (1998) 2059–2106.

B. Dupire, "Pricing with a smile", *RISK* **7** (1996) 18–20.

N. El Karoui and M. C. Quenez, "Dynamic programming and pricing of contingent claims in incomplete markets", *SIAM Journal of Control and Optimization* **33** (1995).

L. P. Hansen and R. Jagannathan, "Assessing specification errors in stochastic factor models", *The Journal of Finance* **52**(2) (1997) 557–90.

J. M. Harrison and D. M. Kreps, "Martingales and arbitrage in multiperiod securities markets", *Journal of Economic Theory* **20** (1979) 381–408.

J.-C. Jackwerth and M. Rubinstein, "Recovering probability distributions form option prices", *The Journal of Finance* **51** (1996) 1611–1631.

R. Jarrow, D. Lando and Turnbull, "A Markov model of the term structure of credit spreads", *The Review of Financial Studies* **10** (1997) 481–523.

H. Kushner and G. Dupuis, *Numerical Methods for Stochastic Control Problems in Continuous Time*, Springer-Verlag, 1992.

E. G. Luttmer, "Asset pricing economics with frictions", *Econometrica* **64**(6) (1997) 1439–76.

R. McCauley and W. R. Melick, "Risk reversal risk", *RISK* **9**(11) (1996) 54–57.

W. Melick and C. Thomas, "Recovering an asset's PDF from option prices: An application to crude oil during the gulf crisis", *Journal of Financial and Quantitative Analysis* **32**(1) (1997) 91–115.

R. C. Merton, "Theory of rational option pricing", *Bell Journal of Economics and Management Science* **4** (1973) 141–183.

R. T. Rockafellar, *Convex Analysis*, Princeton University Press, 1970.

M. Rubinstein, "Implied binomial trees", *The Journal of Finance* **49** (1994) 771–818.

D. Shimko, "Bounds of probability", *RISK* **6** (1993) 33–37.

D. Sondermann, "Currency options: Hedging and social value", *European Economic Review* **31** (1987) 246–256.

——, "Option pricing with bounds on the underlying securities", in *Bankpolitik, finanzielle Unternehmensführung und die Theorie der Finanzmärkte, Festschrift für Hans Krümmel*, (Duncker & Humblot, 1988) pp. 421–442.

P. Zipkin, "Mortgages and Markov chains: A simplified evaluation model", *Management Science* **39** (1993) 683–691.

WEIGHTED MONTE CARLO: A NEW TECHNIQUE FOR CALIBRATING ASSET-PRICING MODELS

MARCO AVELLANEDA, ROBERT BUFF, CRAIG FRIEDMAN
NICOLAS GRANDECHAMP, LUKASZ KRUK and JOSHUA NEWMAN*

A general approach for calibrating Monte Carlo models to the market prices of benchmark securities is presented. Starting from a given model for market dynamics (price diffusion, rate diffusion, etc.), the algorithm corrects for price-misspecifications and finite-sample effects in the simulation by assigning "probability weights" to the simulated paths. The choice of weights is done by minimizing the Kullback–Leibler relative entropy distance of the posterior measure to the empirical measure. The resulting ensemble prices the given set of benchmark instruments exactly or in the sense of least-squares. We discuss pricing and hedging in the context of these weighted Monte Carlo models. A significant reduction of variance is demonstrated theoretically as well as numerically. Concrete applications to the calibration of stochastic volatility models and term-structure models with up to 40 benchmark instruments are presented. The construction of implied volatility surfaces and forward-rate curves and the pricing and hedging of exotic options are investigated through several examples.

1. Introduction

According to Asset-Pricing Theory, security prices should be equal to the expectations of their discounted cash-flows under a suitable probability measure. This "risk-neutral" measure represents the economic value of consuming one unit of account on a given future date and state of the economy. A risk-neutral probability implemented in the context of a specific market is often called a *pricing model*. It is natural to require that a pricing model reproduce correctly the prices of liquid instruments which are actively traded. This ensures that "off-market", less liquid, instruments are realistically priced by the model.[a]

Here, we consider pricing models based on Monte Carlo (MC) simulations of future market scenarios ("paths").[b] Prices are computed by averaging discounted

*Address: Courant Institute of Mathematical Sciences, New York University, 251 Mercer Street, New York, NY, 10012. We are grateful to Graciela Chichilnisky, Freddy Delbaen, Raphael Douady, Darrell Duffie, Nicole El Karoui, David Faucon, Olivier Floris, Helyette Geman, Jonathan Goodman, Robert Kohn, Jean-Pierre Laurent, Jean-Michel Lasry, Jerome Lebuchoux, Marek Musiela, Jens Nonnemacher, and Frank Zhang for their enlightening comments and suggestions. We acknowledge the hospitality and generous support of Banque Paribas, Ecole Polytechnique, Ecole Normale Superieure de Cachan, ETH-Zurich and the Center for Financial Studies (Frankfurt). This work was partially supported by the US National Science Foundation (DMS-9973226).
[a]Throughout this paper, a pricing model refers to a model for pricing less liquid instruments relatively to more liquid ones (the benchmarks) in the context of a particular market. This type of financial model is used by most large investment banks to manage their positions.
[b]See Dupire (1998) for an up-to-date collection of papers by academics and practitioners on Monte Carlo methods in finance.

cashflows over the different paths. We shall be concerned with the *calibration* of such models, i.e., with specifying the statistics of the sample paths in such a way that the model matches the prices of benchmark instruments traded in the market.

Most calibration procedures rely on the existence of explicit formulas for the prices of the benchmark instruments. The unknown parameters of the underlying stochastic process are found by inverting such pricing formulas, either exactly or in the sense of least-squares. Unfortunately, in Monte Carlo simulations, this method may not be sufficiently accurate enough due to sampling errors (the finite sample effect). Furthermore, closed-form solutions for prices may not always be available or easy to code. In the latter case, fitting the model to market prices implies searching the parameter space through direct simulation, a computationally expensive proposition.

This paper consisders an alternative, non-parametric, approach for calibrating Monte Carlo models and applies it to several practical situations. The main idea behind our method is to put the emphasis on determining directly the risk-neutral probabilities of the future states of the market, as opposed to finding the parameters of the differential equations used to generate the paths for the state-variables.

One way to motivate our algorithm is to observe that Monte Carlo simulations can be divided (somewhat arbitrarily) into two categories: those that are *uniformly weighted* and those that are *non-uniformly weighted*. To wit, consider a set of sample paths, denoted by $\omega_1, \ldots, \omega_\nu$, generated according to some simulation procedure. By definition, a uniformy weighted simulation is such that all sample paths are assigned the same probability. Thus, a contingent claim that pays the holder h_i dollars if the path ω_i ocurrs, has model value

$$\Pi_h = \frac{1}{\nu} \sum_{i=1}^{\nu} h_i \,. \tag{1}$$

A non-uniformly weighted simulation is one in which the probabilities are not necessarily equal. Suppose that we assign, respectively, probabilities $p_1, \ldots, p_\nu$ to the different paths. The value of the contingent claim according to the corresponding "non-uniformly weighted" simulation is

$$\Pi_h = \sum_{i=1}^{\nu} h_i p_i \,. \tag{2}$$

Our approach is based on non-uniformly weighted simulations. First, we simulate a large number of paths of a stochastic process followed by the state-variables (prices, rates, etc.) under a *prior distribution*. Second — and here we depart from the conventional Monte Carlo method — we assign a different probability to each path. Probabilities are determined in such a way that (i) the expected values of the discounted cash-flows of benchmark instruments coincide exactly or within tolerance with the market prices of these securities and (ii) they are as close as possible to uniform probabilities ($p_i = 1/\nu$) coresponding to the simulated prior.

This method allows us to incorporate market information in two stages. The first step gives a prior probability measure that corresponds to our best guess for the risk-neutral measure given the information available. This guess may involve real statistics, such as estimates of rates of return, historical volatilities, correlations. It may also use parameters which are implied from market prices (implied volatilities, cost-of-carry, etc). In other words, the path simulation is used to construct a "backbone" or "prior" for the model which incorporates econometric or market-implied data. The second step has two purposes: it reconciles the econometric/prior information with the prices observed at any given time and also corrects finite-sample errors on the prices.

We denote the mid-market prices of the N benchmark instruments by $C_1, \ldots, C_N$ and represent the present values of the cashflows of the jth benchmark along the different paths by

$$g_{1j}, g_{2j}, \ldots, g_{\nu j} \quad j = 1, \ldots, N. \tag{3}$$

The price relations for the benchmark instruments can then be written in the form

$$\sum_{i=1}^{\nu} p_i g_{ij} = C_j, \quad j = 1, \ldots, N, \tag{4}$$

where $(p_1, \ldots, p_\nu)$ are the probabilitites that we need to determine. Generically, this (linear) system of equations admits infinitely many solutions because the number of paths ν is greater than the number of constraints.[c] The criterion that we propose for finding the calibrated probability measure is to minimize the *Kullback–Leibler relative entropy* of the non-uniformly sampled simulation with respect to the prior. Recall that if $p_1, \ldots, p_\nu$ and $q_1, \ldots, q_\nu$ are probability vectors on a probability space with ν states, the relative entropy of p with respect to q is defined as

$$D(p|q) = \sum_{i=1}^{\nu} p_i \, \log\left(\frac{p_i}{q_i}\right). \tag{5}$$

In the case of Monte Carlo simulation with $q_i = 1/\nu \equiv u_i$ we have[d]

$$D(p|u) = \log \nu + \sum_{i=1}^{\nu} p_i \log p_i. \tag{6}$$

We minimize this function under the linear constraints implied by (4). To this effect, we implement a dual, or Lagrangian, formulation which transforms the problem into an unconstrained minimization over N variables. Minimization of the dual objective function is made with L-BFGS (Byrd *et al.* (1994)), a gradient-based quasi-Newton optimization routine.

---

[c]It is also possible that the system of equations admits no solutions if the prior is inadequate or if the prices give rise to an arbitrage opportunity. We shall not dwell on this here.
[d]We shall denote the uniform probability vector by u, i.e., $u = (1/\nu, \ldots, 1/\nu)$.

The use of minimization of relative entropy as a tool for computing Arrow–Debreu probabilities was introduced by Buchen and Kelly (1996) and Gulko (1995, 1996) for single-period models. Other calibration methods based on minimizing a least-squares penalization function were proposed earlier by Rubinstein (1994) and Jackwerth and Rubinstein (1995). Samperi (1997), Avellaneda *et al.* (1997) and Avellaneda (1998) generalized the minimum-entropy method to intertemporal lattice models and diffusions. More recently, Laurent and Leisen (1999) considered the case of Markov chains. These studies suggest that this is a computationally feasible approach that works in several classical settings, such as generalizations of the Black–Scholes model with volatility skew or for one-factor interest rate models.

The use of minimum relative entropy for selecting Arrow–Debreu probabilitites has also been justified on economic grounds. Samperi (1997) shows that there exists a one-to-one correspondence between the calibration of a model starting with a prior probability measure and using a "penalization function" on the space of probabilities and the calculation of state-prices via utility maximization. More precisely, the Arrow–Debreu prices coincide with the marginal utilities for consumption obtained by maximizing the expectation of the utility function $U(x) = -\exp(-\alpha x)$ by investing in a portfolio of benchmark instruments. This correspondence is quite general. It implies, most notably, that other "distances" or "penalization functions" for Arrow–Debreu probabilities of the form

$$\tilde{D}(p|q) = \sum_{i=1}^{\nu} \psi\left(\frac{p_i}{q_i}\right) q_i\,, \quad \psi(x) \quad \text{convex} \tag{7}$$

can be used instead of relative entropy (which corresponds to the special case $\psi(x) = x \log x$). For each such penalization function, there exists a corresponding concave utility $U(x)$, obtained via a Legendre transformation, such that the Arrow–Debreu probabilities are consistent with an agent maximizing his/her expected utility for terminal wealth by investing in a portfolio of benchmarks.[e]

The use of relative entropy also has consequences in terms of price-sensitivity analysis and hedging. Avellaneda (1998) shows that the sensitivities of model values with respect to changes in the benchmark prices are equal to the linear regression coefficients of the payoff of contingent claim under consideration on the linear span of the cashflows of the benchmark instruments. In particular, the price-sensitivities can be computed directly using a single Monte Carlo simulation, i.e., without having to perturb the N input prices and to repeat the calibration procedure each time.

---

[e]While the particular choice of the mathematical distances $\tilde{D}(p|q)$ remains to be justified, the different distances between probabilites which result are, roughly speaking, economically equivalent — except possibly for the particular choice of smooth, increasing, convex utility that might represent the agent's preferences. The Kullback–Leibler distance is convenient because it leads to particularly simple mathematical computations, as we shall see hereafter. Another important feature of relative entropy is that it is invariant under changes of variables and therefore independent of the choice parameterization used to describe the system (Cover and Thomas (1991)). We refer the reader to Samperi (1999) for an in-depth discussion of this correspondence principle.

Thus, we hope that this method may provide an efficient approach for computing hedge-ratios as well.

Practical considerations in terms of model implementation are studied in the last four sections. We show that calibration of Monte Carlo models to the prices of benchmark instruments results in a strong reduction of variance, or simulation noise. This is due to the fact that the model effectively only needs to estimate the *residual* cash-flows (modulo the linear space spanned by the benchmarks). Therefore, instruments which are well-approximated by benchmarks have very small Monte Carlo variance. In particular, the interpolation of implied volatilities and prices of option between strikes and maturities is numerically efficient.

In practice, the success of any calibration method will depend on the characteristics of the market where it is applied. To evaluate the algorithm, we consider a few concrete examples. We study option-pricing models in the foreign-exchange and equity markets, using forwards and liquidly traded options as benchmarks. The models that we use incorporate stochastic volatility and are calibrated to the observed volatility skew. We also discuss the calibration of fixed-income models, and apply the algorithm to the construction of forward-rate curves based on the prices of on-the-run US Treasury securities.

2. Relative Entropy Distance and the Support of the Risk-Neutral Measure

Relative entropy measures the deviation of the calibrated model from the prior. Intiutively, if the relative entropy is small, the model is "close" to the prior and thus is "more desirable" than a model that has large distance from the prior. Let us make this statement more precise in the context of Monte Carlo simulations. The relative entropy distance,

$$D(p|u) = \log \nu + \sum_{i=1}^{\nu} p_i \log p_i \,, \tag{8}$$

takes values in the interval $[0, \log \nu]$. The value zero corresponds to $p_i = 1/\nu$ (the prior) whereas a value of $\log \nu$ is obtained when all the probability is concentrated on a single path. More generally, consider a probability distribution which is supported on a subset of paths of size μ and is uniformly distributed on these paths. If we take $\mu = \nu^\alpha$, with $0 < \alpha < 1$, and substitute the corresponding probabilities in (8), we find that

$$D(p|u) = \log \nu + \log \left(\frac{1}{\nu^\alpha} \right) = (1 - \alpha) \log \nu \,. \tag{9}$$

Within this class of measures, the relative entropy distance counts the number of paths in the support on a logarithmic scale. If $\frac{D(p|u)}{\log \nu} \ll 1$ the support of the calibrated measure is of size ν, whereas $\frac{D(p|u)}{\log \nu} \approx 1$ corresponds to a measure with a "thin support". Thin supports are inefficient from a computational viewpoint. They imply that the calibration algorithm "discards" a large number of simulated paths.

In this case, the *a priori* support of the distribution constructed by simulation will be very different from the the *a posteriori* support. This confirms the intuition whereby calibrations with small relative entropy are desirable.

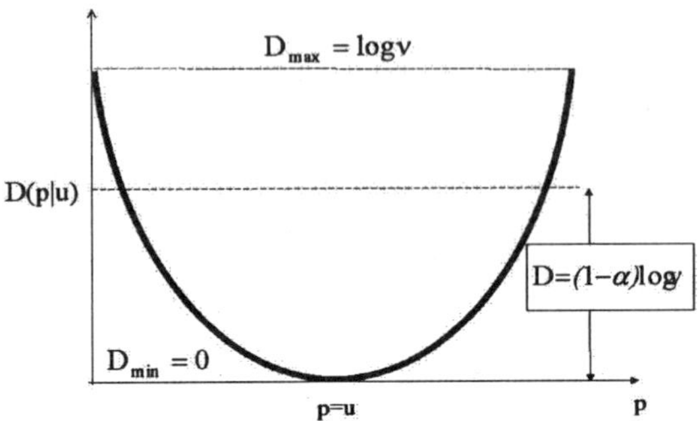

Fig. 1. Schematic graph of the relative entropy function. A probability with $D(p|u) = (1-\alpha)\log\nu$ is supported essentially on a subset of paths of cardinality ν^α. Probabilities with small $D(p|u)$ have large support whereas probabilities supported on a single path have the highest Kullback–Leibler distance, $\log\nu$.

This analysis can be applied to more general probability distributions. Let us write

$$p_i = \frac{1}{\nu^{\alpha_i}}, \quad i = 1, 2, \ldots, \nu. \tag{10}$$

Let N_α represent the number of paths with $\alpha_i = \alpha$, so that we have

$$\sum_\alpha N_\alpha = \nu, \quad \sum_\alpha \frac{N_\alpha}{\nu^\alpha} = 1. \tag{11}$$

Substituting (10) into (8), we find that

$$D(p|u) = \log\nu \left(1 + \sum_\alpha \frac{N_\alpha}{\nu^\alpha} \log\left(\frac{1}{\nu^\alpha}\right) \right)$$

$$= \log\nu \left(1 - \sum_\alpha \frac{N_\alpha}{\nu^\alpha} \alpha \right)$$

$$= \log\nu \left(1 - \mathbf{E}^p(\alpha) \right), \tag{12}$$

which shows that the relative entropy increases if the expected value of α is large. Due to the constraints implied by (11), this is possible only if there is a wide range of exponents α_i. Since probabilities are measured on a logarithmic scale, the measure will be concentrated on those paths which correspond to small values of α. A wide mismatching of probabilities between the calibrated measure and the prior is undesirable because this means that certain state-contingent claims will have very different values under the prior and the posterior measures.

3. Calibration Algorithm

We describe the algorithm for calibrating Monte Carlo simulations under market price constraints. It is a simple adaptation of the classical dual program used for entropy optimization (see Cover and Thomas (1991)). The new idea proposed here is to apply the algorithm to the state-space which consists of a collection of sample paths generated by Monte Carlo simulation of the prior.

To fix ideas, we shall consider a model in which paths are generated as solutions of the stochastic difference equations

$$X_{n+1} = X_n + \sigma(X_n, n) \cdot \xi_{n+1}\sqrt{\Delta T} + \mu(X_n, n)\Delta T, \quad n = 1, 2, \ldots, M \qquad (13)$$

where $M\Delta T = T_{\max}$ is the horizon time. Here $X_n \in \mathbf{R}^d$ is a vector of state variables and as a multi-dimensional process with values and $\xi_n \in \mathbf{R}^{d'}$ is as a vector of independent Gaussian shocks (d, d' are positive integers). The variance-covariance structure is represented by the $\nu \times \nu'$ matrix $\sigma(X, t)$ and the drift is the ν-vector $\mu(X, t)$.[f]

Using a pseudo-random number generator, we construct a set of sample paths of (13) of size ν, which we denote by

$$\omega^{(i)} = \left(X_1(\omega^{(i)}), \ldots, X_M(\omega^{(i)})\right) \quad i = 1, 2, \ldots, \nu. \qquad (14)$$

We assume throughout this paper that the benchmark instruments are such that their *cashflows along each path ω are completely determined by the path itself.* In the case of equities, where the components of the state-vector X generally represent stock prices, instruments satisfying this assumption include forwards, futures and standard European options. It is also possible to use barrier options or average-rate options. American-style derivatives do not satisfy this assumption because the early-exercise premium depends on the value of the option (and hence on the full pricing measure defined on the paths) as well as on the current state of the world. For fixed-income securites, benchmark instruments can include interest rate forwards, futures contracts, bonds, swaps, caps and European swaptions.[g] Under these circumnstances, the price relations can be written in the form (4) where g_{ij} is the present value of the cash-flows of the jth instrument along the ith path. The mathematical problem is to minimize the convex function of p

$$D(p|u) = \log \nu + \sum_{i=1}^{\nu} p_i \log p_i \qquad (15)$$

under linear constraints. This problem has been well-studied (Cover and Thomas, 1991). Introducing Lagrange multipliers $(\lambda_1, \ldots, \lambda_N)$, we can reformulate it as a

---

[f]This formulation extends trivially to the case of jump-diffusions or more general Markov processes and the MRE algorithm applies to these more general stochastic processes.
[g]American-style securities, such as Bermudan swaptions or callable bonds do not satisfy this assumption.

mini-max program (the "dual" formulation of the constrained problem)

$$\min_{\lambda} \max_{p} \left\{ -D(p|u) + \sum_{j=1}^{N} \lambda_j \left(\sum_{i=1}^{\nu} p_i\, g_{ij} - C_j \right) \right\} \tag{16}$$

A straighforward argument shows that probability vector that realizes the supremum for each λ has the Boltzmann–Gibbs form

$$p_i = p(\omega^{(i)}) = \frac{1}{Z(\lambda)} \exp\left(\sum_{j=1}^{N} g_{ij} \lambda_j \right). \tag{17}$$

To determine the Lagrange multipliers, define the "partition function", or normalization factor,

$$Z(\lambda) = \frac{1}{\nu} \sum_{i=1}^{\nu} \exp\left(\sum_{j=1}^{N} g_{ij}\, \lambda_j \right). \tag{18}$$

and consider the function

$$W(\lambda) = \log\left(Z(\lambda) \right) - \sum_{j=1}^{N} \lambda_j C_j$$

$$= \log\left\{ \frac{1}{\nu} \sum_{i=1}^{\nu} \exp\left(\sum_{j=1}^{N} g_{ij}\, \lambda_j \right) \right\} - \sum_{j=1}^{N} \lambda_j\, C_j. \tag{19}$$

We shall denote by $g_j(\omega)$ the present value of the caashflows of the jth instrument along the path ω. (Thus, $g_j(\omega_i) = g_{ij}$). At a critical point of $W(\lambda)$, we have

$$0 = \frac{1}{Z} \frac{\partial Z}{\partial \lambda_k} - C_k$$

$$= \frac{1}{Z(\lambda)} \sum_{i=1}^{\nu} g_{ik} \exp\left(\sum_{j=1}^{N} g_{ij}\, \lambda_j \right) - C_k$$

$$= \mathbf{E}^p \left\{ g_k(\omega) \right\} - C_k. \tag{20}$$

Hence, if λ is a critical point of $W(\lambda)$, the probability vector defined by Eq. (17) is calibrated to the benchmark instruments.

Notice that the function $W(\lambda)$ is convex: differentiating both sides of Eq. (19) with respect to λ yields

$$\frac{\partial^2 W(\lambda)}{\partial \lambda_j\, \partial \lambda_k} = \mathbf{Cov}^p \left\{ g_j(\omega)\, g_k(\omega) \right\}, \tag{21}$$

which is a non-negative definite matrix. In particular, the critical point, if it exists, should correspond to a *minimum* of $W(\lambda)$.

Based on this, we have the following algorithm:

(i) Construct a set of sample paths using the difference equations (1) and a pseudo-random-number generator.

(ii) Compute the cashflow matrix $\{g_{ij}, i = 1, \ldots, \nu, j = 1, 2, \ldots, N, \}$.

(iii) Using a gradient-based optimization routine, minimize the function $W(\lambda)$ in (19).[h]

(iv) Compute the risk-neutral probabilities $p_i, i = 1, 2, \ldots, \nu$ for each path using Eq. (17) and the optimal values of $\lambda_1, \ldots, \lambda_N$.

4. Implementation Using Weighted Least-Squares: An Alternative to Exact Fitting

It may not always be desirable to match model values to the price data exactly due to bid-ask spreads, asynchronous data, and liquidity considerations. Alternatively, we can minimize the sum of the weighted least-squares residuals and the relative entropy. We define the sum of the weighted least-squares residuals as

$$\chi_w^2 = \frac{1}{2} \sum_{j=1}^{N} \frac{1}{w_j} (\mathbf{E}^p \{g_j(\omega)\} - C_j)^2 , \tag{22}$$

where the $w = (w_1, \ldots, w_N)$ is a vector of positive weights.

The proposal is to minimize the quantity

$$\chi_w^2 + D(p|u) \tag{23}$$

over all probability vectors $p = (p_1, \ldots, p_\nu)$. Notice that the limit $w_i \ll 1$ corresponds to exact matching of the constraints. The discrepancy between the model value and market price with a weight w_i is typically of order $\frac{1}{\sqrt{w_i}}$.

We indicate how to modify the previous algorithm to compute the probabilities $(p_1, \ldots, p_\nu)$ that minimize $\chi_w^2 + D(p|u)$. Using the inequality

$$ab \leq \frac{1}{2}a^2 + \frac{1}{2}b^2 , \tag{24}$$

we find that, for all p,

$$\chi_w^2 \geq -\sum_{j=1}^{N} \lambda_j (\mathbf{E}^p \{g_j(\omega)\} - C_j) - \frac{1}{2} \sum_{j=1}^{N} w_j \lambda_j^2 . \tag{25}$$

It follows that

$$\inf_p [D(p|u) + \chi_w^2] \tag{26}$$

$$\geq \sup_\lambda \left\{ \inf_p \left[D(p|u) - \sum_{j=1}^{N} \lambda_j (\mathbf{E}^p \{g_j(\omega)\} - C_j) \right] - \frac{1}{2} \sum_{j=1}^{N} w_j \lambda_j^2 \right\}$$

[h]In our implementation, we use the L-BFGS algorithm.

$$= -\inf_{\lambda} \left[\log(Z(\lambda)) - \sum_{j=1}^{N} \lambda_j \, C_j + \frac{1}{2} \sum_{j=1}^{N} w_j \lambda_j^2 \right]$$

$$= -\inf_{\lambda} \left[W(\lambda) + \frac{1}{2} \sum_{j=1}^{N} w_j \lambda_j^2 \right] . \tag{27}$$

Here, $W(\lambda) = \log(Z(\lambda) - \sum_j \lambda_j \, C_j$ is the function used in the case of exact fitting.

The inequality expressed in (27) is in fact an equality. To see this, observe that the function $D(p|u) + \chi_w^2$ is convex in p and grows quadratically for $p \gg 1$. Therefore, a probability vector realizing the infimum exists and is characterized by the vanishing of the first variation in p. A straightforward calculation shows that if p is a minimum of this function, we have

$$p_i^* = \frac{1}{Z(\lambda^*)} \exp \left(\sum_{j=1}^{N} \lambda_j^* \, g_{ij} \right) \tag{28}$$

with

$$\lambda_j^* = -\frac{1}{w_j} (\mathbf{E}^{p^*} \{ g_j(\omega) \} - C_j) . \tag{29}$$

In particular, notice that this value of λ is such that (25) is an equality. Furthermore, the probability (28) is of exponential type, so we have

$$D(p^*|u) + \chi_w^2 = D(p^*|u) - \sum_{j=1}^{N} \lambda_j^* (\mathbf{E}^{p^*} \{ g_j(\omega) \} - C_j) - \frac{1}{2} \sum_{j=1}^{N} w_j (\lambda_j^*)^2$$

$$= \inf_{p} \left[D(p|u) - \sum_{j=1}^{N} \lambda_j^* (\mathbf{E}^{p} \{ g_j(\omega) \} - C_j) \right] - \frac{1}{2} \sum_{j=1}^{N} w_j \lambda_j^2$$

$$= -\log(Z(\lambda^*)) + \sum_{j=1}^{N} \lambda_j^* C_j - \frac{1}{2} \sum_{j=1}^{N} w_j \lambda_j^2$$

$$\leq -\inf_{\lambda} \left(W(\lambda) + \frac{1}{2} \sum_{j=1}^{N} w_j \lambda_j^2 \right) , \tag{30}$$

so equality must hold. This calculation shows that the pair (λ^*, p^*) is a saddlepoint of the min-max problem and that there is no "duality gap" in (27) and (30).

The algorithm for finding the probabilities that mimimise χ_w^2 under the entropy penalization consists in minimizing

$$\log(Z(\lambda)) - \sum_{j=1}^{N} \lambda_j (\mathbf{E}^{p^*} \{ g_j(\omega) \} - C_j) + \frac{1}{2} \sum_{j=1}^{N} w_j \lambda_j^2 \tag{31}$$

among candidate vectors λ. This algorithm represents a small modification of the one corresponding to the exact fitting of prices and can be implemented in the same way, using L-BFGS.[i]

5. Price Sensitivities and Hedge-Ratios

The MRE setting provides a simple method for computing portfolio price-sensitivities, under the additional assumption that *the prior measure remains fixed* as we perturb the benchmark prices and recalibrate.[j] We show, under this assumption, that sensitivities can be related to regression coefficients of the target contingent claim on the cashflows of the benchmarks. For simplicity, we discuss only the case of exact fitting, but the analysis carries over to the case of least-squares residuals with minor modifications.

Let $F(\omega^{(i)}), i = 1, \ldots, \nu$ represent the discounted cash-flows of a portfolio or contingent claim. To compute the price-sensitivities of the model value of the portfolio we utilize the "chain rule", differentiating first with respect to the Lagrange multipliers. More precisely,

$$\frac{\partial \mathbf{E}^p(F(\omega))}{\partial C_k} = \sum_{j=1}^{N_{\text{sec}}} \frac{\partial \mathbf{E}^p(F(\omega))}{\partial \lambda_j} \frac{\partial \lambda_j}{\partial C_k} \, . \tag{32}$$

We note, using Eq. (17) for the probability p_i, that

$$\frac{\partial \mathbf{E}^p(F(\omega))}{\partial \lambda_j} = \mathbf{Cov}^p\{F(\omega), g_j(\omega)\} \, . \tag{33}$$

Moreover, we have, on account of Eq. (21),

$$\frac{\partial}{\partial \lambda_j}(\mathbf{E}^p(g_k(\omega))) = \frac{\partial}{\partial \lambda_j}\left(\frac{\partial \log(Z(\lambda))}{\partial \lambda_k}\right)$$

$$= \mathbf{Cov}^p\{g_j(\omega), g_k(\omega)\} \tag{34}$$

In particular,

$$\frac{\partial C_k}{\partial \lambda_j} = \mathbf{Cov}^p\{g_j(\omega)\, g_k(\omega)\} \, . \tag{35}$$

Substitution of the expressions in (34) and (35) into Eq. (32) gives

$$\nabla_C \mathbf{E}^p\{F(\omega)\} = \mathbf{Cov}^p\{F(\omega)g_{\cdot}(\omega)\} \cdot [\mathbf{Cov}^p\{g_{\cdot}(\omega)g_{\cdot}(\omega)\}]^{-1} \, , \tag{36}$$

[i]More generally, we can consider the minimization of $W(\lambda) + \sum_{j=1}^{N} w_j \psi(\lambda_j)$, where ψ is a convex function. An argument entirely similar to the one presented above shows that this program corresponds to minimizing the quantity $\sum \frac{1}{w_j} \psi^*(\mathbf{E}^{p^*}\{\Gamma_j(\omega)\} - C_j)$, where ψ^* is the Legendre dual of ψ. The case $\psi(x) = |x|$ can be used to model proportional bid-ask spreads in the prices of the benchmark instruments, for example.

[j]This assumes, implicitly, that the prior probability represents information that "varies slowly" with respect to the observed market prices. For example, the assumption is consistent with interpreting the prior as a historical probability.

with the obvious matrix notation.[k] This implies, in turn, that the sensitivities of the portfolio value with respect to the input prices are the linear regression coefficients of $F(\omega)$ with respect to $g_j(\omega)$. Namely, if we solve

$$\min_{\beta} \left\{ \mathbf{Var}^p \left[F(\omega) - \beta_0 - \sum_{j=1}^{N} \beta_j g_j(\omega) \right] \right\}, \tag{37}$$

we obtain, from (36),

$$\beta_k = \frac{\partial \mathbf{E}^p(F(\omega))}{\partial C_k} \quad k = 1, \ldots, N \tag{38}$$

and[l]

$$\beta_0 = \mathbf{E}^p(F(\omega)) - \sum_{j=1}^{N} \beta_j \mathbf{E}^p(g_j(\omega)). \tag{39}$$

We conclude that a sensitivity analysis with respect to variations of the input prices can be done without the need to perform additional Monte Carlo runs and to perturb the input prices one by one. Instead, the MRE framework allows us to compute prices and hedges with a single Monte Carlo simulation, which is much less costly.[m]

Notice that the characterization of the hedge-ratios as regression coefficients shows that they are "stable" in the sense that they vary continuously with input prices. In practice, the significance of this hedging technique depends on details of the implementation procedure, such as the number of paths used in the simulation. The main issue is whether the support of the probability measure induced by the prior — the basic scenarios of the simulation — is sufficiently "rich in scenarios", for example.

6. Variance Reduction

The calibration of Monte Carlo simulations can significantly reduce pricing errors. Claims that are "well-replicated" by the benchmarks — in the sense that the variance in (37) is small compared to the variance of $F(\omega)$[n] — will benefit from a significant noise reduction in comparison with standard MC evaluation.

---

[k]The invertibility of the covariance matrix presupposes that the cashflow vectors of the benchmark instruments, $g_j(\omega) j = 1, \ldots, N$, are linearly independent. This assumption is discussed, for example, in Avellaneda (1998).

[l]This result can be interpreted as follows. Assume that an agent hedges the initial portfolio by shorting β_j units of the jth benchmark instrument for $j = 1, \ldots, N$. In this case, the model value of the net holdings (initial portfolio + hedge) is β_0. It represents the expected cost of dynamic replication of the residual.

[m]In contrast, a perturbation analysis that uses centered differences to approximate the partial derivatives with respect to input instruments requires $2N + 1$ Monte Carlo simulations.

[n]The ratio of the variances is the statistic $1 - R^2$ in the risk-neutral measure.

In fact, given any vector $\zeta = (\zeta_1, \ldots, \zeta_N)$, we have

$$\mathbf{E}^p(F(\omega)) = \mathbf{E}^p \left\{ F(\omega) - \sum_{j=1}^{N} \zeta_j g_j(\omega) \right\} + \sum_{j=1}^{N} \zeta_j C_j \, . \tag{40}$$

Since the second term on the right-hand is constant, the variance of the Monte Carlo method for pricing the cash-flow F is the same as the one associated with $F - \zeta \cdot g$. This statement is true for any value of the vector ζ and so, in particular, for the regression coefficients $(\beta_1, \ldots, \beta_N)$. Since, by definition, $F - \beta \cdot g$ has the least possible true variance among all choices of ζ, the cash-flow $\beta \cdot \Gamma$ is an "optimal control variate" for the simulation. Our method implicitly uses such control variates.

To measure this variance reduction experimentally in a simple framework, we considered the problem of calibrating a Monte Carlo simulation to the prices of European stock options, assuming a lognormal price with constant volatility.

We considered European options on a stock with a spot price of 100 with no dividends. The interest rate was taken to be zero. Taking a "maximum horizon" for the model of 120 days, we used as benchmarks all European options with maturities of 30, 60 and 90 days and strikes of 90, 100 and 110, as well as forward contracts with maturities of 30, 60 and 90 days. We assumed that the prices of the benchmarks were given by the Black–Scholes formula with a volatility of 25%. The prior was taken to be a geometric Brownian motion with drift zero and volatility 25%.[o]

The test consisted of pricing various options (target options) with strike/maturity distributed along a regular grid (maturities from 20 days to 120 days with 1-day intervals; all integer strikes lying between two standard deviations from the mean of the distribution). For each option, we compared the variances resulting from pricing with the simulated lognormal process with and without calibrating to the "benchmarks". As a matter of general principle, when pricing an option contract, we also include the forward contract corresponding to the option's expiration date in the set of calibration instruments.[P]

All Monte Carlo runs were made with 2000 paths. Each run took roughly half a second of CPU time on a SunOS 5.6. This includes the time required to search for the optimal lambdas. We verified the correctness of the scheme by checking that all model prices fell within three (theoretical) standard deviations of the true price, both with and without the min-entropy adjustment.

We found that there was significant variance reduction for all cases, with the exception of options having strikes far from the money and maturities which did not match the benchmark maturities. As expected, the best results were observed for those options with strikes and maturities near to those of the benchmark options. In particular, options with the "benchmark maturities" (30, 60 and 90 days) yielded

---

[o]We assumed that all benchmark options were correctly priced with the prior to separate the issues of calibration and variance reduction, focussing on the latter.

[P]Doing so guarantees that the mean of the distribution of the asset price at the expiration date is fitted exactly.

some of the best results for most strikes which were not too far away from the money. We also obtained some of the best results of options with strikes at or close to the forward values. The following table gives the factor by which the entropy method improved the variance for selected strikes and maturity dates. Note that the table below includes the benchmark instruments which yield an infinite improvement since the entropy method always prices them correctly (indicated by INF on Fig. 2). Benchmark strikes and maturities are shown in boldface.

The variance reduction from the entropy method translated into some excellent data for the computed standard errors. The figure below contains this information. The data is given in terms of Black–Scholes implied volatility and is obtained by taking the standard error of the option price and dividing by the Black–Scholes value of vega.

Finally, we examined the R^2 statistic given by the entropy method. We found that R^2 was greatest for values with benchmark maturity dates and strikes whose values are close to that of the forward. The results are given in the table below.

Maturity	Strike								
(Days)	80	85	**90**	95	**100**	105	**110**	115	120
20	N/A	N/A	1.03	2.22	8.24	2.78	1.54	N/A	N/A
30	N/A	N/A	INF	13.66	INF	19.11	INF	3.09	N/A
45	N/A	1.25	2.38	5.21	14.83	6.63	3.75	2.25	N/A
60	N/A	5.63	INF	49.83	INF	73.98	INF	10.83	3.03
75	1.54	3.04	6.25	10.61	25.71	14.08	9.12	5.45	3.05
90	2.47	8.79	INF	92.40	INF	150.36	INF	22.96	5.77
120	1.96	2.77	4.09	6.08	13.57	8.34	5.77	4.15	3.03

Fig. 2. Variance improvement ratio.

Maturity	Strike								
(Days)	80	85	**90**	95	**100**	105	**110**	115	120
20	N/A	N/A	0.56	0.37	0.32	0.37	0.51	N/A	N/A
30	N/A	N/A	0	0.16	0	0.15	0	0.38	N/A
45	N/A	0.47	0.35	0.28	0.26	0.28	0.34	0.47	N/A
60	N/A	0.25	0	0.10	0	0.09	0	0.21	0.45
75	0.50	0.32	0.23	0.21	0.21	0.22	0.24	0.30	0.42
90	0.38	0.19	0	0.07	0	0.07	0	0.15	0.32
120	0.40	0.33	0.30	0.29	0.29	0.29	0.31	0.35	0.40

Fig. 3. Standard errors from the entropy method (in percentage of implied volatility).

Maturity	Strike								
(Days)	80	85	**90**	95	**100**	105	**110**	115	120
20	N/A	N/A	0.25	0.57	0.85	0.65	0.44	N/A	N/A
30	N/A	N/A	1	0.93	1	0.95	1	0.75	N/A
45	N/A	0.40	0.63	0.80	0.91	0.83	0.72	0.55	N/A
60	N/A	0.83	1	0.97	1	0.98	1	0.91	0.68
75	0.41	0.68	0.83	0.89	0.95	0.92	0.88	0.81	0.67
90	0.62	0.89	1	0.99	1	0.99	1	0.95	0.80
120	0.51	0.63	0.73	0.79	0.90	0.85	0.80	0.73	0.64

Fig. 4. R^2 statistic from the entropy method.

Our interpretation is that the options with benchmark dates or near-the-money strikes have only a small component of their cashflows which is orthogonal to the benchmark instruments, and conversely. One would expect both greater variance reduction and dependence on the values of the benchmark instruments in these cases. Note that the variance reduction data given in Fig. 2 confirms this interpretation.

exp (days)	type	strike	price	exp(days)	type	strike	price
30	call	1.5421	0.007	180	call	1.6025	0.0141
30	call	1.531	0.0093	180	call	1.5779	0.0191
30	call	1.4872	0.0234	180	call	1.4823	0.0505
30	put	1.4479	0.0092	180	put	1.3902	0.0216
30	put	1.4371	0.0069	180	put	1.3682	0.0162
60	call	1.5621	0.0094	270	call	1.6297	0.0173
60	call	1.5469	0.0126	270	call	1.5988	0.0226
60	call	1.4866	0.0319	270	call	1.4793	0.0598
60	put	1.4312	0.0128	270	put	1.371	0.0254
60	put	1.4178	0.01	270	put	1.3455	0.019
90	call	1.5764	0.0112	30	fwd	0	1.486695
90	call	1.558	0.0149	60	fwd	0	1.484692
90	call	1.4856	0.0378	90	fwd	0	1.482692
90	put	1.4197	0.0153	180	fwd	0	1.476708
90	put	1.4038	0.0114	270	fwd	0	1.470749

Fig. 5. Data used for fitting the implied volatilities of options. The implied volatilities, displayed on the left-hand side of the graph, range from 13% to 14.5%. From Avellaneda and Paras (1996).

7. Example: Fitting a Volatility Skew

We apply the algorithm to calibrate a model using forwards and the prices of European options with different strikes and maturities. This example is take from the interbank foreign exchange market. It is well-known that options with different strikes/maturities trade with different implied volatilities. The goal is to construct a pricing model that incorporates this effect. Notice that such problem has been addressed by many authors in the context of the so-called "volatility surface" (Dupire (1994), Derman and Kani (1994), Rubinstein (1994), Chriss (1996); see Avellaneda *et al.* (1997) for references on this problem up to 1997). The method presented here is completely different since we do not iterpolate option prices or use a parameterization of the local volatility function $\sigma(S, t)$.

We considered a dataset consisting of 25 contemporaneous USD/DEM option prices obtained from a major dealer in the interbank market on August 25, 1995. The maturities are 30, 60, 90, 180 and 270 days. Strikes (quoted in DEM) correspond to 50-, 20- and 25-delta puts and calls. Aside from these options, we introduced five additional "zero-strike options" which correspond to the present value of a dollar in DEM for delivery at the different expiration dates (see Table 5) — the forward prices implied by the interest rates and the spot price. Including these forward prices in the set of benchmark instruments ensures that the model is calibrated to the forward rates and hence that there is no net bias in the forward prices.

As a prior, we considered the system of stochastic differential equations:

$$\frac{dS_t}{S_t} = \sigma_t \, dZ_t + \mu \, dt$$

$$\frac{d\sigma_t}{\sigma_t} = \kappa dW_t + \nu_t \, dt \,,$$

(41)

where Z_t and W_t are Brownian motions such that $\mathbf{E}(dZ_t \, dW_t) = \rho \, dt$. In Eq. (41), S_t represents the value of one US Dollar in DM. The instantaneous volatility is denoted by σ_t. The additional parameters are: μ, the cost-of-carry (interest rate differential), κ, the volatility of volatility, and ν_t is the drift of the volatility. Therefore, we are calibrating a two-factor stochastic volatility model. We assume the following numerical values for the parameters that define the prior dynamics:

1. S_0= midmarket USD/DEM spot exchange rate = 1.4887
2. US rate=5.91%
3. DM rate=4.27%
4. $\mu = -1.64\%$ (for convenience, we take $\mu = $ DM rate $-$ US rate in the prior, i.e., we adjust the model to the standard risk-neutral drift).
5. $\sigma_0 = $ Initial value of the prior volatility of USD/DEM $= 14\%$. (This is essentially the average of the observed implied volatilities.)
6. $\kappa = 50\%$
7. $\rho = - 50\%$

We simulated 5000 paths of Eq. (41), consisting of 2500 paths and their antithetics. The gradient tolerance in the BFGS routine was set to 10^{-7} and we used equal weights $w_i = 10^{-5}$ for the least-squares approximation. We found that the difference between model prices and market prices was typically on the order of 10^{-4}–10^{-5}

exp (days)	type	strike	error	lambda	exp (days)	type	strike	error	lambda
30	c	1.5421	-0.000019	-22.31451	180	c	1.6025	-0.000015	-21.81038
30	c	1.531	0.000045	43.71396	180	c	1.5779	0.000032	24.07218
30	c	1.4872	-0.000042	-27.9719	180	c	1.4823	-0.000006	-10.61863
30	p	1.4479	0.000004	-5.30083	180	p	1.3902	0.000018	19.90037
30	p	1.4371	0.000016	12.82587	180	p	1.3682	-0.000016	-15.64954
60	c	1.5621	-0.000025	-26.09551	270	c	1.6297	-0.000001	6.951896
60	c	1.5469	0.000033	34.28321	270	c	1.5988	-0.000007	-3.65328
60	c	1.4866	-0.000003	-7.354312	270	c	1.4793	0.000009	-3.421416
60	p	1.4312	-0.000029	-31.52781	270	p	1.371	0.000019	10.23301
60	p	1.4178	0.000041	32.31679	270	p	1.3455	-0.000012	-13.96789
90	c	1.5764	-0.000019	-18.9968	30	f	0	0.00002	8.864074
90	c	1.558	0.000034	29.95544	60	f	0	0.000018	-1.028268
90	c	1.4856	0.000004	-18.24796	90	f	0	0.000027	7.661008
90	p	1.4197	0.000029	38.27566	180	f	0	0.000017	5.521524
90	p	1.4038	-0.00005	-36.20459	270	f	0	0.000013	0.71636

Fig. 6. Fitting errors and lambdas for the 30 instruments using 5000 paths. The relative entropy is $D \approx 0.07$.

DM, representing relative errors of 1% in the deep-out-of-the money short-term options and much less 0.1% for at the money options (see Fig. 6).

The algorithm initiated with $\lambda_i = 0, i = 1, \ldots, 30$ converges after approximately 20 iterations of the BGFS routine. The entire calibration procedure takes about four seconds on a desktop PC with a Pentium II 330 Mhz processor. In practice, computation times are much faster because the values of lambdas from the previous runs can be stored and used as better initial guesses.

The relative entropy of the calibrated risk-neutral measure was found to be $D(p|u) = 7.39 \times 10^{-3}$. We can interpret this result in terms of the parameter α of Section 2. We find a value of $\alpha = 1 - \frac{D(p|u)}{\log \nu} = 0.99913$, which would correspond to an "effective number of paths" $\nu^\alpha \approx 4963$ according to the heuristics of Section 2. This represents an excellent fit in terms of the support of the calibrated measure. In Fig. 7, we present descriptive statistics for the calibrated probabilities, in Fig. 8, we plot the probabilities, which appear to be randomly distributed with a small mean about the uniform value $1/5000 = 0.0002$. Figure 9 displays a histogram of the logarithms of the probabilities. These results indicate a relatively small scatter about the mean and a large set of paths which support the posterior measure.

The values of the lambdas were between -0.84 and 0.79, which correspond to moderate variations of the probabilities about their mean. In Fig. 9, we present a histogram of the 5000 probabilities obtained The distibution of probabilites — or, equivalently, the distribution of state-price deflators — is unimodal and strongly peaked about its mean $1/5000 = 0.0002$. This confirms that the risk-neutral measure is supported on the full set of paths generated using the prior.

p	
Mean	0.0002
Standard Error	1.12E-06
Median	0.000188
Mode	0.000162
Standard Deviation	7.95E-05
Sample Variance	6.32E-09
Kurtosis	1.046163
Skewness	0.896427
Range	0.00059
Minimum	0.000039
Maximum	0.000629

Fig. 7. Descriptive statistics for the vector of calibrated probabilities $(p_1, \ldots, p_{5000})$.

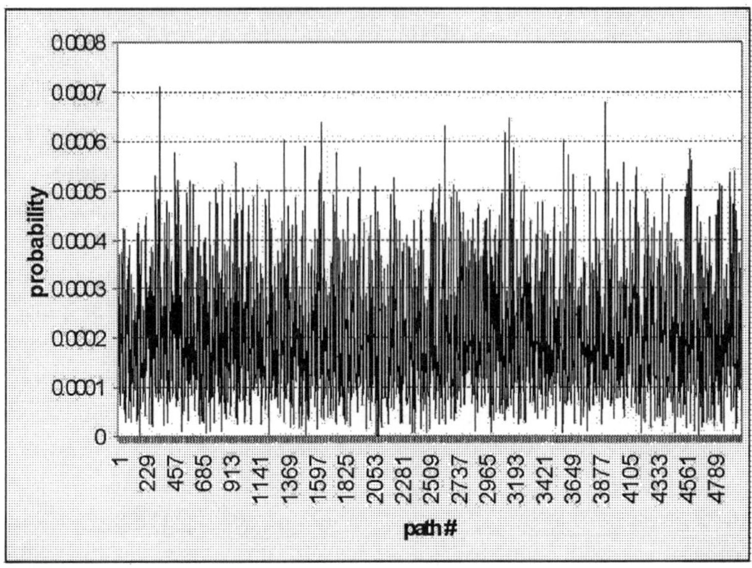

Fig. 8. Snapshot of the 5000 probabilities obtained by the method. The values oscillate about $1/5000 = 0.0002$.

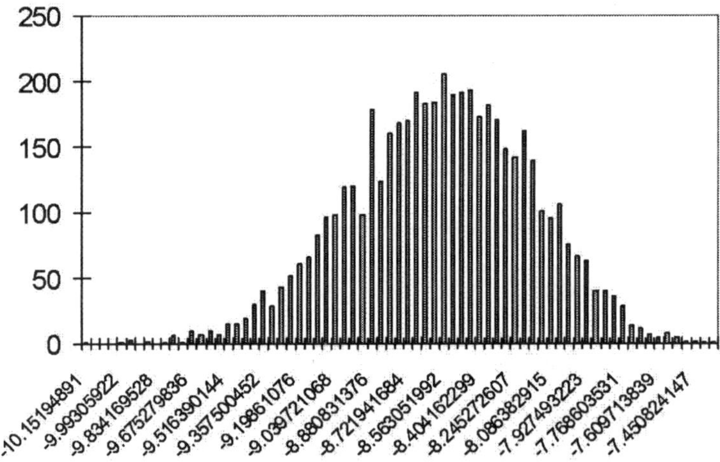

Fig. 9. Histogram of the logarithms of the calibrated probabilities multiplied by 5000. The distribution is unimodal and strongly peaked about $\log(1/5000) = -8.517$. Observe that there are some outliers corresponding to small probabilities.

To gain insight into the calibrated model, we generated an *implied volatility surface*, by repricing a set of options on a fine grid in strike/maturity space. We determined the highest and lowest strikes in the input option, which are, respectively, 1.67 DEM and 1.34 DEM, corresponding to the 20-delta options with the longest maturity (270 days). We then considered strike increments going from the maximum to the minimum value according to the rule

$$k_{\max} = 1.67 = k_0\,, \quad k_i = \frac{k_{i-1}}{1.01}\,, \quad i = 1, 2, \ldots, 20 \tag{42}$$

and maturities

$$t_{\min} = 30 = t_0\,, \quad t_i = t_{i-1} + 10\,, \quad i = 1, 2, \ldots, 24\,. \tag{43}$$

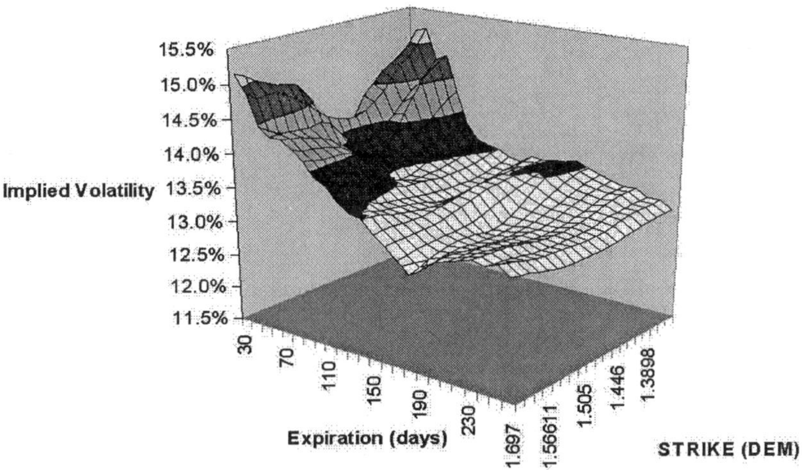

Fig. 10. Implied volatility surface for the USD/DEM options market based on the data in Fig. 5. The prior volatility is 14%.

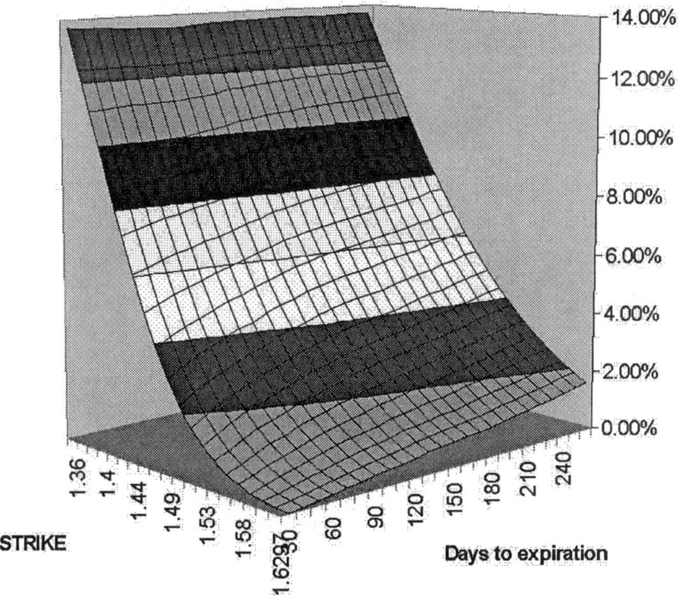

Fig. 11. Call prices associated with the implied volatility surface of Fig. 10.

We repriced these 480 options with the weighted Monte Carlo and generated a
surface by interpolating the implied volatilities linearly between strikes and dates.
The interpolation was done graphically using the Excel 5.0 graphics package. The
results are exhibited in Figs. 10 and 11.

Finally we tested the model by pricing a barrier option with the following characteristics:

1. Option type: USD put, DM call
2. Notional amount: 1 USD
3. Maturity: 180 days
4. Strike: 1.48 DM
5. Knockout barrier: 1.38 DM

and comparing the results with the price and delta hedge obtained using the Black–
Scholes lognormal model.

Closed-form solutions for the price of barrier options in the lognormal setting
were computed by Reiner and Rubinstein (1991). To compare with our model, we
used a Black–Scholes constant volatility of 14% — which corresponds to the implied
volatilities of short-maturity options – as well as with a constant volatility of 13%.
The latter corresponds to the implied volatility of the ATM forward call expiring
in 180 days. The differences in the Black–Scholes prices and deltas by changing the
volatilities from 13% to 14% are quite small. In Fig. 12, we compare the results of

exp.	type	strike	hedge	exp.	type	strike	hedge
30	c	1.5421	-3.95%	180	c	1.6025	-7.35%
30	c	1.531	4.00%	180	c *	1.5779	14.08%
30	c	1.4872	-1.08%	180	c *	1.4823	13.49%
30	p	1.4479	4.19%	180	p *	1.3902	-111.23%
30	p	1.4371	-6.71%	180	p *	1.3682	91.36%
60	c	1.5621	4.41%	270	c	1.6297	-2.91%
60	c	1.5469	-5.89%	270	c	1.5988	4.47%
60	c	1.4866	2.30%	270	c	1.4793	-1.41%
60	p *	1.4312	-17.21%	270	p	1.371	-2.47%
60	p *	1.4178	17.66%	270	p	1.3455	2.41%
90	c	1.5764	-3.50%	30	f		0.48%
90	c	1.558	1.61%	60	f		-0.14%
90	c	1.4856	-4.53%	90	f		5.22%
90	p *	1.4197	-15.59%	180	f		-20.59%
90	p *	1.4038	21.96%	270	f		-0.07%

expiration	options	fwd	net delta
30	0.0%	0.5%	0.4%
60	1.3%	-0.1%	1.2%
90	-3.1%	5.2%	2.2%
180	18.3%	-20.6%	-2.3%
ALL	16.6%	-15.0%	1.5%

MC Value = 0.358%
BS Value = 0.364%
BS Delta =   1%

Fig. 12. Price and hedge report for the reverse-knockout barrier option. Notice that the calibrated
model computes exposures to all the options and forwards entered as reference instruments. Instruments that have an exposure (in notional terms) of more than 10% of the notional amount of
the exotic option are labeled with asterisks. We observe significant exposure (1) at 60 and 90 days
near the knockout barrier and (2) at the expiration date in ATM and low-strike options. Hedges
at th barrier involve a spread in contracts with neighboring strikes, as expected.

applying the closed-form solution with 14% volatility are compared with the Monte Carlo simulation model described above. The values obtained with both models agree to three significant digits.

To evaluate the quality of the hedges, we computed the "net delta" of the exotic option given by the model and compared it with the Black–Scholes delta. For this, we assumed that the input options have the same deltas under BS than under our model. Of course, this a simplification, since the deltas of the individual options may be affected globally by the differences in implied volatilities. However, experience shows that such approximation is reasonable, in the sense that we do not expect the volatility skew in this market to distort significantly the deltas of the plain-vanilla options. The total amount of forward dollars needed to hedge the knockout is obtained by converting each option hedge into an equivalent forward position and summing over all contracts with the same maturity. To this amount, we also add the corresponding sensitivity to the forward contract (modeled here as a call with strike 0).

Adding the "forward deltas" associated with the different maturities, we find a total of 1.57% which is near the Black–Scholes value of 1.80 The results obtained in this example appear reasonable, despite the fact that the computation was done using Monte Carlo simulation with only 5000 paths. We believe that this is due in part to the reduction of variance which results from calibration process.

8. Example 2: Fitting the Smile for America Online Options on May 1999

We calibrated another stochastic volatility model to the mid-market prices of 35 America Online call prices recorded on May 10, 1999 at the close.

To ensure sufficient liquidity of the benchmarks, we took only options with traded volume above 100 for shorter maturities and above 50 for the maturity longer than half a year. It can be seen from the table that the implied volatilities of the benchmark calls vary in the range 78.33–88.52%. These extreme volatilities correspond to deeply in- or out-of-the money short term options. We also included forward prices for the stock at the different delivery dates, namely 12, 40, 68, 159 and 257 days. Thus, we calibrated the simulation to 40 benchmark prices.

We simulated $\nu = 10000$ Monte Carlo paths from this distribution on the time-horizon of 258 days with one time-step per day. We used the following parameters: $S_0 = 128.375$ (the America Online closing price on May 10, 1999), $\sigma_0 = 0.86$, $\rho = -0.5$, $r = 5\%$ and $\kappa = 0.5$. Then, we applied the MRE method described in Subsection 3.2 to calibrate the uniform distribution on the obtained sample paths to the set of 35 European call prices on America Online given in Fig. 13.

The program matched all the given 40 benchmark prices with the predefined accuracy to four decimal places using 181 iterations starting from $\lambda = 0$. The obtained entropy of the calibrated measure on the path space was 0.66 with the maximum possible entropy being $\log 10000 = 9.21$, indicating that the prior distribution its not far in the entropy distance from the calibrated one.

Maturity	Strike	Price	IVOL	Maturity	Strike	Price	IVOL
12	120	12.125	78.33	68	150	12.25	87.99
12	130	7.125	83.73	68	170	7.625	87.60
12	135	5.125	83.29	68	175	6.625	86.80
12	140	3.625	83.40	68	180	6	87.54
12	145	2.625	85.16	68	200	3.75	87.84
40	115	21.75	85.34	159	120	32.625	83.90
40	120	18.875	85.27	159	125	30.25	83.19
40	125	16.25	84.96	159	150	21.5	83.43
40	135	12.25	86.86	159	160	19	84.19
40	140	10.625	87.80	257	100	48.375	81.21
40	160	5.5	87.65	257	110	43.375	80.80
68	100	35.625	88.52	257	120	38.75	80.07
68	110	29	86.98	257	130	35.125	80.76
68	115	25.875	85.55	257	150	28.125	79.73
68	120	23.25	85.61	257	160	25.375	79.77
68	125	20.875	85.78	257	170	23	79.98
68	135	17.125	87.80	257	200	16.625	78.93
68	145	13.625	87.54				

Fig. 13. America Online call (mid-market) prices on May 10, 1999.

Figure 14 displays the implied volatility surface associated with the calibrated model. This surface was obtained by pricing a dense grid of plain-vanilla options with the calibrated Monte Carlo. Figure 15 represents the surface of corresponding call option prices. Notice that the shapes of the implied volatility surfaces in both examples are quite different Of course, the call price surfaces appear to be more similar: from well-know no-arbitrage relations, they are both convex and decreasing in the strike direction and monote-increasing with expiration.

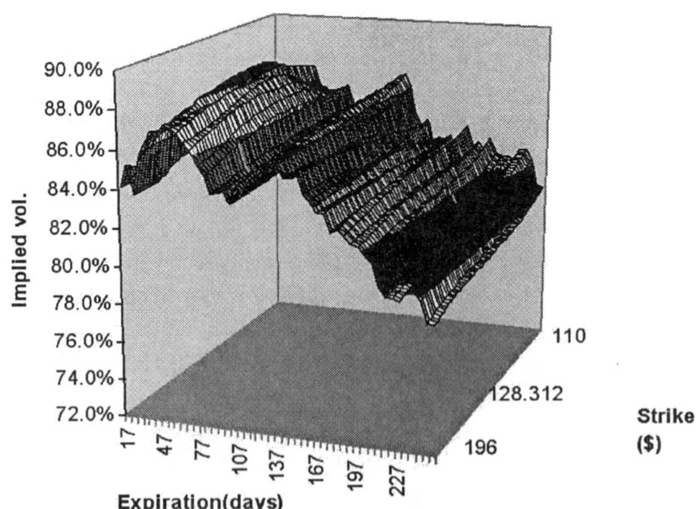

Fig. 14. Implied volatility surface for AOL option closing prices, calibrated to the prices of Fig. 13. The additional parameters are spot price=128.312, $\sigma_0 = 86\%, r = 5\%, \kappa = 50\%$ and $\rho = -50\%$. The relative entropy distance to the prior is $D = 0.66$.

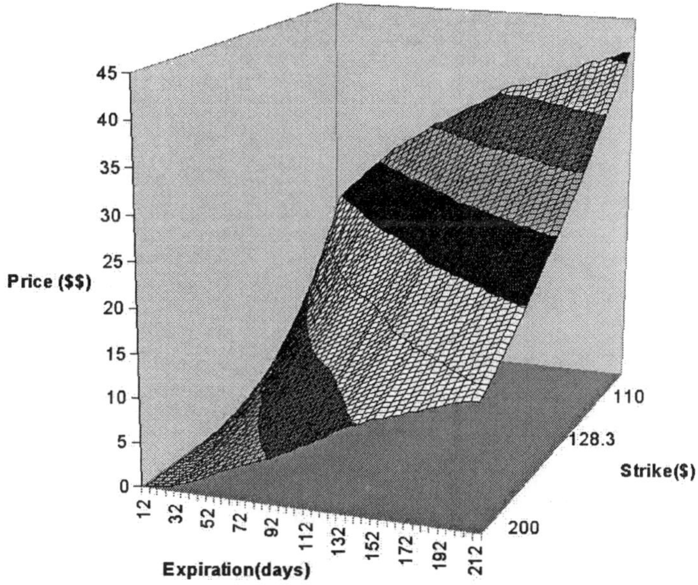

Fig. 15. Surface of call prices corresponding to the implied volatility surface of Fig. 14.

9. Example 3: Constructing a US Treasury Yield Curve

We considered the following calibration problem: given the current prices of on-the-run treasury securities, construct a smooth forward rate curve consistent with these prices.

Figure 16 shows the on-the-run instruments and the prices observed in the morning of Thursday, April 15, 1999.

A stochastic short-rate model was used to discount future cash flows. As a prior, we considered the modified Vasicek model

$$dr = \alpha(m(t) - r)dt + \sigma \, dW \, . \tag{44}$$

Here, $m(t)$ is the possibly time-dependent level of mean reversion, and the constant α controls the rate of mean reversion. We experimented with two types of mean-reversion levels: constant levels and time-dependent levels, where the latter were taken to be $m(t) =$ the piecewise-constant (bootstrapped) instantaneous forward-rate curve.[q]

Figure 17 shows several prior instantiations of the coefficients of (44).

We calibrated the modified Vasicek model with 15000 Monte Carlo paths and 24 time steps per year. Figure 18 shows the calibrated forward-rate curve and

---

[q]These are arbitrary modeling choices. For example, we could start with an econometrically calibrated forward rate curve or with a level of mean-reversion that corresponds to an estimate of forward rates for long maturities. Bootstrapping is standard method for building a forward rate curve: it works by assembling forward rates instrument by instrument, earlier maturities first. Rates are constant between maturities and jump at maturities.

Maturity	Coupon	Price
07/15/99	–	98.955
10/14/99	–	97.823
03/30/00	–	95.725
03/31/01	4.875	99.875
02/15/04	4.75	98.812
11/15/08	4.75	97.219
02/15/29	5.25	96.250

Fig. 16. Seven benchmark US-treasury bills and bonds. Prices are as quoted on 04/15/99 and not adjusted for accrued interest. The alignment of prices reflects the direction of the sensitivities listed below in Fig. 20: left alignment indicates positive sensitivity, right alignment negative sensitivity.

Scenario	α	$m(t)$	σ
I	0.25	0.0426035	**0.01**
II	0.25	0.0426035	**0.05**
III	0.25	bootstrap	**0.01**
IV	0.25	bootstrap	**0.05**

Fig. 17. Various prior instantiations of the coefficients of (44). 0.0426035 is the rate of the first leg of the piecewise constant bootstrapped forward rate curve.

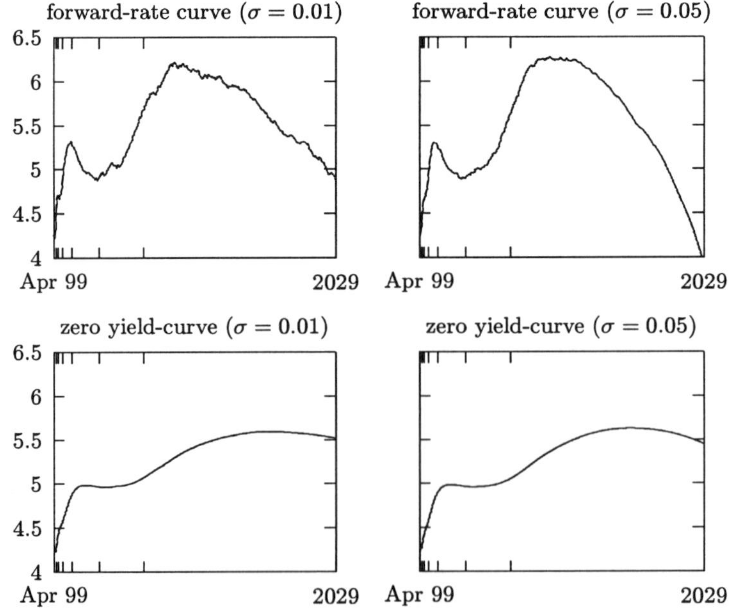

Fig. 18. Forward-rate curve and zero-coupon-bond yield curve for scenario I ($\sigma = 0.01$) on the left side, and scenario II ($\sigma = 0.05$) on the right side.

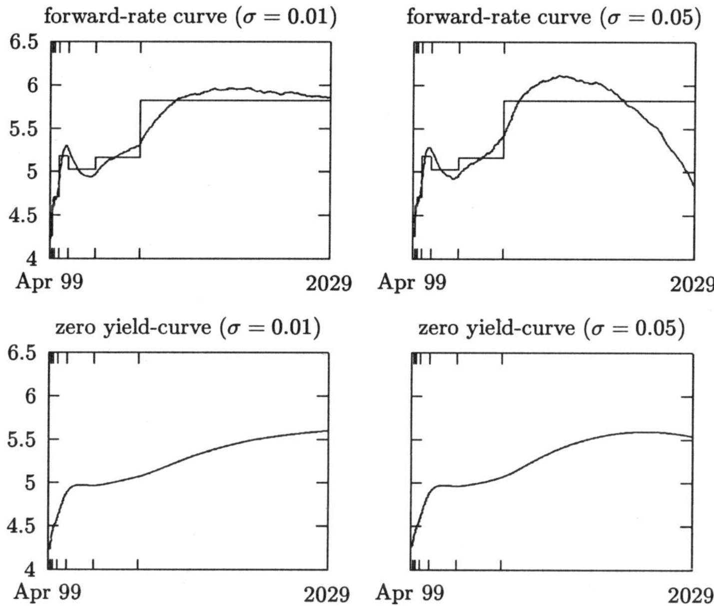

Fig. 19. Forward-rate curve and zero-coupon-bond yield curve for scenario III ($\sigma = 0.01$) on the left side, and scenario IV ($\sigma = 0.05$) on the right side. Both scenarios revert to the piecewise constant bootstrapped prior $m(t)$ superimposed in the top row.

zero-coupon-bond yield curve for scenarios I and II, with constant level of mean reversion. Figure 19 shows the calibrated forward-rate curve and zero-coupon-bond yield curve for scenarios III and IV. In these scenarios, the level of mean reversion $m(t)$ is set to the piecewise constant bootstrapped forward rate curve consistent with the data in Fig. 16.

In accordance with the work of Samperi (1997) and others, the optimal Lagrange multipliers λ_j^* can be interpreted in terms of an optimal investment portfolio. Specifically, consider an expected CARA utility function defined on the space of static portfolios $(\theta_1, \ldots, \theta_N)$ as follows:

$$U(\theta) = -\frac{1}{\nu} \sum_{i=1}^{\nu} e^{-\sum_{j=1}^{N} \theta_j (g_{ij} - C_j)}. \tag{45}$$

Since λ^* minimizes $log(Z(\lambda)) - \lambda \cdot C$, it follows from the analysis of Section 3 that the vector of Lagrange multipliers and the optimal portfolio are in simple correspondence: we have

$$\lambda_j^* = -\theta_j, \quad j = 1, \ldots, N. \tag{46}$$

Sensitivities are the opposites of the optimal portfolio weights for (45). A negative lambda corresponds to a "cheap" instrument (hence $\theta > 0$) and a positive lambda to a "rich" instrument (hence $\theta < 0$). The Lagrange multipliers, or sensitivities, $\lambda_1^*, \ldots, \lambda_7^*$ for all four scenarios are summarized in Fig. 20. Making

Maturity	Sensitivities for scenario...			
	I	II	III	IV
07/15/99	**32.507**	0.582	**29.421**	0.569
10/14/99	**−17.638**	−0.242	**−15.307**	−0.298
03/30/00	**5.359**	0.114	**4.607**	0.123
03/31/01	**−1.667**	−0.056	**−1.692**	−0.055
02/15/04	0.085	0.009	0.211	0.010
11/15/08	0.340	0.013	0.013	0.005
02/15/29	−0.356	−0.021	−0.050	−0.011

Fig. 20. The sensitivities for the seven benchmark instruments in Fig. 16, in each of the four prior scenarios. Sensitivities with absolute value greater than 1 are typeset in boldface.

$m(t$ time-dependend does not change the sensitivities significantly (scenario I versus III and II versus IV). The Lagrange multipliers corresponding to short-term instruments, however, are very high if the prior volatility is low ($\sigma = 0.01$ in scenarios I and III).

This last application of the MRE algorithm has also been implemented by one of the authors (R. Buff) as a prototype software operating remotely via the Internet. The software, which uses periodically updated Treasury securities prices and/or bond prices entered by the users, is accessible in the Courant Finance Server (http://www.courantfinance.cims.nyu.edu).

10. Conclusions

We have presented a simple approach for calibrating Monte Carlo simulations to the price of benchmark instruments. This approach is based on minimizing the Kullback–Leibler relative entropy between the posterior measure and a prior measure. In this context, the prior corresponds to the uniform measure over simulated paths (hence to the "classical" Monte Carlo simulation). This approach is known to be equivalent to finding the Arrow–Debreu prices which are consistent with an investor which maximizes an expected utility of exponential type. The advantage of the minimum-entropy algorithm is that (i) it is non-parametric (and thus not market or model specific) and (ii) it allows the modeler to incorporate econometric information and *a-priori* information on the market dynamics, effectively separating the specification of the dynamics from the issue of price-fitting.

We showed that the algorithm can be implemented as an exact fit to prices or in the sense of least-squares. The notion of entropy distance can be interpreted as a measure of the logarithm of *effective number of paths* which are active in the posterior measure. Large entropy distances correspond therefore to "thin" supports and thus to an ortogonality (in the measure-theoretic sense) between the prior and posterior measures.

The sensitivities produced by the model can be computed via regression, without the need to simulate the market dynamics and to recalibrate each time that we perturb the price of a benchmark instrument. Another interesting feature of the weighted Monte Carlo algorithm is the reduction of variance which results from the exact pricing of benchmark instruments. In fact, the simulation effectively evaluates the "residual risk" obtained after projecting the payoff of interest onto the space of portfolios spanned by the benchmark instruments. Numerical experiments indicate that the reduction of variance can be significant.

We discussed concrete implementations of the algorithm for the case of foreign-exchange and equity options, calibrated the underlying dynamics to two-factor stochastic volatility models. We have exhibited numerical evidence that shows that such an algorithm can be implemented in practice on desktop computers.

References

M. Avellaneda, "Minimum-entropy calibration of asset-pricing models", *International Journal of Theoretical and Applied Finance* **1**(4) (1998) 447.

M. Avellaneda, C. Friedman, R. Holmes and D. Samperi, "Calibrating volatility surfaces via relative-entropy minimization", *Applied Mathematical Finance* **4**(1) (1997) 37–64.

M. Avellaneda and A. Paras, "Managing the volatility risk of portfolios of derivative securities: The Lagrangian uncertain volatility model", *Applied Mathematical Finance* **3** (1996) 21–52.

P. W. Buchen and M. Kelly, "The maximum entropy distribution of an asset inferred from option prices", *Journal of Financial and Quantitative Analysis* **31**(1) (1996) 143–159.

N. Chriss, "Transatlantic trees", *RISK* **9** (1996) 7.

E. Derman and I. Kani, "Riding on a smile", *RISK* **7** (1994) 2.

B. Dupire, "Pricing with a smile", *RISK* **7** (1994) 1.

B. Dupire, *Monte Carlo Methodlogies and Applications for Pricing and Risk Management*, RISK Publications, London, 1998.

T. M. Cover and J. A. Thomas, *Elements of Information Theory*, Wiley, New York, 1991.

L. Gulko, "The Entropy Theory of Option Pricing", Yale University Working Paper, 1995.

———, "The Entropy Theory of Bond Pricing", Working Paper, Yale University, 1996.

J. C. Jackwerth and M. Rubinstein, "Recovering Probability Distributions from Contemporaneous Security Prices", Working Paper, Berkeley University, Hass School of Business, 1995.

J. P. Laurent and D. Leisen, "Building a Consistent Pricing Model from Observed Option Prices", Working Paper, the Hoover Institute, Stanford University, 1999.

D. Samperi, "Inverse Problems, Model Selection and Entropy in Derivative Security Pricing", PhD thesis, New York University, 1997.

M. Rubinstein, "Implied binomial trees", *The Journal of Finance* **69**(3) (1994) 771–818.

M. Rubinstein, and E. Reiner, "Breakind down the barriers", *RISK* **4**(8) (1991) 28–35.

C. Zhu, R. H. Boyd, P. Lu and J. Nocedal, *L-BFGS-B: FORTRAN Subroutines for Large-Scale Bound-Constrained Optimization*, Northwestern University, Department of Electrical Engineering, 1994.

Y. Zhu and M. Avellaneda, "A risk-neutral stochastic volatility model", *International Journal of Theoretical and Applied Finance* **1**(2) (1998) 289.

Reprinted from the Int. Jour. of Theor. and Appl. Finance
2(4) (1999) 441–469

ONE- AND MULTI-FACTOR VALUATION OF MORTGAGES: COMPUTATIONAL PROBLEMS AND SHORTCUTS

ALEXANDER LEVIN

The Dime Bancorp, Inc., Treasury Department, 589 5th Ave.,
New York, NY 10017, USA
E-mail: LevinA@Dime.com

Received 21 November 1998

A new valuation method is proposed that can be mathematically viewed as a numerical shortcut to approximately solve the partial differential equation written for Expected Instantaneous Return. The method derives OAS as "static spread plus cost of convexity" and is based on some simplified parametric assumption about the static spread's time behavior. The modeling and numerical procedure details are disclaimed with proven accuracy and time efficiency in option-adjusted valuation.

The method is especially effective for multi-scenario pricing, portfolio pricing, risk management and reporting, and for quantifying the impact of "non-traded" factors on reward and risk of holding mortgages. A systematic methodology for comprehensive multi-factor analysis is covered.

1. Introduction to Computational Problems of Mortgage Pricing

Pricing, managing and hedging portfolios of mortgages and mortgage-backed securities present several serious problems. Mortgages bear the prepayment option, which, unlike most known interest-rate options, is not exercised efficiently. Mortgages often pay an adjustable rate indexed to some market indicators that have no established term structure and need to be modeled themselves (11th District Cost of Funds, for example). Therefore, even generating mortgage cashflows is an issue itself that can only be solved approximately, using a fair amount of "practical intuition".

Secondly, mortgages are path-dependent instruments. Mathematically, it means that, aside from the interest rate factors, the value may depend on a set of additional state variables. This list includes coupon (for ARMs with periodic reset caps and lookbacks), current ARM index (if not Treasury or LIBOR), current burnout factor (if prepayments are modeled with a burnout), current collateral factor (for modeling sequential CMOs), or even current balances on every CMO tranche.

Thirdly, mortgages are multi-factor instruments as their values depend on inaccurate modeling parameters, which, in turn, appeared to be volatile pricing factors. For example, prepayment speeds can be modeled with certain accuracy and tend to deviate from the model. This modeling error can be treated as a factor that is not correlated to the interest rates (otherwise, we could enhance the explanatory power of the prepayment model), and has measurable volatility and transient characteristics. Many non-interest-rate-related factors are not efficiently traded by the mortgage market but bear a hidden source of additional systematic risk and return, see [10,7].

The existence of additional state variables (path-dependenc) and factors constitutes a very serious obstacle when the most time-efficient pricing structures, trees and grids, are considered. It is not surprising that it has become a common and trivial argument and that the mortgage pricing necessitates time-consuming Monte Carlo simulations. We see the major shortcoming of the Monte Carlo method not in its speed itself, but in the fact that it *is not structurally tailored* for the goals of one- and multi-factor risk measurement and management. Indeed, all (possibly thousands) simulations are solely intended to compute one value. This may satisfy the needs of a trader seeking *one accurate price*, but is reasonably deemed a waste of time by risk managers who need *many* (*possibly, approximate*) prices to assess, hedge and report on each of the risk dimensions. The entire Monte Carlo scheme has to be rerun for each new pricing point (such as a yield curve shock) — in a sharp contrast to the finite difference methods where all prices are sought simultaneously. In addition, securities are typically traded one-at-a-time whereas the risk is routinely assessed for large portfolios with the total number of mortgage loans and securities counting in thousands. The pricing speed will certainly rank higher than the "accuracy" (restricted anyway due to the input approximations made).

The other problem arising in the practical use of Monte Carlo is that such systems are deemed black boxes as they implement purely numeric procedures, require a complete cashflow dynamic model enabled along an arbitrary (random) interest rate path and do not "communicate" in the language of practitioners. To clarify this problem, let us imagine a typical bank's portfolio manager who can easily generate and even download cashflows for any static shock (non-random) scenario using Bloomberg, Oracle Financial Services, Wall Street Analytics, etc. but can never perform option-adjusted valuation (OAV) using those cashflows. Nor can he use a market-based prepayment consensus (normally formulated for deterministic scenarios) or adjust them having some collateral-specific information. A typical Monte Carlo-based OAV system will not allow him to achieve this goal since a random generator produces no shock scenarios. This is not to say that the risk manager is deficient — quite the opposite, his intuitive assumptions deliver almost complete information for option-adjusted valuation, should an appropriate mathematical approach exists.

In this paper, we introduce a new finite difference valuation method that can be mathematically viewed as a numerical shortcut to approximately solve the partial differential equation written for Expected Instantaneous Return (EIR). OAS is easily derived from this equation as EIR spread to a benchmark (EIRS, hence the name of the method) and sought concurrently with pricing the instrument on a grid of interest rate factor values. The only interest rate paths used by the method are those few induced by deterministic initial "shocks" of the factors. It explains the method's computational speed and, at the same time, sets up a natural bridge between the strict and rigorous mathematics of option-adjusted valuation and more traditional (static) practical tests for reward and risk of holding dynamic assets in a volatile environment. The method is ideally tailored for risk, asset/liability and portfolio management, it confidently handles many path-dependencies, can be adapted for the existence of additional early redemption provisions such as a clean-up call or an explicit American call schedule. Complemented with a control variate technique, the EIRS method yields a "very accurate" (yet quick) solution, especially for pass-through MBS including ARMs.

Although the deterministic interest rate paths used by the method are derived from an arbitrage-free model (they therefore differ from those normally used for the static valuation), they still can be understood and assessed by practitioners. Cash-flows required for those scenarios can be generated using a prepayment model, or scenario-specific PSA or CPR assumptions, or even downloaded from a vendor's system. This feature allows us to use the method when pricing CMOs approximately, without having a dynamic access to their cashflow generators.

2. The Instantaneous Return PDE and the Problem of Path-Dependency

Let us consider a hypothetical dynamic asset market price of which $P(t, x)$ depends on time t and one generalized market factor x. The latter can formally be anything and does not necessarily have to be the short market rate or the yield on the security analyzed. We treat $x(t)$ as a random process having a (generally, variable) drift rate μ and a volatility rate σ, and being disturbed by a standard Brownian motion $z(t)$, i.e.,

$$dx = \mu dt + \sigma dz \,. \tag{2.1}$$

Instantaneous Return (IR) is a random return measured over an infinitesimal investment horizon and annualized. The key statement in the Instantaneous Return concept is a partial differential equation that is traditionally derived applying the following mathematical operations:

- The Ito's Lemma (a stochastic differential equation) written for the random dynamics of price $P(t, x)$ given process (1) for $x(t)$;
- Collecting all the cashflow-related components of IR;
- Finding the mathematical expectation and (if needed) variance of the both sides;
- Equating thus obtained expectation to the risk-free rate $r(t, x)$ prevailing on the market plus a return spread (OAS) that investors expect from this type of risky financial instruments.

We assume that the asset continuously pays the $c(t,x)$ coupon rate and its balance B gets amortized at the $\lambda(t,x)$ rate. We formulate the known result in a form convenient for understanding of our approach:

Proposition 2.1 (single-factor case). *The Expected Instantaneous Return (EIR) and the Variance of Instantaneous Return (VIR) equal to*

$$\text{EIR} \equiv r + OAS = IR_0 - D_x\mu + \frac{1}{2}C_x\sigma^2 \tag{2.2}$$

$$\text{VIR} = D_x^2\sigma^2 \tag{2.3}$$

where $D_x = -\frac{1}{P}\frac{\partial P}{\partial x}$ is the "factor-Duration", and $C_x = \frac{1}{P}\frac{\partial^2 P}{\partial x^2}$ is the "factor-Convexity", and IR_0 is Instantaneous Return along the "base scenario":

$$IR_0 = \frac{1}{P}\frac{\partial P}{\partial t} + \frac{1}{P}(c+\lambda) - \lambda. \tag{2.4}$$

Note that the expression (4) for IR_0 can be derived following the above listed steps for the total market value, i.e., P times B, and computing all needed partial derivatives. In particular, $\frac{\partial B}{\partial t} = -\lambda B$, due to the definition of λ, whereas $\frac{\partial B}{\partial x}$ and $\frac{\partial^2 B}{\partial x^2}$ are replaced by zeros because the balance is not an "immediate" function of the factor. Another way to arrive to the formulae (2.2), (2.4) is demonstrated in [6,9] where the expected present value of the principal cashflow is integrated by parts and thus obtained pricing formula is mapped onto the PDE using the "inverse" Feynman–Kac theorem. The square root of the VIR measure presents nothing else than the famed Value-at-Risk (VAR), assessed for the infinitesimal investment horizon at an 84% confidence and annualized. We will use the notions of VIR and VAR interchangeably.

A notable feature of the above written PDE is it does not contain the balance variable, B. The entire effect of possibly random prepayment is represented by the amortization rate function, $\lambda(t,x)$. Although the total cashflow observed for each accrual period does depend on the beginning-period balance, construction of a finite difference scheme and the backward induction will require the knowledge of $\lambda(t,x)$, not the balance. This observation agrees with a trivial practical rule stating that the relative price is generally independent of the investment size.

Another important observation comes as follows. If we transform the economy having shifted all the rates, $r(t,x)$ and $c(t,x)$, by amortization rate $\lambda(t,x)$, formulae (2.2), (2.4) will be reduced to the constant-par asset's pricing formula. It means that a finite difference pricing grid built in the "λ-shifted" economy should, in principle, have as many dimensions as the total number of factors or state variables that affect r, c, and λ. In particular, even if $r(t,x)$ and $c(t,x)$ are functions of time and one factor x, but $\lambda(t,x,\xi)$ depends upon an additional state variable, ξ, the grid will necessarily have all 3 dimensions, for t, x, and ξ.

This "discount-rate-like" role of the λ-variable is in contrast to some other state variables that may affect the asset's value. We have already mentioned that the balance variable drops from the PDE and therefore does not cause any path-dependency. Another class of financial instruments includes "linear" assets where additional state variables linearly affect the coupon rate only (for example, regular floaters without caps and floors). For such instruments, a finite difference scheme can sometimes be built without additional axes. For example, if coupon rate $c(t, x, \xi)$ is linear in ξ, and all the variables, x, r, c, λ, and ξ follow a system of linear SDEs and ODEs, then price $P(t, x, \xi)$ is also going to be linear in ξ, and the diffusion term in the PDE can be computed correctly even without a ξ-axis.[a] Unfortunately, mortgage instruments are prepayable, and the redemption rate λ is a nonlinear function of coupon rate c, the burnout factor and possibly other path-dependent variables. It is therefore not feasible to price mortgages using a "true" low-dimensional finite difference grid. This problem encourages us to look for a descent approximation that would satisfactorily value dynamic assets while retaining all attractiveness of the finite difference methods, from the risk management view.

3. The EIRS Computational Scheme

Let us return to the one-factor setting and, without any loss of generality, define x as a deviation counted off the averaged scenario implied by the model, for which $x(t) = 0$. Indeed, if x was defined in some other way, we could always introduce a new factor by simply subtracting the mean values. This transformation eliminates drift μ from our analysis. We then write Eq. (2.2) twice, for the security in question, and for a benchmark that is expected to earn the risk-free rate, r, and subtract them:

$$OAS = EIR - EIR^b = IR_0 - IR_0^b + \frac{1}{2}(C_x - C_x^b)\sigma^2 \qquad (3.1)$$

where superscript "b" refers to the benchmark. We can simply state that "OAS equals the static return spread plus the relative convexity cost" where both components are clearly specified.

3.1. *The main hypothesis of the EIRS method*

Equation (3.1) above has almost exclusively economical meaning and represents a concept but does not allow for computing OAS itself. Indeed, both terms in the right-hand side of this equation remain unknown because the time- and factor-sensitivity of price $P(t, x)$ have not been determined yet. Note that the only time instance when the path-dependent problem does not exist is the initial moment, $t = 0$. Therefore, if one could approximate the static return spread term, $IR_0 - IR_0^b$, in Eq. (3.1) for $t = 0$ only, it would allow, in principle, to build an efficient computational pricing scheme for path-dependent dynamic assets. Such a simplifying assumption originally proposed by the author concerns with the time behavior of the *static discount spread* (s) over the forward curve, for the cashflows computed

---

[a]All these conditions are listed for illustration only; some of them could be relaxed.

along the corresponding average-rate path. We note as a reminder that the static discount spread is defined from the standard "static" pricing formula:

$$P = \sum_i \frac{CF_i}{\prod_{j=0}^{i-1}(1 + f_j/12 + s/12)} \tag{3.2}$$

where CF_i denotes the ith month cashflow, and f_j is the 1-month benchmark rate, j months forward. Note that the average path for the entire term structure is conditioned upon the observed forward curve.[b]

The simplest ("1st-order") hypothesis has a form of $ds/dt = 0$, therefore, s is simply an unknown constant. This assumption is equivalent to the statement that, for the average-rate scenario, static return spread remains unchanged from $t = 0$ to $t = \delta$. This is to say that the same average-rate scenario's cashflow will be priced at the same static discount spread in the nearest future.[c] It is easy to show that, under the 1st-order assumption, $IR_0 - IR_0^b = s$ and the relationship (3.1) yields the following approximate ordinary differential equation:

$$OAS \approx s(x) + \frac{\sigma^2}{2}\left[\frac{1}{P}\frac{\partial^2 P(0,x)}{\partial x^2} - C_x^{\text{static}}(x)\right] \equiv EIRS \tag{3.3}$$

where C_x^{static} is the static (benchmark) convexity of the average-rate scenario cashflow measured with respect to the same factor x. The OAS approximated according to formula (3.3) was called EIR spread and denoted EIRS. We therefore have transformed the instantaneous return PDE (3.1) to a nonlinear second-order ODE, in which $s(x)$ is a known function of price $P(0,x)$ uniquely defined by the average-rate scenario cashflow conditioned upon the initial value of the factor, $x(0)$. The time variable has disappeared from the model, therefore, the pricing grid need not be propagated along the t-axis.

Equation (3.3) can be viewed as an equation for $P(0,x)$ as well as equation for $s(x)$. To solve it, one needs to know either the base scenario price, $P(0,0)$, or EIRS, and specify two boundary conditions. This step depends on the specific security and an additional information about its behavior. For an MBS, for example, one can consider two extreme scenarios for which the cashflow is practically insensitive to x. The interest rate sensitivity of the prepayment speed is typically ranged between the minimum turnover rate and refinancing credit limitations; in addition, the ARM's coupon is bound by caps and floors. For these boundary scenarios, the convexity costs are assumed to be zero, i.e., $C_x = C_x^{\text{static}}$.

---

[b]Strictly speaking, discount rates do not necessarily have to be the forward rates. Nor does the benchmark have to be "the same-cashflow-Treasury". When pricing ARMs, it sometimes makes sense to use a "perfect floater" as a benchmark. In such a case, the discount rates in formula (6) will be equal to the forward rates till the next coupon reset, and the floating rates thereafter.

[c]As we explain later, in Section 8, the short-rate forward curve will change over time, but the static return spread is still equal to the (unchanged) static discount spread.

3.2. *The finite difference grid*

To solve pricing Eq. (3.3) we apply a finite difference method. Namely, along with the currently observed forward curve ("base case") we consider a grid of scenarios induced by initial shocks of the factor $x(0) = -N_{\mathrm{dn}}\Delta, \ldots, -\Delta, 0, \Delta, \ldots, N_{\mathrm{up}}\Delta$ for sufficiently small step Δ and sufficiently large number of "down" shocks N_{dn} and "up" shocks N_{up}. Then, we replace $\frac{\partial^2 P(0,x)}{\partial x^2}$ by its finite difference approximation, rewrite Eq. (3.3) for every scenario, employ boundary conditions, and solve thus obtained system of $N = N_{dn} + N_{up} + 1$ algebraic equations using the multi-dimensional Newton–Raphson iterations as shown in Appendix A. After the numerical process has converged, prices, effective durations and convexities, static spreads, convexity costs are found for the all "up" and "down" scenarios — along with EIRS (given current price) or current price (given EIRS). Thus, the proposed option-adjusted valuation method delivers a complete set of risk/reward measures that is typically required for portfolio risk management, hedging and reporting. We would like to emphasize the fact that the scenario grid is solely intended to compute components of the differential Eq. (3.3): convexity, static convexity, static discount spread. This equation assumes continuous diffusion of the factor, not the immediate shocks used to construct the pricing grid.

It is worth noting that, unlike the Monte Carlo method or a true finite-difference scheme, Eq. (3.3) yields absolutely accurate prices for option-free claims regardless the number of scenarios composing the pricing grid and the grid's step.

3.3. *The EIRS equation for the 2nd-order hypothesis*

The 2nd-order assumption such as $\frac{\partial^2 s}{\partial t^2} = 0$ leads to a more accurate, but more time-consuming analysis. First, formula (3.3) has to be modified in order to include an additional instantaneous return component, $D_s \frac{\partial s}{\partial t}$ where D_s denotes the "static spread duration", i.e., $D_s = -d(LnP)/ds$. Thus,

$$EIRS = s(x) + \frac{\sigma^2}{2}\left[\frac{1}{P}\frac{\partial^2 P(0,x)}{\partial x^2} - C_x^{\mathrm{static}}(x)\right] - D_s(x)\frac{\partial s}{\partial t}(x). \tag{3.4}$$

Let us equate the time derivatives of the both sides of (3.4) keeping in mind that $\frac{\partial^2 s}{\partial t^2} = 0$:

$$0 = \frac{\partial s}{\partial t}(x)\left[1 - \frac{\partial}{\partial t}D_s(x)\right] + \frac{\sigma^2}{2}\frac{\partial}{\partial t}\left[\frac{1}{P}\frac{\partial^2 P(0,x)}{\partial x^2} - C_x^{\mathrm{static}}(x)\right]. \tag{3.5}$$

Systems (3.4) and (3.5) contains two unknown functions, $s(x)$ and $\frac{\partial s}{\partial t}(x)$. Indeed, given a cashflow, price is uniquely determined by static spread $s(x)$, therefore $\frac{\partial P}{\partial t}$ can be expressed via $s(x)$ and $\frac{\partial s}{\partial t}(x)$. We use the same grid and the same numerical approach as in Appendix A involving now 2 differential equations and 2 unknown functions.

4. The Method's Implementation for the Hull–White Term Structure

Having introduced the EIRS method let us turn our attention to the interest rate process. Although the model's choice is certainly up to the user's taste, one ultimately has to construct the scenario grid, i.e., to find the base and the shocked paths for all points of the yield curve that are needed for MBS cashflow generation and discounting. For example, mortgage prepayments are typically driven by a long Treasury rate whereas ARM's coupon can be indexed to a relatively short or intermediate point of the curve.

The averaged paths and the relationships between different rates are determined by an x-factor stochastic model. In a general case, one cannot derive closed-form analytics, but for a linear model, we can. The Hull–White model for the short rate $r(t)$ is as follows:

$$dr(t) = a[\theta(t) - r(t)]dr + \sigma dz(t) \tag{4.1}$$

where $\theta(t)$ is the long-term equilibrium rate, $dz(t)$ is a standard Brownian motion, a is the mean reversion, σ is volatility. It can be shown[d] that model (4.1) is made arbitrage-free [i.e., consistent with the observed short-rate forward curve, $f(t)$] if $\theta(t) = f(t) + \frac{1}{a}\frac{df}{dt} + \frac{\sigma^2}{2a^2}(1 - e^{-2at})$.

4.1. *Mean rate paths*

The above given choice of the long-term equilibrium leads to the following mean path equations for the short spot rate $r(t)$, the long T-maturity spot rate $r_T(t)$, and the short forward rate $f(t, T)$ that is in effect for instance $t + T$ as seen at instance t (called sometimes "forward forward rate"):

$$E[r(t)] = f(t) + \frac{\sigma^2}{2a^2}(1 - e^{-at})^2 \tag{4.2}$$

$$E[r_T(t)] = f_T(t) + \frac{\sigma^2}{4a^2}B(T)[2(1 - e^{-at})^2 + (1 - e^{-aT})(1 - e^{-2at})] \tag{4.3}$$

$$E[f(t,T)] = f(t+T) + \frac{\sigma^2}{2a^2}e^{-aT}[(1 - e^{-at})^2 + aTB(T)(1 - e^{-2at})] \tag{4.4}$$

where $f_T(t)$ is the rate for a T-maturity zero, t years forward, and $B(T) = (1 - e^{-aT})/aT$.

One can notice that, in the presence of volatility, the mean rate paths are not centered around the currently observed forward curves. The spread between the mean rates and the forward rates is volatility-dependent and always positive as it is caused by the nonlinear, positively convex nature of discounting. That is why, curves (4.2)–(4.4) define what is often called "convexity-adjusted forward rates". For a relatively small mean reversion and a limited time horizon, these adjustments

---

[d]Most derivations for the 1- and 2-factor term structure models can be found in [6] and [9].

can be approximated by $\frac{1}{2}T\sigma^2 t$ for the spot rates, and by $T\sigma^2 t$ for the "forward forward rates".

4.2. *Volatility term structure and adjusted duration and convexity*

We can think of the yield curve diffusion in the 1-factor Hull–White model as a set of continuous, perfectly correlated deviations of each rate from its base (i.e., mean) path. At any time t, a deviation of the short rate $x(t) = r(t) - Er(t)$ from its base path causes proportional, $B(T)$-reduced deviations of all other spot rates: $x_T(t) = r_T(t) - Er_T = B(T)x(t)$. That is why, function $B(T) < 1$ uniquely determines the volatility term structure in the Hull–White model. Each point of the curve has a $B(T)$-reduced volatility, therefore, for a T-maturity zero-coupon bond, the traditional duration and convexity measured with respect to the bond's own yield have to be adjusted in order to get factor-specific measures: $D_x = TB(T)$, $C_x = D_x^2$.

4.3. *"Shocked" paths*

We have already established a term structure feature: each deviation x of the short rate (counted from its base path) immediately causes a $B(T)x$ deviation of the T-maturity spot rate (also counted from its base path). This implies an immediate change in the entire spot curve shape for any initial shock $x(0) = \Delta \neq 0$. After the shock occurred, the spot, the forward, and the mean curves are shifted and these shifts are implied by the term structure model and generally are non-parallel. It is easy to prove that the shocked path for the short

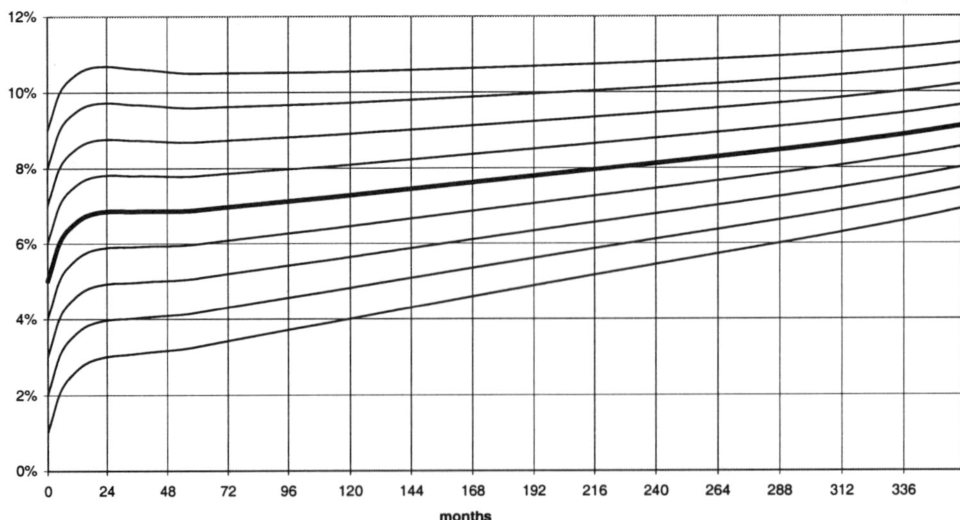

Fig. 1. The base path (bold) and the shocked paths of the short rate under 1-factor Hull–White term structure.

Security	Type	Price	Exact Solution[*]				EIRS Method				Same-OAS Pricing Errors		
			OAS bp	Duration	Convexity	Option Cost bp	OAS bp	Duration	Convexity	Option Cost bp	Base Case bp	MSE bp	"Time Ratio"[**]
G2AR 5.5	1yr CMT ARM	98-17	60.2	2.78	-0.67	22.4	57.8	2.63	-0.57	24.7	8.5	25.0	16 : 1
G2AR 6.0	1yr CMT ARM	99-23	60.8	2.36	-0.66	18.8	56.5	2.26	-0.54	23.1	15.8	23.1	16 : 1
G2AR 6.5	1yr CMT ARM	100-28	56.2	1.96	-0.63	15.7	51.0	1.94	-0.49	20.9	17.5	18.4	17 : 1
G2AR 7.0	1yr CMT ARM	101-26	53.6	1.59	-0.52	12.3	48.4	1.62	-0.40	17.4	15.6	15.4	17 : 1
TBA COFI	COFI ARM	98-24+	64.5	1.34	0.00	9.5	65.4	1.39	-0.01	8.6	4.4	7.9	14 : 1
FNCL 6.5	Fixed rate PT	94-6	40.8	5.59	-0.29	21.3	36.0	5.63	-0.23	26.1	28.7	17.1	24 : 1
FNCL 7.0	Fixed rate PT	96-26	46.2	5.06	-0.60	29.3	43.2	5.18	-0.47	32.3	17.3	15.0	28 : 1
FNCL 7.5	Fixed rate PT	99-4	52.1	4.45	-0.84	39.0	51.3	4.61	-0.76	39.8	4.8	11.7	28 : 1
FNCL 8.0	Fixed rate PT	101-6	56.0	3.80	-0.98	49.5	58.4	3.94	-1.07	47.1	11.5	10.6	32 : 1
FNCL 8.5	Fixed rate PT	102-31	57.1	3.19	-0.88	54.4	62.4	3.18	-1.25	49.1	4.3	9.5	32 : 1
8% A-Class	Fixed rate CMO	101	61.3	1.11	-0.78	22.1	58.2	1.13	-0.75	25.1	4.7	2.8	17 : 1
7% B-Class	Fixed rate CMO	93	74.9	6.55	-0.71	32.3	66.8	6.73	-0.42	39.1	55.5	34.4	29 : 1

[*] Based on a large number of Monte Carlo runs

[**] "Time Ratio" is the **same-accuracy-based** measure of the processing time required by antithetic Monte Carlo related to that of EIRS method

Fig. 2. EIRS Method vs. Exact Solution for most frequently traded agency pass-throughs (as of May 8, 1997, Volatility = 80 hp/yr, MeanRev = 2%).

rate will differ from its mean path by Δe^{-at}. Due to the term structure paradigm, the shocked path for the T-maturity rate will differ from its mean path by $\Delta e^{-at} B(T)$. For the short rate, Fig. 1 illustrates the base path along with 4 "up" and 4 "down" scenarios using a 2% mean reversion. We clearly see a non-parallel pattern of the shocks, the size of which diminishes over time with a $1/a = 50$-year time constant.

4.4. *Effectiveness and accuracy of the EIRS method*

For 10 actively traded agency pass-throughs and 2 unstable CMO tranches, Fig. 2 demonstrates that even the 1st-order hypothesis EIRS method provides a reasonable accuracy in computing grid prices, OAS (within 5 basis points), duration, convexity, and option cost. The processing time is significantly smaller than that of the antithetic Monte Carlo algorithm if the latter is set up to reach just the same accuracy when pricing the same set of 9 term structures.

Considering a broad range of MBS (fixed rates vs. ARMs, premium-priced vs. discount-priced, pass-throughs vs. CMOs) we observe that the EIRS method tends to understate OAS slightly, on average. As an MBS rolls to maturity, the convexity cost ultimately vanishes forcing the market to reduce the static discount spread — according to the underlying Eq. (3.3). Therefore, the MBS will get an additional price return component ignored by the 1st-order hypothesis. For some MBS, however, convexity cost may temporarily rise, along the mean path. Thus, the refinancing rate for high-premium fixed-rate MBS (FNCL 8.5 in the example) is likely to be saturated by credit and other limitations. As we follow the convexity-adjusted forward curve (sloped positively), "refinancing spread" is tightening thereby inducing prepayment variability and increasing convexity cost. The 1st-order hypothesis EIRS method will not quantify this future loss as the same static discount spread is applied over time, and, for such an MBS, the OAS is overstated.

5. The Multi-Factor Case: A Theoretical Excerpt

If n factors disturb the market conditions and security's cashflow, $x(t)$ should be interpreted as an n-dimensional Brownian motion (hereinafter, vectors and matrices are bolded). Duration $(\boldsymbol{D_x})$ becomes an n-dimensional vector $(D_i = -\frac{1}{P}\frac{\partial P}{\partial x_i})$ whereas convexity $(\boldsymbol{C_x})$ becomes a symmetric matrix with $C_{ij} = \frac{1}{P}\frac{\partial^2 P}{\partial x_i \partial x_j}$. The "base case" refers to a scenario in which all factors remain unchanged (i.e., $\boldsymbol{x} =$ Const). We assume that vector $\boldsymbol{x}$ of Brownian motions has a drift vector $\boldsymbol{m} = [\mu_1, \ldots, \mu_n]^T$ and a volatility (covariance) matrix $\boldsymbol{V} : V_{ij} = \rho_{ij}\sigma_i\sigma_j$; ρ_{ij} denotes correlation between dx_i and $dx_j (\rho_{ii} = 1)$, σ_i is the ith factor's volatility. We can reformulate Proposition 1.1 for the multi-factor case.

Proposition 5.1 (multi-factor case). *The Expected Instantaneous Return (EIR) and the Variance of Instantaneous Return (VIR) equal:*

$$\text{EIR} = IR_0 - \boldsymbol{D}_{\boldsymbol{x}}^T \boldsymbol{m} + \frac{1}{2} \operatorname{tr}(\boldsymbol{V} \boldsymbol{C}_{\boldsymbol{x}}) = IR_0 - \sum_{i=1}^{n} D_i \mu_i + \frac{1}{2} \sum_{ij=1}^{n} C_{ij} \rho_{ij} \sigma_i \sigma_j \qquad (5.1)$$

$$\text{VIR} = \boldsymbol{D}_{\boldsymbol{x}}^T \boldsymbol{V} D_{\boldsymbol{x}} = \sum_{i,j=1}^{n} D_i D_j \rho_{ij} \sigma_i \sigma_j \qquad (5.2)$$

where the time return IR_0 *is still determined by formula* (2.4).

Formula (5.1) involves the drift vector m. To exclude one, we assume again that the factors are counted off their mean paths. Subtracting both sides of formula (5.1) written for the benchmark from the ones written for the asset in question we derive the arbitrage-free OAS measure:

$$OAS = \text{EIR} - \text{EIR}^b = IR_0 - IR_0^b + \operatorname{tr}[\boldsymbol{V}(\boldsymbol{C}_{\boldsymbol{x}} - \boldsymbol{C}_{\boldsymbol{x}}^b)]$$

$$\equiv IR_0 - IR_0^b + \frac{1}{2} \sum_{i,j=1}^{n} (C_{ij} - C_{ij}^b) \rho_{ij} \sigma_i \sigma_j \,. \qquad (5.3)$$

Formula (5.3) clearly states that OAS consists of the static return spread and generally $n(n+1)/2$ "convexity" costs including the mixed term costs. Selecting independent factors ($\rho_{ij} = 0, i \neq j$) would certainly be beneficial from the numerical point of view; only n convexities, not the mixed terms, need to be estimated in this case.

We would like to address the problem of creating a benchmark portfolio to determine the mean path for each factor. To clarify this issue, we first start with the interest-only factors (say, Treasury rates). Since all market participants are well versed on their values, the mean paths for these rates are immediately derived from an appropriate arbitrage-free rate model. Let us consider another factor, a prepayment uncertainty, that is a noise in the prepayment model that would otherwise perfectly explain prepayment rates via interest rates changes. Any Treasury portfolio has no exposure to this factor and therefore should be complemented by a "benchmark" MBS. Then, a security in question can be formally priced relative to such a defined benchmark, but the latter does not fit our standard understanding of the option-adjusted valuation. Indeed, a "benchmark" MBS usually needs to be priced itself, and there may exists no price that perfectly reflects the prepayment uncertainty of mortgages.

If a benchmark does not exist or is not suitable for us, we can simplify the problem by making some assumption about the drifts of "inconvenient" factors. Thus, a zero drift is not a frivolous choice for the prepayment factor if the prepayment model is built as an unconstrained mean-least-squared regression. The noise of such a model (i.e., the prepayment error) is centered and uncorrelated with the other ("explanatory") factors. This is a very convenient decomposition of OAS into "off-Treasury" (traded) OAS and "hidden" convexity costs of "non-Treasury" (non-traded) factors. We will list some non-traded factors further in Section 7 and discuss their impact on the total valuation process.

6. Introducing a Second Factor into the Gaussian Interest Rate Model

Although a mean-reverting one-factor model implies a certain change in the yield curve shape (Fig. 1), this change is perfectly correlated with the short rate, the only factor of the model. The other limitation of the one-factor model (3.5) is that its volatility term structure always declines from the short end to the long end. As observed historically, the curve can twist independently from its vertical move and a long volatility can exceed the short volatility.

6.1. *A two-factor Gaussian term structure*

A two-factor Gaussian term structure model closest to our specification was first proposed by Hull and White [3]. Levin [6,9] developed a universal procedure for constructing any 2-factor Gaussian model (with real eigen-values) as the following four-step mathematical derivation:

- Present the short rate process in the form of $r(t) = x_1(t) + x_2(t)$ where x_1 and x_2 are two Gaussian variables satisfying first-order stochastic differential equations similar to (3.4):

$$dx_1(t) = a_1[\theta(t) - x_1(t)]dt + \sigma_1 dz_1(t)$$
$$dx_2(t) = -a_2 x_2(t)dt + \sigma_2 dx_2(t) \tag{6.1}$$

  where two Brownian motion increments $dz_1(t)$ and $dz_2(t)$ have a correlation of ρ;
- Find $\theta(t)$ such to insure arbitrage-free conditions, i.e., correct pricing of all default-free and option-free claims (Treasury bills, notes, and bonds, or LIBOR contracts). Find the average and the shocked paths for the short rate.
- At any point of time t and any maturity T, establish the linear relationships between long spot rates $r_T(t)$ and the factors $x_1(t)$ and $x_2(t)$. Find the average and the shocked paths for $r_T(t)$.
- Fit the 5 model's parameters (2 mean reversions, 2 volatilities, and the correlation) in order to approximate the observed volatility term structure by that implied by the model.

Factors x_1 and x_2 have no financial meaning. In fact, the transient solution of any second-order system with real eigen-values can be presented as a linear combination of exponents, i.e., the transient solutions of model (6.1). Therefore, $r(t) = x_1(t) + x_2(t)$ is a generic mathematical statement that holds true for any 2-factor Gaussian term structure with real eigen-values, $-a_1$ and $-a_2$, regardless the form it is written in. In addition, volatilities for long-rates can exceed the short-rate volatility only if $\rho < 0$. Then, the model can be rewritten in a unique equivalent "(r, v)-form" introducing a new "slope" variable $v = x_1 + \beta x_2$ with $\beta = -\sigma_1(\sigma_1 + \rho\sigma_2)/\sigma_2(\rho\sigma_1 + \sigma_2) \neq 1$. In the transformed model, factors r and

v are independent, for an infinitesimal time horizon. Factor v determines a term structure move net of that propagated by the short rate r. Thus, an immediate change in the initial value $v(0)$ causes a change in the entire term structure, except for its short end driven by $r(0)$.

All needed formulae for the (r,v)-model are listed in Appendix B. A method of finding the 5 model parameters, $\sigma's, a's$ and ρ that would fit the observed volatility term structure is introduced in [8,9]. We suggest selecting 2 long-rate maturity points and equating their volatilities and correlations with the short rate to those values implied by the model. Many different volatility term structures can be fitted including "humped" and rising, but not all. An alternative to fitting $\sigma's, a's$, and ρ would be using the best approximations of historical day-to-day yield curve diffusion. Parameters are selected to minimize a norm of errors (measured across maturities) between the actual daily increments and those simulated by the 2-factor term structure.

6.2. *EIRS method for the 2-factor model*

The EIRS Eq. (3.3) is generalized for the 2-factor (r,v)-model as:

$$EIRS = s(r,v) + \frac{\sigma_r^2}{2}\left[\frac{1}{P}\frac{\sigma^2 P(0,r,v)}{\partial r^2} - C_r^{\text{static}}(r,v)\right]$$

$$+ \frac{\sigma_v^2}{2}\left[\frac{1}{P}\frac{\partial^2 P(0,r,v)}{\partial v^2} - C_v^{\text{static}}(r,v)\right] \tag{6.2}$$

where the last 2 terms are convexity costs associated with the r-factor and the v-factor, correspondingly. In the derivation of (6.2), we have employed again the 1st-order assumption and replaced the static return spread $IR_0 - IR_0^b$ by the static discount spread, s. This approximation automatically eliminates the time variable and converts (6.2) into a 2-factor PDE.

Let us consider constructing the two-dimensional (r,v)-grid of scenarios required by the EIRS method. We start building the grid from its base path using Eqs. (B.2), (B.4) listed in Appendix B. Immediate changes ("shocks") in initial conditions, Δr_0 or Δv_0, are then considered as the "shocked" trajectories for our factors and for the long rates will be counted off their mean paths, see formulae (B.9). One can construct some 11–15 "r-scenarios" (Δr_0's are equally spaced, $\Delta v_0 = 0$) and, independently, 11–15 "v-scenarios" (Δv_0's are equally spaced, $\Delta r_0 = 0$) keeping the total number of paths (and therefore unknown prices) within some reasonable range (up to 150–200). For example, possible basic "$r-$" and "$v-$" scenarios are plotted in Fig. 3; all others are proportional. Each scenario sets up a unique price-static discount spread relationship. The rest of our technique generally repeats that of the one-factor case.

If our goal was to find only OAS, the time required to solve Eq. (6.2) would approach that of the 2-factor Monte Carlo method as doubling the number of factors will only slightly delay the Monte Carlo runs. The relative effectiveness of the EIRS

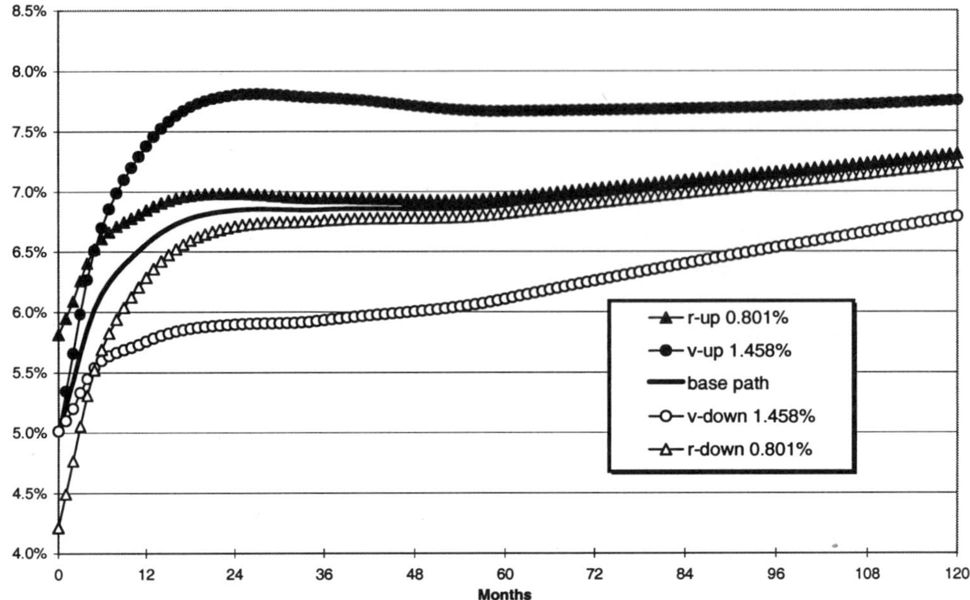

Fig. 3. The short rate mean paths in the 2-factor term structure.

method increases as the risk management task involves deriving price sensitivity measures for different "up", "down", "slope-up", and "slope-down" scenarios. The EIRS method generates these measures concurrently whereas Monte Carlo runs have to be repeated every time the initial yield curve is shocked. Thus, if all the grid points we use to solve Eq. (6.2) are indeed "points of pricing interest", the relative performance of the Monte Carlo approach will be even worse than that for the one-factor model.

6.3. *MBS valuation under the 2-factor model*

The full-scale analysis "2-factor model vs. 1-factor model" could be quite a vast research, results of which essentially depend on 5 parameters used for the 2-factor term structure. Accordingly, we restrict our presentation to the major (sometimes, trivial) practical observations assuming that the short rate volatility is given; the major conclusions follow below:

- Given correlations with the short rate, increments in the intermediate- and long-rate volatilities magnify the total price volatility (VAR). They also cause higher convexity costs and typically reduce OAS.
- Given volatilities of all the rates, a reduction of correlations (i.e., "activating" the v-factor) gradually reduces the prepayment option costs, as well as sensitivity to the short rate factor (D_r and C_r) and increases OAS, until certain point. Then, the reverse becomes true.

Security	One-factor model		Two-factor model						Two-factor hedge	
	Duration	VAR	OAS added	r-Duration	r-VAR	v-Duration	v-VAR	Total VAR	2-yr TSY	10-yr TSY
G2AR 5.5	2.63	2.10%	1.4	0.83	0.66%	1.36	1.97%	2.08%	0.79	0.17
G2AR 6.0	2.26	1.81%	0.2	0.79	0.63%	1.16	1.70%	1.81%	0.83	0.11
G2AR 6.5	1.94	1.55%	-0.4	0.74	0.59%	0.97	1.41%	1.53%	0.83	0.05
G2AR 7.0	1.62	1.30%	-0.7	0.70	0.56%	0.79	1.19%	1.32%	0.88	0.00
TBA COFI	1.39	1.11%	-3.6	0.40	0.32%	0.82	1.21%	1.25%	0.28	0.15
FNCL 6.5	5.63	4.50%	5.5	1.12	0.89%	2.86	4.12%	4.22%	0.46	0.66
FNCL 7.0	5.18	4.14%	5.4	1.08	0.86%	2.66	3.78%	3.88%	0.50	0.60
FNCL 7.5	4.61	3.69%	5.2	1.03	0.83%	2.34	3.36%	3.46%	0.61	0.48
FNCL 8.0	3.94	3.15%	4.1	0.96	0.77%	1.96	2.83%	2.93%	0.69	0.36
FNCL 8.5	3.18	2.54%	2.5	0.90	0.72%	1.53	2.25%	2.36%	0.82	0.21
8% A-Class	1.13	0.90%	-3.6	0.68	0.54%	0.48	0.68%	0.87%	1.00	-0.12
7% B-Class	6.73	5.38%	10.9	1.26	1.00%	3.17	5.00%	5.10%	0.54	0.73

Assumed Volatility Term Structure		
Maturity	Volatility	Correlation
1-mo	16.0%	100.0%
2-yr	12.4%	43.4%
10-yr	10.8%	16.5%

Implied Parameters					
$B_1(T)$	$B_2(T)$	$B_r(T)$	$B_v(T)$	$\sigma_1=1.272\%$	$\sigma_r=0.801\%$
1.000	1.000	1.000	0.000	$\sigma_2=1.446\%$	$\sigma_v=1.458\%$
0.909	0.352	0.425	0.485	$a_1=0.097$	$\rho=-83.4\%$
0.641	0.076	0.150	0.491	$a_2=1.319$	$\beta=-0.15$

Fig. 4. 1-factor model vs. same-volatility 2-factor model: Risk/Reward comparison.

- For the same volatility of all the rates, the total interest-rate-related VAR for fixed-rate MBS will be lower under the 2-factor model. This is not necessarily true for ARMs.

Results presented in Fig. 4 are obtained for a 2-factor term structure that has the same short (1-mo), intermediate (2-yr), and long (10-yr) rate volatilities as those implied by the 1-factor Hull–White model. The intermediate rate and the long rate were assumed to have 43.4% and 16.5% correlations with the short rate, correspondingly. Although the total VAR may look comparable for both models, the 2-factor term structure substantially "re-distributes" risk between the factors reducing the r-component and introducing the v-component. The perfect 2-factor Δ-hedges for each MBS are non-trivial and also listed in Fig. 4. Note that a trivial 1-factor Δ-hedge may even induce an additional risk when used in the 2-factor economy.

7. Non-Traded Factors: A Systematic Methodology and Practical Results

If a factor is not traded in the sense introduced in the Section 5 above, computations of the corresponding convexity cost and the contribution to VIR are rather simple. "Traded OAS" is first found for the model that involves traded factors only. The non-traded factor is then shocked "up" and "down" and new prices are computed. Since the factor is non-traded, the market would not change "traded OAS". Using thus calculated 3 prices we estimate factor-Duration and factor-Convexity. The product of factor-Duration and factor-Volatility geometrically contributes to VIR whereas the product of a half of factor-Convexity and factor-Volatility squared is Convexity Cost (a positive number reduces the total OAS). Generally, no "static convexity" exists since no benchmark sensitive to the factor is considered. The only exception to this rule is a benchmark-related factor (therefore, a traded factor) that, for the sake of simplicity, is treated as a non-traded one. This means that, while computing "traded OAS", we ignore this factor, but while computing its Convexity Cost, we should subtract the corresponding static convexity (the interest rate volatility is an example considered below). In this section, we list the major non-traded factors, and, for each one, a brief commentary is presented on how the factor effects risk and reward of holding mortgages. For a more comprehensive coverage of this subject we refer the reader to paper [7]. The results are summarized and combined in Fig. 5.

7.1. *Prepayment model error*

There exists no ideal prepayment model. The deviations of actual prepayment speeds from those computed by a model can be treated as a factor.[e] The prepayment

---

[e]A. Sparks and F. Feiken Sung [10] conclude that a positive (negative) "prepayment convexity" will increase (decrease) OAS if this factor is included in the OAS model, but, they do not provide any analytical tools to quantify this OAS addition.

A. Prepayment speed multiplier

Security	Base path Avg PSA	Volatility[*] % of speed / yr	Duration yr / % of speed	VAR[**] %	Convexity (yr / % of speed)^2	Convexity Cost bp
G2AR 5.5	361	20%	0.016	0.31%	0.029	-5.9
G2AR 6.0	315	20%	0.015	0.30%	0.029	-5.7
G2AR 6.5	348	24%	0.018	0.44%	0.033	-9.6
G2AR 7.0	391	26%	0.021	0.55%	0.037	-12.6
TBA COFI	250	20%	-0.002	-0.04%	-0.003	0.5
FNCL 6.5	106	20%	-0.012	-0.43%	-0.005	1.0
FNCL 7.0	120	20%	-0.012	-0.25%	0.002	-0.4
FNCL 7.5	142	20%	-0.003	-0.07%	0.011	-2.1
FNCL 8.0	182	20%	0.006	0.13%	0.021	-4.2
FNCL 8.5	280	20%	0.017	0.34%	0.035	-7.1
A-Class	216	20%	0.002	0.05%	0.002	-0.5
B-Class	118	20%	-0.035	-0.07%	-0.006	1.2

B. Index model error

Security	Volatility bp / yr	Duration yr / bp	VAR %	Convexity (yr / bp)^2	Convexity Cost bp
TBA COFI	15	-0.034	-0.51%	-0.020	2.2

C. Interest rate volatility

Security	Volatility % of Vol / yr	Duration yr / % of Vol	VAR %	Convexity (yr / % of Vol)^2	Convexity Cost bp
G2AR 5.5	15	0.014	0.22%	-0.110	1.3
G2AR 6.0	15	0.012	0.18%	-0.008	0.9
G2AR 6.5	15	0.009	0.13%	-0.005	0.6
G2AR 7.0	15	0.006	0.09%	-0.004	0.4
TBA COFI	15	0.002	0.03%	0.000	0.0
FNCL 6.5	15	0.033	0.50%	-0.010	1.2
FNCL 7.0	15	0.034	0.51%	-0.008	0.9
FNCL 7.5	15	0.034	0.51%	-0.003	0.4
FNCL 8.0	15	0.031	0.46%	-0.003	0.3
FNCL 8.5	15	0.025	0.38%	-0.005	0.5
A-Class	15	0.006	0.09%	-0.002	0.2
B-Class	15	0.059	0.88%	-0.012	1.4

D. Discount spread

Security	Volatility bp / yr	Duration yr	VAR %	Convexity / 100 (yr)^2	Convexity Cost bp
G2AR 5.5	15	3.89	0.58%	0.25	-0.3
G2AR 6.0	15	3.62	0.54%	0.24	-0.3
G2AR 6.5	15	3.37	0.50%	0.21	-0.2
G2AR 7.0	15	3.13	0.47%	0.19	-0.2
TBA COFI	15	4.27	0.64%	0.32	-0.4
FNCL 6.5	15	6.03	0.90%	0.63	0.7
FNCL 7.0	15	5.66	0.85%	0.54	-0.6
FNCL 7.5	15	5.21	0.78%	0.48	-0.5
FNCL 8.0	15	4.67	0.70%	0.39	-0.4
FNCL 8.5	15	4.08	0.61%	0.35	-0.4
A-Class	15	1.49	0.22%	0.03	0.0
B-Class	15	7.18	1.08%	0.67	-0.8

E. Cumulative effect of non-traded factors

Security	Traded VAR %	Non-traded addition %	Total VAR %	VAR increase %	Convexity Cost bp
G2AR 5.5	2.10%	0.68%	2.21%	5.1%	-4.1
G2AR 6.0	1.81%	0.65%	1.92%	6.3%	-5.1
G2AR 6.5	1.55%	0.68%	1.69%	9.2%	-9.2
G2AR 7.0	1.30%	0.73%	1.49%	14.8%	-12.4
TBA COFI	1.11%	0.82%	1.38%	24.2%	2.4
FNCL 6.5	4.50%	1.12%	4.64%	3.0%	1.4
FNCL 7.0	4.14%	1.02%	4.27%	3.0%	-0.1
FNCL 7.5	3.69%	0.94%	3.81%	3.2%	-2.3
FNCL 8.0	3.15%	0.85%	3.26%	3.6%	-4.3
FNCL 8.5	2.54%	0.79%	2.66%	4.7%	-6.9
A-Class	0.91%	0.25%	0.94%	3.3%	
B-Class	5.39%	1.56%	5.61%	4.1%	

[*] Volatility assumptions are illustrative and averaged for specific models used

[**] All VARs are at 84% confidence

Fig. 5. Multi-factor analysis.

errors are strongly serially negatively correlated, and one should find a "filter" (a system of differential equations) that is disturbed by a Wiener "white" process (i.e., a factor) and generates the observable colorful noise.[f] We assume that the prepayment error factor can be isolated as a "speed multiplier" satisfying the strongly mean reverting diffusion model but uncorrelated with other factors. The reason why we suggest using the speed multiplier instead of, say, an additive speed error, is that prepayment uncertainties usually grow when prepayment speed grows itself [7]. A prepayment model can be built much more accurately for discount MBS' than for premium ones (the same is generally true about brokers' prepayment forecasts). Using the speed multiplier we smooth out the difference in prepayment error volatilities.

It is a simple but notable finding that both prepayment duration and prepayment convexity grow as one moves from discount mortgages to premium ones. In a very simplified form, this conclusion can be explained by the generalized Gordon's stock pricing formula:

$$P = (c + \lambda)/(r + \lambda)$$

in which parameters c, r, and λ have the same meanings as above. A simple exercise in calculus proves that both sensitivity measures are proportional to $(c - r)$. Since contributions to VIR from independent factors are always positive, both discount and premium assets are riskier than par-priced ones. However, convexity cost has sign and a premium (discount) fixed-rate asset will have a positive (negative) prepayment convexity. For actual mortgages, this relationship is more complicated since prepayment speeds affect the cost of prepayment option itself as well as the interest income and the values of life and periodic caps and floors, for ARMs.

7.2. *ARM index model error*

If an ARM is indexed to a non-Treasury and non-LIBOR index (COFI, NMCR, Prime, etc.) that has no established term structure, it needs to be modeled. An error arising in such a model can be treated as an additional factor. For the sake of demonstration, we will assume that a 1st-order autoregressive model provides suitable accuracy, and the residual is directly proportional to a standard Wiener process I^{res}:

$$I = I^{\text{reg}} + \sigma_I I^{\text{res}} \tag{7.1}$$

$$dI^{\text{reg}} = a_I(\alpha + \beta r_T - I^{\text{reg}})dt \tag{7.2}$$

where I is an ARM index's value, r_T is a T-maturity market rate, a_I is the mean reversion, σ_I is the error's volatility, α and β are the "long-term" regression coefficients. If the best model is sought as the least-mean-squared model, then the

---

[f]The filter is formally constructed using factorization of the spectral density function $F(\omega) = W(i\omega)W(-i\omega)$ of the noise. Then, $W(s)$ is the transfer function of the filter, where s is the Laplace operator provided that all poles of $W(s)$ belong to the left half-plane.

residual is uncorrelated with the explanatory variable (r_T), centered, and its volatility can be naturally estimated. Following the standard technique described in the beginning of this section, we can compute factor-Duration (therefore, the appropriate VIR contribution) and factor-Convexity (therefore, cost of convexity). Due to the nature of the prepayment option, Convexity Cost with respect to the ARM index model error is always positive. Indeed, should the index be above the modeled values, potential interest income benefits for investors will be limited by a rising prepayment activity. If the index is below the model's projections, the loss of interest will be extended. In the above example (see Fig. 5), the investor loses 2.2 basis points due to this "hidden", non-traded factor. At the same time, he takes an additional risk (0.51% of VAR, compare to 1.08% of the "traded VAR").

7.3. *Interest rate volatility*

Practically all parameters in interest rate, prepayment rate, and ARM index models considered in this paper are known with certain accuracy and could be treated as factors themselves. From this wide universe of potential factors we should distinguish those having the major impact on the OAV. Interest rate volatility is an obvious candidate as "volatility of volatility" is a known market phenomenon. This is to say that process (2.1) has to be heteroskedastic with σ being described by some stochastic differential equation which would be comprised of an "explanatory" part complemented by a residual term:

$$d\sigma = \mu_\sigma dt + \sigma_\sigma dz_\sigma . \tag{7.3}$$

We assume that μ_σ and σ_σ may depend on time and all other factors; however, the standard Wiener process $z_\sigma(t)$ is assumed to be uncorrelated with those factors.

Formally, interest rate volatility is a traded factor since there exists an established market for interest rate options. As we could see above, the introduction of an additional traded factor makes the finite difference grid much more complicated. This observation, and our aversion to having options in the benchmark portfolio, necessitate the use of a simpler approach. In fact, we propose to treat the interest rate volatility as a non-traded factor and employ our general method to find the required sensitivity measures. As one can see from Fig. 5, "volatility of volatility" results in some positive convexity cost not exceeding 1 basis points. Contributions to VAR can reach as much as 15% of the traded VAR.

7.4. *"Spread" factor*

Loans and mortgage-backed securities are traded at an option-adjusted spread above the Treasury market or LIBOR market. This spread can vary over time thus becoming a pricing factor, its volatility is easily observed. It might seem that the spread factor is one of the easiest to analyze because spread-Duration and spread-Convexity are, in essence, the well-known (positive) static measures. However, this straightforward approach can serve as an approximation only. Indeed, any change in the static discount spread is not evenly propagated across the interest rate "up" and "down" scenarios (the reason is the difference in the scenarios' cashflows and therefore their price sensitivities to the discount spread). In order to find the correct price-to-OAS sensitivity, one needs to re-run the entire traded OAS model. In addition, the OAS is a highly mean-reverting variable, and a stable filter has to be constructed first similarly to that discussed for the prepayment model error.

For most mortgages, spread-related cost of convexity is found within 1 basis point range. Contributions to VAR can be more significant, comparable to the traded VAR, for COFI ARMs.

7.5. *Summary*

Undoubtedly, prepayment error volatility represents the major source of non-traded ("hidden") convexity cost for all MBS' considered, except for the COFI ARM. Our analysis shows that an additional expected return grows as the price rises. The only exception is the COFI ARM having a positive convexity cost related to the COFI modeling error. Thus, an investor should expect a higher traded OAS for COFI ARMs just to compensate for this difference.

Total VAR contributions by non-traded factors are within 0.62% to 1.12% range, but have different dominant components, for different securities. The total resultant VAR is certainly larger than just traded VAR. This "risk increment" is 3%–4% for fixed-rate MBS', 5%–15% for CMT ARMs, and 26% for the COFI ARM. Thus, the overall risk of holding fixed-rate MBS' seems to be almost the same as the interest rate risk whereas ARMs suffer from more substantial additional exposure. However, it is important to keep in mind that we have been considering *unhedged* mortgages. Should these instruments be Delta-hedged, the traded VAR will be eliminated, and the remaining risk is entirely driven by the non-traded factors.

8. EIRS Method Under Additional Early Termination Provisions

In addition to the prepayment option, many MBS are subject to early calls according to some schedule or to a cleanup condition. Even if an asset does not have any legal early termination provision, it may have to be modeled that way. For example, unlike residential loans, high-balance commercial mortgages are usually managed by financial experts. The prepayment option is exercised very efficiently forcing us to view such assets as amortizing (possibly, adjustable-rate) loans with an embedded American call schedule.[g]

Valuation of such securities becomes a real challenge, as neither Monte Carlo approach (cannot price American options), nor finite difference schemes or trees (do not handle path-dependency) may be used. We show below that at the cost of accuracy and processing time, the EIRS method can be adapted to pricing MBS with additional American options. Although this can be viewed as a significant achievement itself, the adjustments we have to make in the EIRS algorithm are

---

[g]The author thanks Kenneth Schmidt for pointing this out.

rather simple and straightforward. First, we assume that the prepayment activity of mortgagors (if any) is totally independent from the additional embedded options. This natural assumption enables the cashflow generation along all scenarios of the grid under the assumption that no early termination is possible.

From the option pricing theory, we know that the optimal exercise of an American option maximizes its value [5]. This theoretical fact justifies the following practical valuation rule: when working backward along a tree or a grid, the obtained prices have to be capped by call prices and floored by put prices. In application to the EIRS scheme it means that the method has to be complemented by a backward induction procedure that would correct the cashflow under appropriate economic conditions.

To implement this idea, we consider the same grid of scenarios as was used in the regular EIRS method, but, in addition, introduce a time grid (i.e., a discrete time sequence $0, \delta, 2\delta, \ldots, M\delta$) that covers, at least, the interval of possible early redemption. For each scenario, we first use the MBS cashflow that ignores early termination. Starting from the maturity date or balloon date we derive prices at all previous time instances that belong to the grid. An important technical detail one has to take into account is changing the entire forward curve when moving from one pricing node to another. Indeed, along a chosen scenario, each rate follows its mean curve (convexity-adjusted forward curve), not the original forward curve. For example, under the Hull–White term structure, formula (4.4) for the "forward forward rates" has to be applied when assessing price at time t and discounting the value from $t + T + \delta$ back to $t + T$. Because the EIRS scheme is iterative, it makes sense to pre-compute the convexity adjustment term for all t and T and store it in a 2-dimensional array. As in the regular EIRS, the same static spread s (different for different scenarios) is added to all forward discount rates. If, for any scenario, the assessed price for time t appears to exceed the applicable call price (of falls below the put price), the prices are capped (or floored) and the cashflow "tail" is cut. The early-terminated cashflow is then used to compute price at $t - \delta$, etc.

When a previous approximation for static spreads $s_{N_{dn}}, \ldots, s_0, \ldots, s_{N_{up}}$ is known, we do not rush with computing the Newton–Raphson's corrections as in the regular EIRS method. Rather we test if the cashflows can be early terminated and modify them if so. For thus modified cashflows, we compute all needed static measures: static factor convexity C_x^{static} (required by the EIRS equation), and spread static duration D_s (used by the Newton–Raphson algorithm, Appendix A). Only then, the Newton–Raphson correction is applied again to find the next static spreads approximation.

The processing time for this algorithm is substantially longer than that for the regular EIRS scheme because the cashflow can vary from iteration to iteration, and future prices have to be computed. On the other hand, the constancy of the static spread over time may appear to be a too rough assumption for estimating the future prices. However, the problem we address here cannot be solved by the traditional methods at all, and it is almost impossible to find a pricing benchmark, as we were able to do for the regular EIRS version. For non-amortizing callable corporate bonds, our numerical experiments reveal an OAS error ranging from moderate 5–7 basis points for out-of-the-money case to 10–12 basis points for in-the-money case.[h] The errors for amortizing MBS are, on average, certainly smaller.

9. Possible Accuracy Enhancements of the EIRS Computational Scheme

The EIRS scheme is ideally suited for asset-liability, portfolio, and risk management computational tasks and has a considerable speed advantage over the more traditional Monte Carlo method. It can be used for pricing many path-dependent assets, assets with additional (American) termination provisions, assets for which dynamic cashflow generation mechanism is not even known. But, the EIRS method is an approximation and its accuracy cannot be substantially enhanced by simply constructing a finer scenario grid. In fact, the method draws the main error from the PDE transformation into ODE, not from the discretization.

Accounting for quickness of the EIRS scheme one can consider the use of a *control variate* technique, i.e., to employ the method twice, for the asset in question and for an auxiliary asset ("control variate"), for which the exact or very accurate price can be computed. For some mortgage products the closed-form solution can be derived as shown in [9]. In order to implement this approach, the principal amortization rate has to be a Gaussian variable. Since most prepayment models are essentially nonlinear, we suggest using a more traditional finite-difference scheme rather than the closed-form solution in order to price the control variate. The latter, therefore, should bear no path-dependency. An obvious advantage of such an approach is that both the EIRS method and a traditional finite-difference method solve the same problem of finding prices for the same set of initial term structures.

Thus, when pricing mortgage instruments with a burnout, it is reasonable to create a control variate without a burnout. In order to have the control variate to be as close to the original MBS as possible, one can first model the burnout factors for the base scenario and then apply them for each scenario of the pricing grid. Such a transformation will eliminate the existing path-dependency, but replicate prepayment speeds for the base scenario. Using the EIRS scheme for the control variate, and comparing it to the solution achieved via a true finite difference method, we find the price corrections for all the scenarios concurrently. As seen from Fig. 6, there is a considerable accuracy improvement for the agency MBS. The base case error is remarkably corrected whereas the grid pricing errors are generally smoothed out among scenarios.

---

[h]A standard backward inducting Crank–Nicholson finite difference scheme was used for comparison.

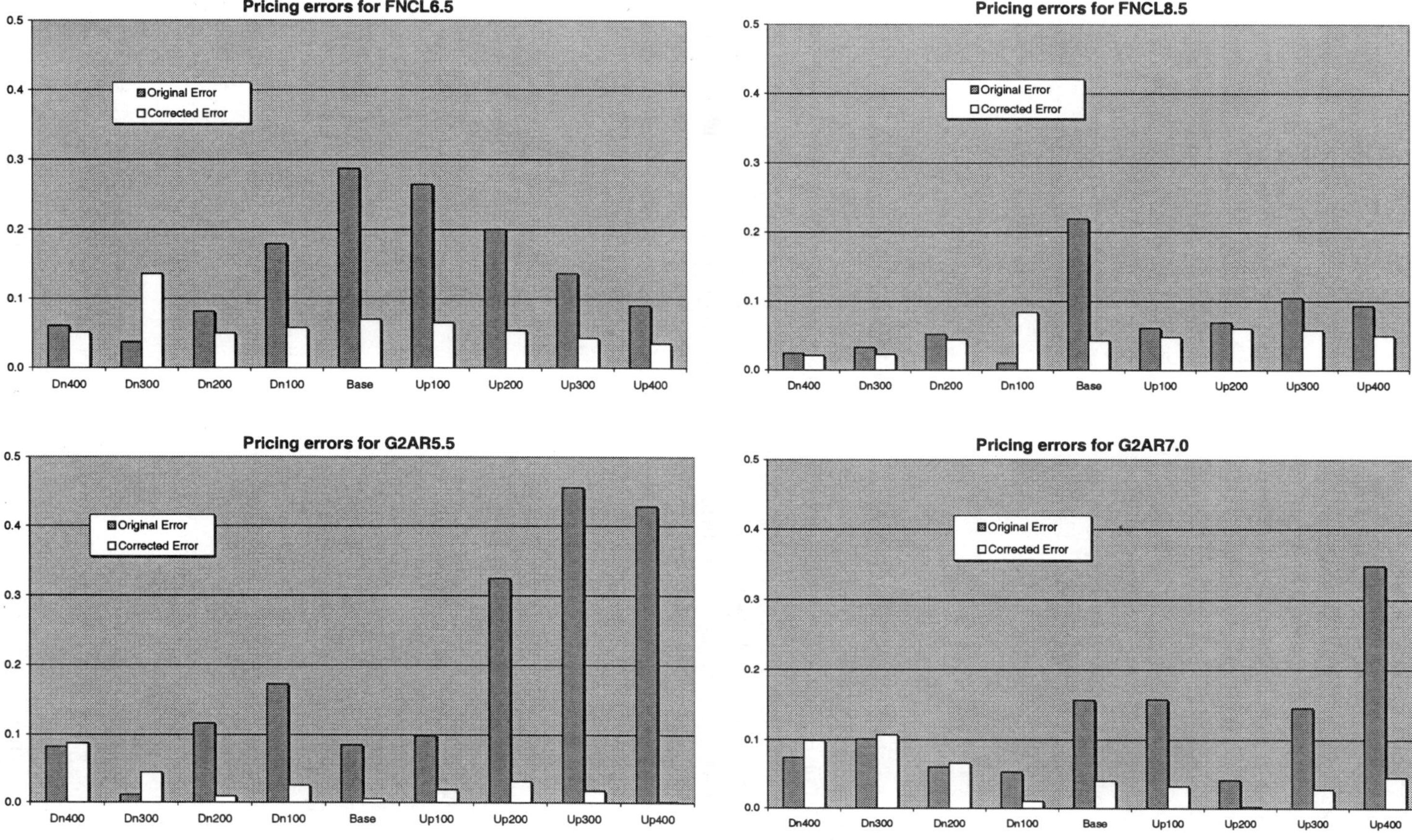

Fig. 6. EIRS pricing errors before and after control variate correction.

10. Conclusions

The EIRS method we have presented in the paper combines mathematical rigor (starting from the partial differential equation of instantaneous return) with an "ad-hoc" practical assumption regarding the behavior of static discount spread along relatively few deterministic scenarios. This approximation removes the time grid from the numerical procedure and explains the time efficiency of the method. OAS, duration, convexity and option cost are found concurrently with pricing an asset on the grid of interest rate scenarios. The method "communicates" with users in practical terms of "up" and "down" scenarios and retains a traditional ("static-like") level of control over prepayment assumptions and cashflow generation that most practitioners love to have. These features make the method especially valuable for:

- multi-scenario option-adjusted valuation of mortgage securities and portfolios, an essential task in risk, asset/liability, and portfolio management;
- historical OAS analysis; methodological consistency and computational speed provided by the EIRS method throughout historical observations rank higher than finding "true" OAS;
- pricing complex securities (such as CMOs) while having limited ability to generate their cashflows dynamically;
- advanced research (such as a multi-factor or sensitivity analysis) which includes OAS pricing as a routine sub-task to be done quickly and consistently.

The method's drawback is that it is an approximation that does not converge to the exact solution via simply using more scenarios (i.e., constructing a "finer grid"). Any improvements should either involve a relatively universal control variate technique, or will interfere with the method's major assumptions. In addition, the method will successfully work only for those path-dependent embedded provisions that are well reflected in the grid scenarios.

Our approach to the multi-factor analysis distinguishes "traded" and "non-traded" factors. We suggest using the underlying EIRS numerical loop for arbitrage-free pricing on a grid of very few (one or two) interest rate traded factors. Then, one measures the exposure (duration and convexity) to uncorrelated non-traded factors such as prepayment model error, ARM index model error, pricing spread, or interest rate volatility. This step immediately quantifies additional convexity costs and VAR contributions creating a comprehensive picture of reward and risk of investing in a mortgage loan or MBS.

Acknowledgments

The author is grateful to the participants of NYU/Courant Institute Mathematical Finance Seminar led by Marco Avellaneda. The paper benefits from a long-term collaboration with James Daras and Kenneth Schmidt. Arthur Beaudoin, Ibrahim Sadiq, and Michael Di Brienza helped significantly with the information system design, and Diahann Rothstein, Krystn Paternostro, Robert Wyle, and Paul Barrett have been excellent users who stimulated development enhancements. Comments made at various stages of this work by Robert Jarrow, John Hull, Andrew Davidson, Yung Lim, Michael Purvis, Kevin Berger, and Douglas Love are greatly appreciated.

Appendix A. A Newton–Raphson Scheme for the EIRS Equation

We consider here a numerical Newton–Raphson scheme intended to solve the pricing Eq. (3.3). First we assume that one knows an initial approximation for the static spread as a function of the factor x, let us denote it $\tilde{s}(x)$. Thus, to start iterations, one can set, for example, $\tilde{s}(x) = s(0) \equiv$ Const if price $P(0)$ is known, or $\tilde{s}(x) = $ EIRS $\equiv$ Const if EIRS is known. The corresponding price, spread duration, and static convexity will be denoted $\tilde{P}(x)$, $\tilde{D}_s(x)$, and $\tilde{C}_x^{\text{static}}(x)$. If the next approximation, $s(x)$, is close to the previous one, we can write down:

$$s(x) \approx \tilde{s}(x) + \frac{\tilde{P}(x) - P(x)}{\tilde{P}(x)\tilde{D}_s(x)} \equiv a(x) - b(x)P(x) \tag{A.1}$$

where $a(x) = \tilde{s}(x) + 1/\tilde{D}_s(x), b(x) = 1/\tilde{P}(x)\tilde{D}_s(x)$.

Substituting (A.1) into Eq. (3.3) and multiplying it by $2\tilde{P}(x)/\sigma^2$ we get the following linearized equation for $P(x)$ and EIRS:

$$P(x)\beta(x) + \frac{\partial^2 P}{\partial x^2} - \delta(x)\text{EIRS} = \gamma(x) \tag{A.2}$$

where $\beta(x) = -2/\tilde{D}_s(x)\sigma^2, \gamma(x) = \tilde{P}(x)[\tilde{C}_x^{\text{static}}(x) - 2a(x)/\sigma^2]$, and $\delta(x) = 2\tilde{P}(x)/\sigma^2$. We have to complement Eq. (A.2) with 2 boundary conditions. For example, we may require a zero convexity cost as factor x approaches some extreme values, x_1 and x_N:

$$\begin{aligned}
P(x_1)\beta(x_1) - \delta(x_1)\text{EIRS} &= -\delta(x_1)a(x_1) \\
P(x_N)\beta(x_N) - \delta(x_N)\text{EIRS} &= -\delta(x_N)a(x_N).
\end{aligned} \tag{A.3}$$

We then use the difference approximation for the second derivative in Eq. (A.2) arriving at a sparse system of linear algebraic equations:

$$A^*P + B^*\text{EIRS} = \Gamma \tag{A.4}$$

where $P = [P_1, \ldots, P_N]$, A is an N by N matrix, B and Γ are N-dimensional columns; only non-zero elements of A, B, and C are listed below for the case of uniform grid, $x_k - x_{k-1} = \Delta$:

$$A_{i,i} = \beta(x_i) - 2/\Delta^2, A_{i,i-1} = A_{i,i+1} = 1/\Delta^2, \Gamma_i = \gamma(x_i), \quad i = 2, \ldots, N-1;$$

$$A_{1,1} = \beta(x_1), \ A_{N,N} = \beta(x_N), \ \Gamma_1 = -a(x_1)\delta(x_1), \ \Gamma_N = -a(x_N)\delta(x_N);$$

$$B_i = -\delta(x_i), \quad i = 1, \ldots, N.$$

Since either EIRS or the base price is known, system (A.4) has exactly N unknowns. The algorithm can be formulated as follows.

Step 1. Select the N-point x-factor grid, build N scenarios and generate N cashflows.

Step 2. Take an initial guess about $\tilde{s}(x)$, i.e., assume some known $\tilde{s}(x_1), \ldots,$ $\tilde{s}(x_N)$ (if there exists no specific preference, use the suggestion above).

Step 3. Compute the required coefficients at the grid points, $a(x_i)$, $b(x_i)$, $\beta(x_i)$, $\delta(x_i)$, and $\gamma(x_i)$, $i = 1, \ldots, N$. Then build matrix A and vectors B and Γ for linear system (A.4).

Step 4. Solve linear system (A.4) for all prices and EIRS (if unknown) using factorization technique.[i] Compute static spreads $s(x_i), \ldots, s(x_N)$ again.

Step 5. If the norm $\|s(x) - \tilde{s}(x)\|$ exceeds tolerance, denote $\tilde{s}(x) = s(x)$ and GOTO Step 3. Otherwise, the solution is found.

This procedure converges surely and quickly for all loans and MBS' the author analyzed.

Appendix B. Mathematics of the 2-Factor Gaussian Term Structure

The author has derived all statistics for linear system (6.1), along with the arbitrage-free choice of $\theta(t)$ function:

$$\theta(t) = f(t) + \frac{dL(t)}{dt} + \frac{1}{2}\frac{d\mathrm{Var}[y(t)]}{dt} + \frac{1}{a_1}\frac{d}{dt}\left\{f(t) + \frac{dL(t)}{dt} + \frac{1}{2}\frac{d\mathrm{Var}[y(t)]}{dt}\right\} \quad \text{(B.1)}$$

where $dy(t) = -r(t)dt$ and $L(t) = -x_1(0)(1 - e^{-a_1 t})/a_1 - x_2(0)(1 - e^{-a_2 t})/a_2$. Note that the first two statistical moments, including the variance, can be explicitly derived for $y(t)$, see [6,9].

The short rate's mean path will be defined by the following formula:

$$E[r(t)] = f(t) + \frac{1}{2}\sum_{i=1}^{2}\frac{\sigma_i^2}{a_i^2}(1 - e^{-a_i t})^2 + \rho\frac{\sigma_1\sigma_2}{a_1 a_2}(1 - e^{-a_1 t})(1 - e^{-a_2 t}). \quad \text{(B.2)}$$

Any T-maturity long rate is a linear function of the factors:

$$r_T(t) = B_1(T)x_1(t) + B_2(T)x_2(t) - A(t,T), \quad \text{(B.3)}$$

where $B_i(T) = (1 - e^{-a_i T})/a_i T, i = 1, 2;$

$$A(t,T) = B_1(T)f(t) - f_T(t) - x_2(0)e^{-a_2 t}[B_1(T) - B_2(T)]$$

$$+ \sum_{i=1}^{2}\frac{\sigma_i^2}{2a_i^2}[B_1(T)(1 - e^{-a_i t})^2 - 2B_i(T)(1 - e^{-a_i t}) + B_i(2T)(1 - e^{-2a_i t})]$$

$$+ \rho\frac{\sigma_1\sigma_2}{a_1 a_2}\{-B_1(T)(1 - e^{-a_1 t})e^{-a_2 t} - B_2(T)(1 - e^{-a_2 t})$$

$$+ B_\Sigma(T)[1 - e^{-(a_1 + a_2)t}]\}.$$

[i]If EIRS is given, then system (A.4) is already tri-diagonal. If the price is given, then we first solve (A.4) for two arbitrarily chosen, distinct values of EIRS. Using the apparent linearity we then recover the true solution connecting thus found two base prices with a straight line until it crosses the given value, $P(0)$.

The expected long rates and "forward forward" short rates are computed as

$$E[r_T(t)] = f_T(t) + \frac{1}{2}\sum_{i=1}^{2}\frac{\sigma_i^2}{a_i^2}[2B_i(T)(1 - e^{-a_i t}) - B_i(2T)(1 - e^{-2a_i t})]$$

$$+ \rho\frac{\sigma_1\sigma_2}{a_1 a_2}\{B_1(T)(1 - e^{-a_1 t}) + B_2(T)(1 - e^{-a_2 t})$$

$$- B_\Sigma(T)[1 - e^{-(a_1 + a_2)t}]\} \tag{B.4}$$

$$E[f(t,T)] = f(t+T) + \frac{1}{2}\sum_{i=1}^{2}\frac{\sigma_i^2}{a_i^2}[2e^{-a_i T}(1 - e^{-a_i t}) - e^{-2a_i T}(1 - e^{-2a_i t})]$$

$$+ \rho\frac{\sigma_1\sigma_2}{a_1 a_2}\{e^{-a_1 T}(1 - e^{-a_1 t}) + e^{-a_2 T}(1 - e^{-a_2 t})$$

$$- e^{-(a_1 + a_2)T}[1 - e^{-(a_1 + a_2)t}]\} \tag{B.5}$$

where $B_\Sigma(T) = [1 - e^{-(a_1 + a_2)T}]/(a_1 + a_2)T$.

Once again, the mean paths always evolve above the corresponding forward curves. Since function $A(t,T)$ is deterministic, the first two terms in the right-hand side of formula (B.3) defines the volatility term structure:

$$\sigma^2 = \sigma_1^2 + \sigma_2^2 + 2\rho\sigma_1\sigma_2$$

$$\sigma_T^2 = B_1^2(T)\sigma_1^2 + B_2^2(T)\sigma_2^2 + 2B_1(T)B_2(T)\rho\sigma_1\sigma_2 \tag{B.6}$$

$$\text{Cov}[r, r_T] = B_1(T)\sigma_1^2 + B_2(T)\sigma_2^2 + [B_1(T) + B_2(T)]\rho\sigma_1\sigma_2\,.$$

The underlying 2-factor model (6.1) is transformed to its "$(r - v)$-form" as follows:

$$dr(t) = \left[a_1\theta(t) + \frac{a_1\beta - a_2}{1 - \beta}r(t) - \frac{a_1 - a_2}{1 - \beta}v(t)\right]dt + \sigma dz_r$$

$$dv(t) = \left[a_1\theta(t) + \frac{a_1 - a_2}{1 - \beta}\beta r(t) + \frac{a_2\beta - a_1}{1 - \beta}v(t)\right]dt + \sigma_v dz_v \tag{B.7}$$

where the short rate volatility, σ, is defined in formula (B.6) above, $\sigma_v^2 = \sigma_1^2 + \beta^2\sigma_2^2 + 2\rho\beta\sigma_1\sigma_2$, and two Brownian motion increments, dz_r and dz_v are uncorrelated, as are the factors r and v at $t = 0$ (i.e., $\text{corr}[r(t), v(t)] \to 0$ as $t \to 0$).

Factor $v(t)$ determines the term structure move net of that propagated by the short rate shock; any long rate is linear in factors:

$$r_T(t) = \frac{B_2(T) - \beta B_1(T)}{1 - \beta}r(t) + \frac{B_1(T) - B_2(T)}{1 - \beta}v(t) - A(t,T)\,. \tag{B.8}$$

The "shocked" trajectories for the factors and the long rates will be counted off their mean paths as follows:

$$\Delta r(t) = [(\Delta v_0 - \beta \Delta r_0)e^{-a_1 t} + (\Delta r_0 - \Delta v_0)e^{-a_2 t}]/(1 - \beta)$$

$$\Delta v(t) = [(\Delta v_0 - \beta \Delta r_0)e^{-a_1 t} + \beta(\Delta r_0 - \Delta v_0)e^{-a_2 t}]/(1 - \beta) \qquad \text{(B.9)}$$

$$\Delta r_T(t) = \{[B_2(T) - \beta B_1(T)]\Delta r(t) + [B_1(T) - B_2(T)]\Delta v(t)\}/(1 - \beta).$$

Note that the average path for $v(t)$ is not essential for constructing the grid.

References

[1] J. C. Hull, *Options, Futures and Other Derivative Securities*, 3rd ed., Prentice Hall, Englewood Cliffs, NJ, 1997.

[2] J. C. Hull and A. D. White, "Pricing interest rate derivatives securities", *Rev. Financial Studies* **3** (1990) 573–592.

[3] J. C. Hull and A. D. White, "Numerical procedures for implementing term structure model II: Two-factor models", J. Derivatives **2**(2) (Winter 1994) 37–48.

[4] B. Ryan, *The CMS Option Model: An HJM Application*, Capital Management Sciences, 1995.

[5] D. Duffie, *Dynamic Asset Pricing Theory*, 2nd ed., Princeton Univ. Press, Princeton, NJ, 1996.

[6] A. Levin, "Linear system theory in stochastic pricing models", in *Yield Curve Dynamics*, R. Ryan, ed., Glenlake, Chicago, 1997.

[7] A. Levin and J. Daras, "Non-traded factors in MBS portfolio management", in *Advances in the Valuation and Management of Mortgage-Backed Securities*, F. Fabozzi, ed., Frank J. Fabozzi Associates, 1998.

[8] A. Levin, "A new approach to option-adjusted valuation of MBS on a multi-scenario grid", in *Advances in the Valuation and Management of Mortgage-Backed Securities*, F. Fabozzi, ed., Frank J. Fabozzi Associates, 1998.

[9] A. Levin, "Deriving closed-form solutions for Gaussian pricing models: A systematic time-domain approach", *Int. J. Theoretical and Appl, Finance* **1**(3) (1998) 349–376.

[10] A. Sparks and F. Feiken Sung, "Prepayment duration and convexity", *J. Fixed Income* (March 1995) 7–11.

SIMULATING BERMUDAN INTEREST RATE DERIVATIVES

PETER CARR

Banc of America Securities,
9 West 57th Street, 40th Floor,
New York, NY 10019, USA
(212) 583-8529
E-mail: pcarr@bofasecurities.com

GUANG YANG

NumeriX LLC,
546 Fifth Avenue, 17th Floor,
New York, NY 11553, USA
(212) 302-2220
E-mail: gyang@numerix.com

Received 21 July 1999

We use simulation to develop a Markov chain approximation for the value of caplets and Bermudan interest rate derivatives in the Market Model developed by Brace, Gatarek, and Musiela (1995) and Jamshidian (1996a, b). One and two factor versions of the Market Model were numerically studied. Our approach yields numerical values for caplets which are in close agreement with analytical solutions. We also provide numerical solutions for several Bermudan swaptions.

1. Introduction

Term structure modeling is one of the most challenging problems in asset pricing theory. The modern approach has focused on the restrictions imposed by the absence of arbitrage and emanates from the seminal paper of Merton (1973). Significant contributions to the valuation and hedging of interest rate derivatives such as bond options were made by Black (1976), Vasicek (1977), Brennan and Schwartz (1982), Courtadon (1982), Ball and Torous (1983), Cox, Ingersoll, and Ross (CIR, 1985), Ho and Lee (1986), Schaefer and Schwartz (1987), Black, Derman, and Toy (BDT, 1990), Black and Karasinski (BK, 1991), Heath, Jarrow, and Morton (HJM, 1988), Jamshidian (1989), Hull and White (1990), and more recently by Brace, Gatarek, and Musiela (BGM, 1995), Jamshidian (1996a, b), and Flesaker and Hughston (1996).

The papers by Black (1976), Ball and Torous (1983) and by Schaefer and Schwartz (1987) all mimicked the original Black-Scholes stock option theory in specifying the underlying bond price dynamics directly. The other papers prior to

Ho and Lee took off from Merton (1973) and instead specified the dynamics of the spot rate and possibly another factor such as the long-term rate. In these models, the bond price dynamics arise as a consequence of the absence of arbitrage and a specification of the market price of interest rate risk. While these models provide valuable insights into the determinants of bond prices and into the relationship between bond prices of different maturities, they have difficulty in matching the initial term structure exactly. An "inversion of the term structure" is required to eliminate the market prices of risk from interest rate derivative values.

Ho and Lee (1986) pioneered a new approach to term structure modeling. They took the initial discount curve as given and modelled the arbitrage-free movement of the entire discount curve in a binomial setting. Inspired by the work of Ho and Lee (1986) and the martingale methods of Harrison and Pliska (1981), Heath, Jarrow, and Morton (1988) developed an elegant mathematical framework for the term structure of interest rates. The HJM approach takes the initial forward rate curve as given and develops a drift restriction for the stochastic evolution of the forward rate curve under the equivalent martingale measure. In common with Ho and Lee, this approach has the advantage of providing arbitrage-free interest rate derivative prices that do not explicitly depend on the "market price of risk". The HJM framework is very general and most of the work before 1988 can be viewed as special cases of it. Research on interest rate derivatives pricing after HJM has focused on specializing the framework to tractable models so that it can be efficiently calibrated to market prices of observable instruments and can efficiently price other interest rate derivatives.

Despite the efforts of many academics and practitioners, there is still no universal model for pricing interest rate derivatives. The challenges of interest rate derivative pricing are more computational than methodological. In order to obtain tractable sub-cases of the HJM model, the volatility structure has to be restricted. For realistic volatility structures, the HJM model is very expensive to calibrate. Furthermore, the valuation of Bermudan and American interest rate derivatives is computationally intensive. For such options, the computation time for even a single factor model is too slow to meet industry requirements. Although there are published results on certain versions of HJM, the issues of convergence, accuracy and efficiency have never been precisely documented. Realizing the theoretical advantages of the HJM approach, researchers have looked for an "ideal" special case of the general HJM model that possesses the following properties:

a. Arbitrage-free:

There must be no arbitrage in the price dynamics of bonds and other interest rate derivatives. This is the most fundamental requirement for any term structure model. As mentioned before, the HJM model solves the above problem and most of the current term structure models do enforce the no arbitrage condition in the above sense.

b. Versatility:

A good model should be consistent with a broad range of possible term structure shapes, volatility profiles, and correlations among yields of different maturity. The empirical studies of Canabarro (1995) demonstrate the inability of certain one-factor models to account for observed yield correlations. In contrast, multifactor models such as HJM are consistent with arbitrary yield correlations.

c. Positive Interest Rates:

Although rare cases may require negative interest rates, it is widely accepted that a model that guarantees positive interest rates will be preferred over ones that do not, ceteris paribus. Despite the claims that the likelihood of negative interest rates is small by Gaussian modelers, Rogers (1995a, b) demonstrated that under certain circumstances, large negative interests can arise and cause non-negligible distortions of derivative prices. The positive term structure models in wide use are CIR, BDT, BK, nearly proportional HJM, BGM, and Flesaker and Hughston.

d. Computational Efficiency:

Whether for model calibration, deal pricing, book revaluation, dynamic hedging, or firmwide risk management, the pricing model will be used many times during the day. Consequently, it is very critical that a model be able to evaluate derivatives efficiently. Closed form solutions for realistic problems are highly desired, although extremely rare in practice. Consequently, efficient numerical schemes are necessary for a valuation model to make the transition from theory to practice. In particular, one of the most computationally demanding problems is the pricing of Bermudan interest rate derivatives in a non-Markovian multi-factor model.

In general, the goal of term structure modeling is to find an "ideal" model that possesses the four features above. Recently, a special case of the HJM framework has emerged which satisfies most of the above criteria. This approach, termed the Market Model by its developers BGM (1995) and Jamshidian (1996a, b), is arbitrage-free, provides strictly positive interest rates, and generates closed form solutions for caps or European swaptions. Since models are usually calibrated to these instruments, this feature speeds up the calibration process considerably. Unfortunately, for Bermudan derivatives, extant applications of the Market Model are computationally demanding.

In this paper, we use simulation to develop a Markov Chain Approximation (MCA) to value Bermudan derivatives in the Market Model. We show that this approach allows such derivatives to be evaluated more efficiently than existing approaches such as non-recombining trees. The structure of this paper is as follows. The next section reviews the Market Model, while the following section reviews our MCA approach. The fourth section applies MCA to the Market Model, while the following section presents our numerical implementation. The final section concludes.

2. The Market Model

A significant advance in the search for an "ideal" term structure model was made by Brace, Gatarek, and Musiela (1995) and Jamshidian(1996a, b). Conscious of structures trading in the market, they developed the so-called Market Model, which takes forward Libor rates or swap rates as inputs and directly models their evolution. Although the Market Model can be viewed as a special case of the HJM model, it differs from traditional approaches in its use of the so-called "numeraire induced measure", instead of the conventional risk-neutral measure. Assuming a deterministic forward rate volatility, the Market Model prices cap or European swaptions by the standard Black formula. Besides being consistent with the industry standard, the assumption of deterministic volatility guarantees positive interest rates.

Since we will price Bermudan derivatives using the Market Model, we now give a brief introduction to the model following Jamshidian's (1996a) approach. We will not go into technical details such as how to change measure etc. but instead refer the reader to Jamshidian (1996a, b) for the mathematical foundations of the Market Model.

We denote the sequence of Libor payment dates by and the sequence of spot Libor reset dates by T_n. Assuming that the reset dates match the payment dates $(t_n = T_n)$, we have a tenor structure:

$$0 < T_1 < T_2 < \cdots < T_{N+1} = T^* \, .$$

The day count factors are defined as:

$$\delta_n = T_{n+1} - T_n \, , \quad (\leq n \leq N) \, .$$

For later use, we define $n(t)$ as one plus the number of payments as of date t:

$$n(t) = \{m : T_{m+1} < t \leq T_m\} \, .$$

Let $B_n(t)$ be the price at of a zero-coupon bond maturing at T_n for $n \geq 1$ and $0 \leq t \leq T_n$. Let $L_n(t)$ be the forward Libor rate for accrual period $[T_n, T_{n+1}]$ and let $S_{n,N}(t)$ the forward swap rate for a swap starting at T_n and maturing at T_{N+1}. Then we have:

$$L_n(t) = \delta_n^{-1} \left(\frac{B_n(t)}{B_{n+1}(t)} - 1 \right) \, ,$$

$$B_{n+1}(t) = B_{n(t)}(t) \prod_{i=n(t)}^{n} \frac{1}{1 + \delta_i L_i(t)} \, ,$$

$$S_{n,N}(t) = \frac{B_n(t) - B_{N+1}(t)}{B_{n,N}(t)},$$

$$B_{n,N}(t) = \sum_{i=n}^{N} \delta_i B_{i+1}(t), \quad (t \leq T_{n+1}).$$

The Market Model can be generically represented by:

$$dK_n(t) = D_{\text{num}}(t, K_n(t))dt + \sum_{i=1}^{L} V_i(t, K_n(t))dz_{\text{num}}^i(t).$$

Here, $K_n(t)$ can be either the Libor rate $L_n(t)$ or the swap rate $S_{n,N}$, $D_{\text{num}}(t, K_n(t))$ is the drift term, $V^i(t, K_n(t))$ is the absolute volatility, and z_{num}^i is the Brownian motion associated with the measure induced by the numeraire $B_{\text{num}}(t)$. By specifying the rate $K_n(t)$ and the numeraire $B_{\text{num}}(t)$, we can get the following three versions of the Market Model:

(1) Libor market model in spot Libor measure:

Introducing a "rolling zero-coupon bond"

$$B(t) = \frac{B_{n(t)}(t)}{B_1(0)} \prod_{n=1}^{n(t)-1} (1 + \delta_n L_n(T_n)) = B_{n(t)}(t) \prod_{n=2}^{n(t)} \frac{1}{B_n(T_{n-1})}$$

and taking $K_n(t) = L_n(t)$ and $B_{\text{num}}(t) = B(t)$, we have the following general equations governing the stochastic evolution of the forward Libor rate:

$$dL_n(t) = \sum_{i=n(t)}^{n} \frac{\delta_i \beta_i(t)\beta_i(t)^t}{1 + \delta_i L_i(t)} dt + \beta_n(t)dz_B(t).$$

The value of any Libor rate derivative satisfies:

$$C(t) = B(t)E_t^B \left[\frac{C(T)}{B(T)}\right] \quad (t \leq T).$$

Here, E^B is the expectation under the measure P^B induced by $B(t)$ (the spot Libor measure), and $z_B(t)$ is the associated Brownian motion (the spot Libor Brownian motion).

If $\lambda_n(t) = \frac{\beta_n(t)}{L_n(t)}$ is a bounded deterministic function, then the general equation reduces to the following "Libor market model in spot Libor measure":

$$\frac{dL_n(t)}{L_n(t)} = \sum_{i=n(t)}^{n} \frac{\delta_i \lambda_i(t)\lambda_n(t)L_i(t)}{1 + \delta_i L_i(t)} dt + \lambda_n(t)dz_B(t).$$

There exists a unique positive solution for $L_n(t)$ and caplets are priced by the Black formula.

(2) Libor market model in terminal measure:

Taking $K_n(t) = L_n(t)$ and $B_{\text{num}}(t) = B_n(t)$, we have:

$$dL_n(t) = \beta_n(t)dz_{n+1}(t) \quad (t \le T_n)$$

and

$$C(t) = B_n(t)E_t^n\left[\frac{C(T)}{B_n(T)}\right] \quad (t \le T \le T_n).$$

Fixing $z_*(t) = z_{N+1}(t)$, we have

$$dL_n(t) = \sum_{i=n+1}^{N} \frac{\delta_i\beta_i(t)\beta_i(t)^t}{1+\delta_iL_i(t)}dt + \beta_n(t)dz_*(t)$$

and

$$C(t) = B_*(t)E_t^*\left[\frac{C(T)}{B_*(T)}\right] \quad (t \le T \le T_*).$$

Under the lognormal volatility assumption, we have

$$\frac{dL_n(t)}{L_n(t)} = -\sum_{i=n+1}^{N} \frac{\delta_i\lambda_i(t)\lambda_n(t)L_i(t)}{1+\delta_iL_i(t)}dt + \lambda_n(t)dz_*(t).$$

This is the Libor market model in terminal measure which BMG (1995) developed. There exists a unique positive solution and caplets are priced by the Black formula.

(3) Swap market model:

Taking $K_n(t) = S_{n,N}(t)$ and $B_{\text{num}}(t) = B_{n,N}(t)$, we have:

$$dS_{n,N}(t) = \phi_n(t)dz_{n,N}(t) \quad (t \le T_n) \quad \text{and}$$

$$C(t) = B_{n,N}(t)E_t^{n,N}\left[\frac{C(T)}{B_{n,N}(T)}\right] \quad (t \le T \le T_n).$$

Under the lognormal volatility assumption, we have the following swap market model:

$$\frac{dS_{n,N}(t)}{S_{n,N}(t)} = \theta_n(t)dz_{n,N}(t) \quad (t \le T_n).$$

There exists a unique positive solution and now European swaptions are priced by the Black formula.

As with the general HJM model, the numeraire (the money market account in HJM) plays a very important role. Jamshidian (1996b) demonstrates that under appropriate measurability assumptions, payoffs which are a function of the path of Libor rates or swap rates can be attained by a self-financing trading strategy involving only the finite number of zero-coupon bonds that define the rates, even when the market is incomplete. This means:

$$C(t) = E_t^{\text{num}}\left(\sum_{i=1}^{N+1} N_{i_i}(T)B_i(T) \Big/ B_{\text{num}}(T)\right)B_{\text{num}}(t).$$

Here $B_{\text{num}}(t)$ is the numeraire, which can be either the spot Libor numeraire $B(t)$ or any of the zero-coupon bonds $B_i(t)$, or the forward swap rate numeraire $B_{n,N}(t)$.

In the special case of the pure discount bond, we get:

$$B_i(T) = E_T^{\text{num}}(1/B_{\text{num}}(T_i))B_{\text{num}}(T).$$

Thus the numeraire value is the only state variable that appears explicitly in the above formula, if one knows the dynamics of the numeraire $B_{\text{num}}(T)$ in the appropriate measure, then the value of the derivative, $C(t)$, is determined. It is important to note that the dynamics of the numeraire $B_{\text{num}}(T)$ can depend on several state variables. This dependency shows up implicitly through the transitional probability of the numeraire processes. In the Market Model, one needs to know the dynamics of the numeraire only at reset dates. This is fortunate because in the Market Model, the dynamics of the numeraire are not known inbetween reset dates.

The above property of the numeraire demonstrates a very general principle: the dynamics of the numeraire in the appropriate measure are all we need to price interest rate derivatives whose payoff can be attained by a self-financing trading strategy in the appropriate bonds. The focus on the risk-neutral dynamics of the spot rate when the money market account is the numeraire can be attributed to the above principle. We will see that this principle is the key to pricing Libor derivatives efficiently.

As mentioned previously, the Market Model is efficient in pricing caps or swaptions. There also exist other closed form solutions for some simple derivatives. However, for Bermudan derivatives, the market models face the same difficulties as the general HJM model. Although Bermudan derivatives can be evaluated by a non-recombining tree, the computational inefficiency of the method makes it almost useless in practice. For this reason, the next section discusses Markov Chain Approximation which will be used to efficiently price Bermudan derivatives.

3. Markov Chain Approximation (MCA)

MCA is a method for approximating a continuous time stochastic process. A graphical representation of the MCA used in this paper is given in Fig. 1. MCA includes, recombining (binomial, trinomial, and multinomial) lattices, non-recombining (binomial, trinomial, and multinomial) trees and the finite difference (explicit and implicit) schemes, spectral methods, and finite elements as special cases. Derivatives are valued in a MCA by the usual backward induction method.

A MCA can be fully described by the node values of the underlying and the transition probability matrix. For $t \in \{0, t_1, t_2, \ldots, t_{n-1}, t_n = T\}$, we represent the values of the underlying variable by

$$\mathbf{r}_{i(t)}(t) \ (\text{or more briefly } \mathbf{r}_i(t))$$

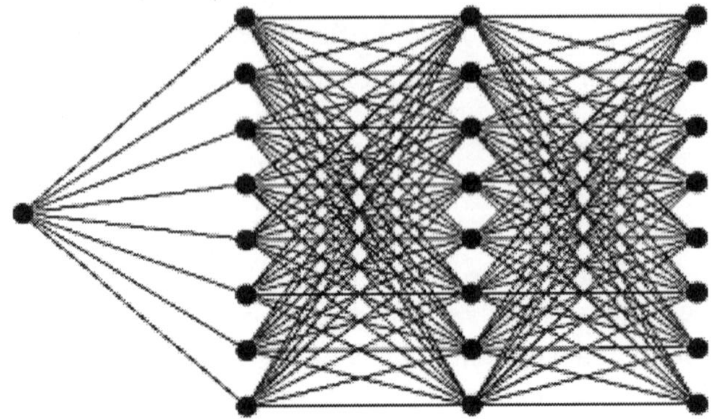

Fig. 1. A Markov Chain Approximation (MCA).

and the transition probability matrix by

$$p_{i(t),j(t+1)}(t) \text{ (or more briefly } p_{i,j}(t)) \text{ with } \sum_{j=0}^{I(t+1)} p_{i,j}(t) = 1,$$

where $i(t) = 0, 1, \ldots, I(t)$, $j(t+1) = 0, 1, \ldots, I(t+1)$, $I(t) \geq 0$,

$$\mathbf{r}_i(t) = (r_i^l(t) \in R), \ l = 0, 1, \ldots, L(i,t), \ L(i,t) \leq 0 \text{ and } p_{ij}(t) \geq 0.$$

In its most general form, a MCA with n steps has $\sum_{k=1}^{n} \sum_{l=0}^{I(k)} L(l, t_k) + \sum_{k=0}^{n-1}(I(k) + 1) * (I(k + 1))$ degrees of freedom and the degrees of freedom go to infinity as n goes to infinity. In practice, we usually set $I(t)$ and $L(i, t)$ to constant or simple linear functions.

The conditional expectation of a function, $f(t, \mathbf{r})$, in terms of a MCA is represented by:

$$E(f(t + \Delta t, \mathbf{r}(t + \Delta t))|\mathbf{r} = \mathbf{r}_i(t)) = \sum_{j=0}^{I(t+\Delta t)} f(t + \Delta t, \mathbf{r}_j) p_{i,j}(t)$$

Traditional numerical methods such as non-recombining trees have the problem of exponential growth in the number of nodes. The advantage of our MCA method lies in the fact that by having a transition matrix which is not sparse, we can control the growth of the number of nodes. The ability to control the growth of the number of nodes results in significant savings in computation time. While MCA also has a cost in terms of the extra computational effort required to get the transition matrix, this cost is very small compared to the resulting reduction in computational effort.

4. Approximating Market Model Values by MCA

In this section, we show how Monte Carlo simulation can be used to generate a MCA of derivative security values in the Market Model. Our focus will be on

Bermudan derivatives since these are not efficiently valued by other techniques. The main difficulty in implementing the Market Model for Bermudan derivatives is that the model has a very high dimension, which is equal to the number of rates being modeled. In order to compute Bermudan prices efficiently, we collapse the dimensionality of the state space down to a single variable plus time. The single variable chosen is the numeraire, since as mentioned in Sec. 2, one only needs the dynamics of this variable in the appropriate measure to price any interest rate derivative and the numeraire is the only state variable that shows up explicitly in the derivative pricing formula. In collapsing the dimensionality, the joint distribution of the rates and the numeraire is replaced with just the marginal distribution of just the numeraire. This marginal distribution is obtained by Monte Carlo simulation as detailed below. Since the exercise strategy is only allowed to depend on the level of the numeraire, our method is an approximation for the true value of the Bermudan derivative. Thus, our approach is similar in spirit to the approach taken by Barraquand and Martineau (1995) for the valuation of multivariate American equity derivatives.

We first approximate the (forward or swap) rate diffusions using an explicit Euler scheme:

$$\Delta K_n(t) = D_{\text{num}}(t, K_n(t))\Delta t + \sum_{i=1}^{L} V_i(t, K_n(t))\Delta z^i_{\text{num}}(t)$$

with

$$\Delta K_n(t) = K_n(t + \Delta t) - K_n(t), \ \Delta z^i_{\text{num}}(t) = z^i_{\text{num}}(t + \Delta t) - z^i_{\text{num}}(t)$$

and

$$\forall t \in \{0, \Delta t, \ldots, T\}\,.$$

Next, we discretize the numeraire space. More specifically, $\forall t \in \{0, t_1, \ldots, T_{N+1}\}$ choose a finite partition, $R(t) = (R_1(t), \ldots, R_{k(t)}(t))$, of the numeraire state space R, i.e. a set of $k(t)$ subsets of R satisfying:

$$\bigcup_{i \in [1,k(t)]} = R \text{ and } \forall\, (i, j) \in [1, k(t)]^2, i \neq j, R_i(t) \cap R_j(t) = \varnothing\,.$$

The time grid for the MCA differs from that for the simulated rate equations since for Bermudan derivatives, we only need a MCA at the reset dates. We assume that the partition $R(0)$ has only two cells:

$$R_1(0) = \{B_{\text{num}}(0)\}, \text{ and } R_2(0) = R \backslash \{B_{\text{num}}(0)\}\,.$$

Once the sample paths $k^1(t), \ldots, k^M(t)$ are computed, $a_i(t)$, the number of samples entering $R_i(t)$, $b_{i,j}(t)$, the number of samples moving from $R_i(t)$ to $R_j(t + \Delta t)$, and $B^{\text{sum}}_{\text{num}}(t)$, the sums of sample numeraire values within the cell $R_i(t)$ are easily computed. The entries in the transition probability matrix can be calculated as:

$$p_{i,j}(t) = \frac{b_{i,j}(t)}{a_i(t)}$$

the average value of the numeraire within $R_i(t)$, $B^i_{\text{num}}(t)$, is

$$B^i_{\text{num}}(t) = \frac{B^{\text{sum}}_{\text{num}}(t)}{a_i(t)} \, .$$

With the calculated values of $p_{i,j}(t)$ and $B^i_{\text{num}}(t)$, a well-defined MCA is generated. The conditional expectation in terms of a MCA is represented by:

$$E^{\text{num}}(X(t + \Delta t, B_{\text{num}}(t - \Delta t))|B_{\text{num}} = B^i_{\text{num}}(t)) = \sum_{j=0}^{I(t+\Delta t)} X(t + \Delta t, B^j_{\text{num}})p_{i,j}(t) \, .$$

Contingent claims can be evaluated from the following formula:

$$C(t, B^i_{\text{num}}(t)) = B_{\text{num}}(t)^* E^{\text{num}} \left(\frac{C(t + \Delta t, B_{\text{num}}(t + \Delta t))}{B_{\text{num}}(t + \Delta t)} \bigg| B_{\text{num}} = B^i_{\text{num}}(t) \right)$$

for European securities, and

$$C(t, B^i_{\text{num}}(t)) = \max \left(C(t)_{ex}, B_{\text{num}}(t)^* \right.$$

$$\times \left. E^{\text{num}} \left(\frac{C(t + \Delta t, B_{\text{num}}(t + \Delta t))}{B_{\text{num}}(t + \Delta t)} \bigg| B_{\text{num}} = B^i_{\text{num}}(t) \right) \right)$$

for American securities, where $C(t)_{ex}$ is the exercise value of the American security.

We use one of two approaches for valuing a derivative depending on the nature of the claim. If the derivative security's payoff is insensitive to the variation in term structures, then we obtain the average yield curve conditional on the numeraire being in a specified cell by averaging over all term structures that happen to have the numeraire value in the given cell. The yield curve corresponding to $B^i_{\text{num}}(t)$ can be obtained by:

$$f^n_i(t) = \frac{f^n_{\text{sum}}(t; B^i_{\text{num}}(t))}{a_i(t)} \, ,$$

where $f^n_{\text{sum}}(t; B^i_{\text{num}}(t))$ is the sum of all term structures with the numeraire in cell $B^i_{\text{num}}(t)$.

The above method for constructing the yield curve is efficient in obtaining the average yield curve and in pricing several derivatives using one simulation. However, for derivatives that are sensitive to variation in the term structure, the above method can cause errors. For one-factor models and for derivatives on rates which are highly correlated with the numeraire, our numerical results show that averaging over term structures is very accurate. However, for claims whose payoff is tied to a specified rate quite different from those which drive the numeraire, we use a different approach. Specifically, for each cell describing a small range in value of the numeraire, we average the payoff of the security over all the term structures whose corresponding numeraire was in the specified cell. We benchmarked our numerical results by comparing them with the traditional Monte Carlo method for European derivatives which are sensitive to variation in the shape of the term structure, e.g., swaptions and caps in a two-factor model.

5. Numerical Results

We applied our MCA method to price caps and swaptions in both a one-factor and a two-factor settings. The payer swaptions we value are (1) European swaptions, (2) Bermudan swaptions, (3) Fixed tail Bermudan swaptions, (4) Constant maturity Bermudan swaptions, and (5) Bermudan forward swaptions. Since the terminology in (2) to (5) may not be standard, we now give a precise definition of each swaption.

Bermudan swaptions:

A Bermudan swaption is an option on a swap that can be exercised on every reset date of the underlying swap up to and including the option maturity. When exercised, the swap one gets is one that starts at the exercise date and matures at a fixed date which is independent of the time of exercise. A Bermudan swaption is equivalent to a Bermudan option on a coupon bond with strike that is the par value of the bond. As an option on a coupon bond, a Bermudan swaption has positive probability of early exercise.

Fixed tail Bermudan swaptions:

A fixed tail Bermudan swaption is a Bermudan swaption with a maturity date equal to the last reset date of the underlying swap and which has an initial lockout period during which exercise is prohibited. A fixed tail Bermudan swaption is equivalent to a Bermudan option on a coupon bond with strike that is the par value of the bond. As an option on a coupon bond, a Bermudan fixed tail swaption has positive probability of early exercise.

Constant maturity Bermudan swaptions:

A constant maturity swaption is an option on a swap that can be exercised on every reset date from now up to and including the swaption maturity. When exercised, the swap one gets is one that starts at the exercise date and has a time to maturity which is independent of the time of exercise. A constant maturity Bermudan swaption is equivalent to a Bermudan option on a constant maturity coupon bond with a strike that is the par value of the bond. Since the underlying bond pays coupons, there is positive probability of early exercise.

Bermudan forward swaptions:

A Bermudan forward swaption is an option on a swap that can be exercised on every reset date up to and including the swaption maturity. When exercised, the swap which one gets is a forward swap that starts at the swaption maturity date and ends at a specified period of time later. A Bermudan forward swaption is equivalent to an option to exchange a forward floating rate bond for a forward fixed rate bond. Since the underlying pays no cash flows between initiation and the swaption maturity, a Bermudan forward swaption should not be exercised before maturity.

The confidence intervals of our results are calculated using the central limit theorem. Based on the theorem, the error must be less than four times the observed standard deviation in order to obtain a 99.95% confidence interval. The observed standard deviations are obtained by doing 100 runs of the option price calculations with each run having a sample size of 100,000.

Our numerical results for caps are compared with closed-form solutions. For European swaptions, our results are very close to the Black formula and to BMG's approximation formula in the one-factor case. For the two-factor model, we can only compare our European swaption values to those obtained by traditional Monte Carlo simulation. For Bermudan swaptions, even traditional Monte Carlo simulation cannot be used to check the validity of our calculations. However, we compared our results with traditional Monte Carlo simulation of European swaption valuations and found that Bermudan swaption valuations are never smaller than European values as expected.

The first test is for a one-factor model. The initial term structure is flat at 10%. The volatility structure is also flat at $\lambda_1(t) = 20\%$. This is the same one-factor model used by BMG (1995) to test their approximation formula. The numerical results for caps are shown in Table 1. As we can see, our MCA results agree with the closed form solution to 4–5 digits depending on the cap maturity. This close agreement with the closed form solution is a strong indication that the MCA can approximate the Market Model accurately.

Table 2 shows our results for European swaptions. As shown in BMG (1995), the Black formula is a very accurate approximation for European swaptions in this one-factor model. Our MCA results are in good agreement with both the Black formula and the BMG (1995) approximation formula.

The results for Bermudan forward swaptions are given in Table 3. As we stated in the description of Bermudan forward swaptions, the value of a Bermudan forward swaption should be the same as that of the corresponding European swaption. Our numerical results confirmed this prediction. The differences between values of the European swaptions and Bermudan forward swaptions are within the Monte Carlo simulation errors.

The Bermudan constant maturity swaption calculations are listed in Table 4. As we expected, constant maturity Bermudan swaptions have larger value than European swaptions. The early exercise premium for constant maturity Bermudan swaptions increases with the swaption maturity, but decreases with the strike rate. For the values tested, the maturity of the underlying swap does not seem to have a significant effect on the exercise premium.

Table 1. Cap price.

| Option maturity | Strike | Cap Price (one-factor) | | Black price |
| | | MC price | | |
		value	4* st. dv.	
	8%	0.004684	0.000002	0.004681
1	10%	0.001762	0.000008	0.001760
	12%	0.000473	0.000004	0.000474
	8%	0.004625	0.000004	0.004621
2	10%	0.002253	0.000009	0.002251
	12%	0.000967	0.000004	0.000967
	8%	0.004506	0.000006	0.004501
3	10%	0.002497	0.000011	0.002494
	12%	0.001297	0.000005	0.001296
	8%	0.004342	0.000009	0.004336
4	10%	0.002609	0.000012	0.002604
	12%	0.001512	0.000007	0.001510
	8%	0.004149	0.000010	0.004142
5	10%	0.002640	0.000014	0.002634
	12%	0.001345	0.000009	0.001641
	8%	0.003941	0.000012	0.003930
6	10%	0.002618	0.000015	0.002609
	12%	0.001721	0.000011	0.001714
	8%	0.003726	0.000011	0.003711
7	10%	0.002562	0.000015	0.002549
	12%	0.001753	0.000012	0.001743
	8%	0.003510	0.000010	0.003490
8	10%	0.002482	0.000014	0.002465
	12%	0.001755	0.000015	0.001740
	8%	0.003297	0.000012	0.003271
9	10%	0.002388	0.000015	0.002365
	12%	0.001734	0.000014	0.001714

The Bermudan swaption calculations are given in Table 5. As we expected, Bermudan swaptions have larger value than European swaptions. The early exercise premium for Bermudan swaptions increases with the swaption maturity and decreases with the strike rate and with the maturity of the underlying swap. The early exercise premium is about 55% of the corresponding European swaption for a three-year at-the-money Bermudan swaption on a three-year swap. For the values tested, we found that a Bermudan swaption always has greater value than a constant maturity Bermudan swaption with the same swaption maturity date.

Table 6 shows our results for Bermudan fixed tail swaptions. As expected, Bermudan fixed tail swaptions have greater value than European swaptions maturing at the end of the lockout period. We term this difference the later exercise

Table 2. European swaption price.

| Option maturity × Swap length | Strike | European Swaption Price (one-factor) | | Black price | BMG price |
| | | MC price | | | |
		value	4* st. dv.		
	8%	0.018391	0.000004	0.018388	0.018388
.25 × 1	10%	0.003660	0.000017	0.003659	0.003659
	12%	0.000131	0.000009	0.000135	0.000135
	8%	0.034426	0.000018	0.034405	0.034405
1 × 2	10%	0.012946	0.000062	0.012936	0.012935
	12%	0.003477	0.000029	0.003487	0.003487
	8%	0.074844	0.000043	0.074802	0.074797
1 × 5	10%	0.028138	0.000140	0.028124	0.028114
	12%	0.007553	0.000058	0.007582	0.007573
	8%	0.120499	0.000078	0.120452	0.120419
1 × 10	10%	0.045258	0.000227	0.045288	0.045220
	12%	0.012120	0.000085	0.012208	0.012160
	8%	0.047377	0.000073	0.047329	0.047321
3 × 3	10%	0.026246	0.000130	0.026220	0.026209
	12%	0.013629	0.000075	0.013627	0.013617

Table 3. Bermudan forward swaption price.

| Option maturity × Swap length | Strike | Bermudan Forward Swaption Price (one-factor) | | | European price | |
| | | Bermudan forward price | | | | |
		value	normalized value	4* st. dv.	value	4* st. dv.
	8%	0.018391	1.000000	0.000004	0.018391	0.000004
.25 × 1	10%	0.003660	1.000000	0.000017	0.003660	0.000017
	12%	0.000131	1.000000	0.000009	0.000131	0.000009
	8%	0.034631	1.005963	0.000080	0.034426	0.000018
1 × 2	10%	0.013033	1.006699	0.000085	0.012946	0.000062
	12%	0.003509	1.009205	0.000035	0.003477	0.000029
	8%	0.075274	1.005742	0.000172	0.074844	0.000043
1 × 5	10%	0.028319	1.006439	0.000183	0.028138	0.000140
	12%	0.007619	1.008696	0.000074	0.007553	0.000058
	8%	0.121158	1.005474	0.000276	0.120499	0.000078
1 × 10	10%	0.045536	1.006142	0.000283	0.045258	0.000227
	12%	0.012218	1.008107	0.000113	0.012120	0.000085
	8%	0.047786	1.008631	0.000098	0.047377	0.000073
3 × 3	10%	0.026513	1.010182	0.000166	0.026246	0.000130
	12%	0.013797	1.012311	0.000117	0.013629	0.000075

Table 4. Constant maturity Bermudan swaption.

| Option maturity | Strike | Bermudan Constant Maturity Swaption Price (one-factor) | | | | |
| × Swap length | | Bermudan constant maturity price | | | European price | |
		value	normalized value	4* st. dv.	value	4* st. dv.
	8%	0.018810	1.022767	0.000000	0.018391	0.000004
.25 × 1	10%	0.003660	1.000000	0.000017	0.003660	0.000017
	12%	0.000131	1.000000	0.000009	0.000131	0.000009
	8%	0.036330	1.055312	0.000093	0.034426	0.000018
1 × 2	10%	0.013291	1.026647	0.000094	0.012946	0.000062
	12%	0.003542	1.018631	0.000039	0.003477	0.000029
	8%	0.079003	1.055572	0.000198	0.074844	0.000043
1 × 5	10%	0.028893	1.026856	0.000198	0.028138	0.000140
	12%	0.007691	1.018296	0.000084	0.007553	0.000058
	8%	0.127244	1.055982	0.000306	0.120499	0.000078
1 × 10	10%	0.046491	1.027234	0.000301	0.045258	0.000227
	12%	0.012337	1.017953	0.000132	0.012120	0.000085
	8%	0.056640	1.195515	0.000137	0.047377	0.000073
3 × 3	10%	0.029765	1.134058	0.000172	0.026246	0.000130
	12%	0.015022	1.102187	0.000164	0.013629	0.000075

Table 5. Bermudan swaption.

| Option maturity | Strike | Bermudan Swaption Price (one-factor) | | | | |
| × Swap length | | Bermudan price | | | European price | |
		value	normalized value	4* st. dv.	value	4* st. dv.
	8%	0.023229	1.263061	0.000000	0.018391	0.000004
.25 × 1	10%	0.003660	1.000000	0.000017	0.003660	0.000017
	12%	0.000131	1.000000	0.000009	0.000131	0.000009
	8%	0.051289	1.489841	0.000000	0.034426	0.000018
1 × 2	10%	0.015257	1.178543	0.000088	0.012946	0.000062
	12%	0.003753	1.079286	0.000049	0.003477	0.000029
	8%	0.089425	1.194818	0.000000	0.074844	0.000043
1 × 5	10%	0.030104	1.069877	0.000189	0.028138	0.000140
	12%	0.007812	1.034346	0.000095	0.007553	0.000058
	8%	0.132519	1.099758	0.000000	0.120499	0.000078
1 × 10	10%	0.047133	1.041425	0.000297	0.045258	0.000227
	12%	0.012399	1.023020	0.000139	0.012120	0.000085
	8%	0.089425	1.887509	0.000000	0.047377	0.000073
3 × 3	10%	0.040717	1.551336	0.000186	0.026246	0.000130
	12%	0.018742	1.375147	0.000174	0.013629	0.000075

Table 6. Bermudan fixed tail swaption.

Lockout Period × Swap length	Strike	Fixed Tail Bermudan Swaption Price (one-factor)				
		Bermudan fixed tail price			European price	
		value	normalized value	4* st. dv.	value	4* st. dv.
	8%	0.018471	1.004364	0.000008	0.018391	0.000004
.25 × 1	10%	0.004918	1.343553	0.000019	0.003660	0.000017
	12%	0.000874	6.657649	0.000012	0.000131	0.000009
	8%	0.035589	1.033785	0.000043	0.034426	0.000018
1 × 2	10%	0.015750	1.216607	0.000055	0.012946	0.000062
	12%	0.006181	1.777506	0.000062	0.003477	0.000029
	8%	0.081121	1.083870	0.000185	0.074844	0.000043
1 × 5	10%	0.041987	1.492176	0.000170	0.028138	0.000140
	12%	0.021581	2.857282	0.000187	0.007553	0.000058
	8%	0.139524	1.157886	0.000325	0.120499	0.000078
1 × 10	10%	0.083011	1.834168	0.000285	0.045258	0.000227
	12%	0.051185	4.223304	0.000358	0.012120	0.000085
	8%	0.049547	1.045802	0.000106	0.047377	0.000073
3 × 3	10%	0.029514	1.124512	0.000100	0.026246	0.000130
	12%	0.017110	1.255417	0.000106	0.013629	0.000075

premium of a Bermudan fixed tail swaption. This premium increases with the strike rate and the underlying swap maturity. This is in contrast to a Bermudan swaption whose early exercise premium decreases with the strike rate and swap maturity. The later exercise premium for a one-year Bermudan fixed tail swaption on a ten-year swap struck at 12% is about 325% of the corresponding European swaption value.

As we have seen, our numerical results for cap and European swaptions are very accurate. For Bermudan type swaptions, we have no data to compare with, but our calculations are consistent with known bounds. These results indicate that our MCA is a viable technique in the one-factor Market Model.

Having successfully simulated the one-factor case, we now extend our MCA method to the two-factor Market Model. The initial term structure is still flat at 10%. The two-factor volatility structure we tested is defined by:

$$\lambda_1(t) = .15$$

$$\lambda_2(t) = .15 - \frac{.3}{\sqrt{10}}\sqrt{T-t}, \ T \geq t$$

The first factor is a constant shock for all maturities, while the second factor adds a twist to the yield term structure.

The numerical results for caps are shown in Table 7. As in the one-factor case, our MCA results agree with the Black formula to 4–5 digits. This close agreement with the closed-form solution in the two-factor case is another indication that the

Table 7. Cap price (two-factor case).

| Option maturity | Strike | Cap Price (two-factor) | | Black price |
| | | MC price | | |
		value	4* st. dv.	
	8%	0.004588	0.000002	0.004584
1	10%	0.001529	0.000006	0.001538
	12%	0.000311	0.000003	0.000323
	8%	0.004385	0.000003	0.004379
2	10%	0.001853	0.000008	0.001857
	12%	0.000622	0.000004	0.000631
	8%	0.004168	0.000003	0.004160
3	10%	0.002000	0.000010	0.002000
	12%	0.000836	0.000004	0.000841
	8%	0.003950	0.000005	0.003940
4	10%	0.002069	0.000009	0.002066
	12%	0.000990	0.000005	0.000993
	8%	0.003736	0.000005	0.003727
5	10%	0.002096	0.000010	0.002092
	12%	0.001107	0.000005	0.001107
	8%	0.003531	0.000007	0.003524
6	10%	0.002096	0.000010	0.002092
	12%	0.001195	0.000007	0.001194
	8%	0.003336	0.000007	0.003331
7	10%	0.002078	0.000011	0.002075
	12%	0.001261	0.000007	0.001260
	8%	0.003150	0.000008	0.003149
8	10%	0.002047	0.000012	0.002046
	12%	0.001309	0.000008	0.001309
	8%	0.002974	0.000009	0.002977
9	10%	0.002005	0.000013	0.002007
	12%	0.001341	0.000010	0.001343

MCA can approximate the Market Model accurately. The values of caps in the two-factor case are all smaller than that of the one-factor case. This indicates that the overall volatility of the two-factor case is smaller.

Table 8 shows our results for European swaptions. Unlike the one-factor case, the Black formula is not a good approximation for European swaptions. This is consistent with Jamshidian's conclusion that we cannot generally have a lognormal Libor rate and a lognormal swap rate at the same time. The Black formula can be made more accurate by adjusting the volatility accordingly as shown by Andersen and Andreasen (1998). Our numerical results for European swaptions show that the Black formula overprices European swaptions. As a consistency check, we found that our MCA results for European swaptions are the same as the traditional Monte

Table 8. European swaption price (two-factor case).

Option maturity × Swap length	Strike	MC price value	MC price 4* st. dv.	Black price
	8%	0.018370	0.000002	0.018377
.25 × 1	10%	0.003157	0.000014	0.003502
	12%	0.000046	0.000005	0.000105
	8%	0.033313	0.000016	0.033688
1 × 2	10%	0.010146	0.000049	0.011305
	12%	0.001652	0.000025	0.002376
	8%	0.072140	0.000040	0.073245
1 × 5	10%	0.021185	0.000111	0.024579
	12%	0.003121	0.000053	0.005166
	8%	0.116589	0.000066	0.117944
1 × 10	10%	0.035351	0.000179	0.039579
	12%	0.005584	0.000070	0.008319
	8%	0.043184	0.000030	0.043744
3 × 3	10%	0.020065	0.000097	0.021029
	12%	0.007957	0.000033	0.008843

Table 9. Bermudan forward swaption price (two-factor case).

Option maturity × Swap length	Strike	Bermudan forward price value	Bermudan forward price normalized value	Bermudan forward price 4* st. dv.	European price value	European price 4* st. dv.
	8%	0.018370	1.000000	0.000002	0.018370	0.000002
.25 × 1	10%	0.003157	1.000000	0.000014	0.003157	0.000014
	12%	0.000046	1.000000	0.000005	0.000046	0.000005
	8%	0.033510	1.005908	0.000065	0.033313	0.000016
1 × 2	10%	0.010206	1.005974	0.000067	0.010146	0.000049
	12%	0.001665	1.007784	0.000026	0.001652	0.000025
	8%	0.072411	1.003764	0.000128	0.072140	0.000040
1 × 5	10%	0.021249	1.003004	0.000131	0.021185	0.000111
	12%	0.003135	1.004647	0.000055	0.003121	0.000053
	8%	0.116909	1.002739	0.000196	0.116589	0.000066
1 × 10	10%	0.035400	1.001371	0.000192	0.035351	0.000179
	12%	0.005598	1.002565	0.000075	0.005584	0.000070
	8%	0.043512	1.007606	0.000096	0.043184	0.000030
3 × 3	10%	0.020245	1.008935	0.000126	0.020065	0.000097
	12%	0.008051	1.011753	0.000066	0.007957	0.000033

Table 10. Bermudan constant maturity swaption (two-factor case).

| Option maturity × Swap length | Strike | Bermudan Constant Maturity Swaption Price (two-factor) | | | | |
| | | Bermudan constant maturity price | | | European price | |
		value	normalized value	4* st. dv.	value	4* st. dv.
	8%	0.018810	1.023971	0.000000	0.018370	0.000002
.25 × 1	10%	0.003157	1.000000	0.000014	0.003157	0.000014
	12%	0.000046	1.000000	0.000005	0.000046	0.000005
	8%	0.036027	1.081476	0.000076	0.033313	0.000016
1 × 2	10%	0.010792	1.063704	0.000070	0.010146	0.000049
	12%	0.001719	1.040550	0.000032	0.001652	0.000025
	8%	0.077946	1.080483	0.000000	0.072140	0.000040
1 × 5	10%	0.021865	1.032077	0.000150	0.021185	0.000111
	12%	0.003169	1.015335	0.000058	0.003121	0.000053
	8%	0.125514	1.076546	0.000000	0.116589	0.000066
1 × 10	10%	0.035715	1.010297	0.000223	0.035351	0.000179
	12%	0.005615	1.005531	0.000079	0.005584	0.000070
	8%	0.054716	1.267052	0.000119	0.043184	0.000030
3 × 3	10%	0.024056	1.198881	0.000133	0.020065	0.000097
	12%	0.009135	1.147994	0.000101	0.007957	0.000033

Carlo results (not reported here). As was true for caps, European swaption values in the two-factor case are all smaller than in the one-factor case.

The Bermudan forward swaption results are given in Table 9. As in the one-factor case, the value of Bermudan forward swaptions should be the same as that of the European swaptions. Our numerical results confirmed this result. The differences between values of the European swaptions and Bermudan forward swaptions are within the Monte Carlo simulation errors.

Constant maturity Bermudan swaption calculations are listed in Table 10. As expected, the values of constant maturity Bermudan swaptions are larger than that of European swaptions. The early exercise premium for constant maturity Bermudan swaptions increases with the option maturity but decreases with the strike rate. The early exercise premium in the two-factor case is more sensitive to the strike rate than in the one-factor case and is generally larger especially when the option maturity is longer.

Bermudan swaption calculations are given in Table 11. As in the one-factor case, Bermudan swaptions have larger values than European swaptions. The early exercise premium for Bermudan swaptions increases with the option maturity and decreases with the strike rate and the swap maturity. The early exercise premium is about 51% of the corresponding European swaption value for a three-year at-the-money Bermudan swaption on three-year swap. A Bermudan swaption has greater value than a constant maturity Bermudan swaption. Again, the early exercise

Table 11. Bermudan swaption (two-factor case).

| Option maturity × Swap length | Strike | Bermudan Swaption Price (two-factor) | | | | |
| | | Bermudan price | | | European price | |
		value	normalized value	4* st. dv.	value	4* st. dv.
	8%	0.023229	1.234547	0.000000	0.018370	0.000002
.25 × 1	10%	0.003157	1.000000	0.000014	0.003157	0.000014
	12%	0.000046	1.000000	0.000005	0.000046	0.000005
	8%	0.051289	1.539600	0.000000	0.033313	0.000016
1 × 2	10%	0.012079	1.190541	0.000079	0.010146	0.000049
	12%	0.001767	1.069185	0.000036	0.001652	0.000025
	8%	0.089425	1.239606	0.000000	0.072140	0.000040
1 × 5	10%	0.022315	1.053328	0.000154	0.021185	0.000111
	12%	0.003174	1.017086	0.000060	0.003121	0.000053
	8%	0.132519	1.136632	0.000000	0.116589	0.000066
1 × 10	10%	0.035788	1.012356	0.000223	0.035351	0.000179
	12%	0.005615	1.005595	0.000079	0.005584	0.000070
	8%	0.089425	2.070799	0.000000	0.043184	0.000030
3 × 3	10%	0.030301	1.510138	0.000171	0.020065	0.000097
	12%	0.010303	1.294841	0.000135	0.007957	0.000033

Table 12. Bermudan fixed tail swaption price (two-factor case).

| Lockout period × Swap length | Strike | Fixed Tail Bermudan Swaption Price (two-factor) | | | | |
| | | Bermudan fixed tail price | | | European price | |
		value	normalized value	4* st. dv.	value	4* st. dv.
	8%	0.018416	1.002517	0.000007	0.018370	0.000002
.25 × 1	10%	0.004327	1.370468	0.000019	0.003157	0.000014
	12%	0.000550	11.930164	0.000010	0.000046	0.000005
	8%	0.034097	1.023539	0.000036	0.033313	0.000016
1 × 2	10%	0.012695	1.251323	0.000053	0.010146	0.000049
	12%	0.003710	2.245603	0.000045	0.001652	0.000025
	8%	0.075004	1.039703	0.000115	0.072140	0.000040
1 × 5	10%	0.031667	1.494727	0.000144	0.021185	0.000111
	12%	0.012792	4.098874	0.000117	0.003121	0.000053
	8%	0.122095	1.047221	0.000330	0.116589	0.000066
1 × 10	10%	0.060220	1.703479	0.000377	0.035351	0.000179
	12%	0.032336	5.791074	0.000305	0.005584	0.000070
	8%	0.044723	1.035652	0.000074	0.043184	0.000030
3 × 3	10%	0.022819	1.137231	0.000098	0.020065	0.000097
	12%	0.010765	1.352899	0.000081	0.007957	0.000033

premium in the two-factor case is more sensitive to the strike rate. For our values, Bermudan swaptions were all exercised at a strike rate of 8% or more in both the one- and two-factor cases.

Table 12 shows the results for Bermudan fixed tail swaptions. As expected, Bermudan fixed tail swaptions have greater value than European swaptions maturing at the lockout date. The later exercise premium of Bermudan fixed tail swaptions increases with the strike rate and swap maturity. This is in contrast to Bermudan swaptions whose early exercise premium decreases with the strike rate. The later exercise premium for a one-year Bermudan fixed tail swaption on a ten-year swap struck at 12% is about 170.3% of the corresponding European swaption. As with other Bermudan type swaptions, Bermudan fixed tail swaptions in the two-factor case are also more sensitive to the strike rate than in the one-factor case. The later exercise premium in the two-factor case can be more than twice as big as in the one-factor case for a strike rate of 12%.

6. Summary

The Market Model by BMG and Jamshidian is a widely used model in interest rate derivative markets. However, its application to Bermudan derivatives has been problematic due to the explosive growth in computation time and memory of non-recombining trees. In this paper, we applied the MCA method for pricing Bermudan interest rate derivatives using Monte Carlo simulation. One- and two-factor volatility structures were studied numerically. Numerical data for several types of Bermudan swaptions were presented as a benchmark for comparison with alternative implementations of the Market model. In particular, the efficacy of the MCA method for long maturity derivatives was demonstrated.

Acknowledgment

We thank Leif Andersen, Mark Broadie, Joseph Langsam, Dilip Madan, Walter Muller and Bill Shen for comments. Any errors are our own.

References

L. Andersen and J. Andreasen, "Volatility Skews and Extensions of the Libor Market Model", Working Paper, General Re Financial Products, 1998.

J. Barraquand and D. Martineau, "Numerical valuation of high dimensional multivariate American securities", *Journal of Financial and Quantitative Analysis* **30** (1995) 3.

F. Black, E. Derman and W. Toy, "A one-factor model of interest rates and its applications to treasury bond options", *Financial Analyst Journal* (1990) 33–39.

F. Black and P. Karasinski, "Bond and option pricing when short rates are lognormal", *Financial Analyst Journal* (1991) 52–59.

M. J. Brennan and E. Schwartz, "A continuous time approach to the pricing of bonds", *Journal of Banking and Finance* **3**(2) (1979) 133–56.

M. J. Brennan and E. Schwartz, "An equilibrium model of bond pricing and a test of market efficiency", *Journal of Financial and Quantitative Analysis* **17**(3) (1982) 301–329.

C. Ball and W. Torous, "Bond price dynamics and options", *Journal of Financial and Quantitative Analysis* **18**(4) (1983) 517–32.

F. Black, "The pricing of commodity contracts", *Journal of Financial Economics* **3**(1) (1976) 167–179.

A. Brace, D. Gatarek and M. Musiela, "The Market Model of Interest Rate Dynamics", *Mathematical Finance* **7**(2) April (1995) 127–155.

G. Courtadon, "The pricing of options on default free bonds", *Journal of Financial and Quantitative Analysis* **17** (1982) 75–100.

E. Canabarro, "Where do one-factor interest rate models fail?" *Journal of Fixed Income* **5**(2) (1995) 31–52.

J. C. Cox, J. E. Ingersoll and S. A. Ross, "A theory of the term structure of interest rates", *Econometrica* **53** (1985) 385–407.

B. Flesaker and L. Hughston, "Positive interest", *Risk Magazine* **9**(1) January (1996) 46–49.

D. Heath, R. Jarrow and A. Morton, "Bond Pricing and Term Structure of the Interest Rates: A New Methodology", Working Paper, Cornell University, 1988.

R. J. Harrison and S. Pliska, "Martingales and stochastic integral in the theory of continuous trading", *Stochastic Processes and Their Applications* **11** (1981) 215–260.

J. Hull and A. White, "Pricing interest-rate derivative securities", *The Review of Financial Studies* **3** (1990) 573–592.

T. S. Y. Ho and S. B. Lee, "Term structure movements and pricing interest rate contingent claims", *Journal of Finance* **41** (1986) 1011–1029.

F. Jamshidian, "The Preference-Free Determination of Bond Prices from the Spot Interest Rate", Working Paper, Merrill Lynch Capital Market, World Financial Center, New York, 1989.

F. Jamshidian, "Libor and Swap Market Models and Measures", Working Paper, Sakura Global Capital, 1996.

F. Jamshidian, "Libor and Swap Market Models and Measures II", Working Paper, Sakura Global Capital, 1996.

R. Merton, "The theory of rational optional pricing", *Bell Journal of Economics and Management Science* **4** (1973) 141–183.

L. C. Rogers, "Which model for the term structure of interest rates should one use ?", *Proc. IMA Workshop on Math. Finance*, June 93, D. Duffie and S. E. Shreve, eds., (Springer Verlag, 1995) 93–116.

L. C. Rogers, "Gaussian Errors", preprint, Bath University, 1995.

S. Schaefer and E. Schwartz, "Time-dependent variance and the pricing of bond options", *Journal of Finance* **42**(5) (1987) 1113–1128.

O. A. Vasicek, "An equilibrium characterization of the term structure", *Journal of Financial Economics* **5** (1977) 177–188.

HOW TO USE SELF-SIMILARITIES TO DISCOVER SIMILARITIES OF PATH-DEPENDENT OPTIONS

ALEXANDER LIPTON

Deutsche Bank

Received 11 August 1999

It is well-known that self-similarity is a powerful concept which reveals the fundamental laws of physic. This concept has important financial applications as well. It allows one to study the properties of derivatives from a unified prospective and significantly simplifies their mathematical modelling and hedging. In the present paper, I show how the concept of self-similarity can be used in order to find similarities between various types of path-dependent options and price them in the unified framework. Specifically, I consider lookback, passport, Asian, and imperfectly hedged European options. I present some new important valuation formulas and rederive a few known ones by elementary means.

1. Introduction

It is customary to begin papers on mathematical finance with a reference to the celebrated work of Black-Scholes (1973) and Merton (1973) on rational pricing of options. Even though I cannot break this rule, as a point of departure for the present paper I use the equally celebrated paper by Taylor (1950) in which he analyzes the nuclear explosion of 1945. After processing data from a series of the (unclassified) photographs of the expansion of a fireball Taylor concludes that the (highly classified) energy of the explosion is approximately 10^{21} erg. The main tool Taylor uses to arrive at his remarkable results is the concept of self-similarity. In this and many other cases a skillful use of this concept allows one to reveal the fundamental properties of physical phenomena and to simplify their mathematical treatment. The standard source of information on self-similarities and their physical applications in the well-known book by Barenblatt (1996) to which I refer the interested reader.

In the opinion of the present author, the concept of self-similarity is very important for the purposes of financial engineering. Merton successfully used self-similarity (specifically, homogeneity) early on in order to price derivatives in the markets without risk-free instruments and for other purposes. Subsequently, this concept (in various disguises) has been used by many financial engineers, including Margrabe, Ingersol, Wilmott *et al.*, Rogers and Shi, and Derman. In this paper,

I use self-similarity in order to develop a unified description of various path-dependent options frequently occurring in practice. The concept itself can be used in many other situations. It can be used in order to establish relations between different instruments such as put-call symmetry, and to value a variety of financial products stretching from outperformance options, to options on assets constrained to a band, to basket options.

In order to fix the notation (and to reflect my current area of responsibility), below I consider options on foreign exchange within the Garman-Kohlhagen framework even though my results are equally applicable for equity markets. The main results are as follows. (A) Self-similarities are used in order to find the relations between lookback, Asian, passport, and imperfectly hedged European options and to establish the fact that the valuation formulas for all these options are remarkably similar in nature. (B) It is shown that in the case when the foreign interest rate is small compared to the squared volatility (which is approximately the case for the USD/JPY and EUR/JPY currency pairs) the price of a passport option on FX can be found by elementary means which are much simpler than any of the methods used for this purpose so far (including the original method of Green's functions developed by the present author). Moreover, if the domestic interest rate is small as well, then this price is equal to one-half the price of a lookback put (this special relation was noted by Lipton, 1999 and, independently, by Henderson and Hobson, 1999). (C) An elementary derivation of the valuation formula for Asian options via a partial differential equation (PDE) method is given (this formula was originally derived by Geman and Yor, 1993 by a complicated probabilistic method). (D) The same idea is used in order to obtain a new valuation formula for the passport option in the case when the foreign interest rate is non-zero. (E) Some unexpected relations between Asian and imperfectly hedged European options are described and the so-called stop-loss start-gain hedge is analyzed in some detail.

The structure of the paper is as follows. In Sec. 2, I show how to use the concept of self-similarity in the standard Black-Scholes framework. In Sec. 3, I price lookback puts by using self-similarity and the Laplace transform. In Sec. 4, I give a PDE based derivation of the standard valuation formula for Asian calls. This derivation combines self-similarity, splitting, and Laplace transform. In Sec. 5, I present a new valuation formula for passport options. In Sec. 6, I establish a useful relation between lookback puts and passport options when the foreign interest rate is zero. In Sec. 7, I describe relations between passport, Asian and imperfectly hedged European options. Finally, in Sec. 7, I draw some conclusions.

A short version of this paper will soon appear in The Risk Magazine (Lipton, 1999). After this short version was submitted for publication, I became aware of an interesting paper by Delbaen and Yor (1999) in which the authors establish certain relations between Asian and passport options via probabilistic methods. In general, there is an interesting duality between the PDE approach which uses the concept of self-similarity and the probabilistic approach which uses the idea of the Levy's local time.

2. Valuation of Call Options Via Self-Similarity

Consider call options on foreign exchange (FX). Denote the FX rate, i.e., the number of units of domestic currency (DC) per unit of foreign currency (FC) by S, the domestic and foreign interest rates by r_d and r_f, respectively, and the volatility of S by σ. The European call option (EC) gives the buyer the right to exchange M units of foreign currency for MK units of domestic currency at maturity T. Its payoff at maturity is $M(S - K)$.

The pricing problem for this option has the standard form

$$V_t^{(EC)} + (r_d - r_f)SV_S^{(EC)} + \frac{1}{2}\sigma^2 S^2 V_{SS}^{(EC)} - r_d V^{(EC)} = 0,$$

$$V^{(EC)}(S,T) = M(S - K)_+, \tag{1}$$

plus boundary conditions at $S = 0$, $S = \infty$. Without going into its detailed analysis, I want to show what information about the price of this option can be obtained by virtue of self-similarity. It is clear that the dimensional price as a function of seven-dimensional variables

$$V^{(EC)} = V^{(EC)}(S, K, M, T - t, \sigma, r_d, r_f), \tag{2}$$

where the dimensions of $V, S, K, M, T - t, \sigma, r_d, r_f$ are DC, DC/FC, DC/FC, FC, time, $1/\sqrt{\text{time}}$, 1/time, 1/time, respectively. Without loss of generality one can assume that variables K, M, σ^2 have independent dimensions, and express dimensions of $V^{(EC)}$, S, $T - t$, r_d, r_f as products of powers of the dimensions of K, M, σ^2. Thus, by virtue of the well-known Π-theorem (see, e.g., Barenblatt, 1996) one can represent the price $V^{(EC)}$ in the form

$$V^{(EC)} = MK\Phi^{(EC)}(\xi, \tau, \hat{r}_d, \hat{r}_f), \tag{3}$$

where $\xi = S/K$ is the non-dimensional moneyness, $\tau = \sigma^2(T - t)$ is the non-dimensional time to maturity, and $\hat{r}_d = r_d/\sigma^2$, $\hat{r}_f = r_f/\sigma^2$ are the non-dimensional interest rates. In other words, the "non-trivial" information contained in the price is a function of four variables. This observation allows one to reduce the computational effort to a very large degree.

The non-dimensional pricing problem has the form

$$\Phi_\tau^{(EC)} - (\hat{r}_d - \hat{r}_f)\xi\Phi_\xi^{(EC)} - \frac{1}{2}\xi^2\Phi_{\xi\xi}^{(EC)} + \hat{r}_d\Phi^{(EC)} = 0, \quad 0 \leq \xi < \infty,$$

$$\Phi^{(EC)}(\xi, 0) = (\xi - 1)_+,$$

$$\Phi^{(EC)}(0, \tau) = 0, \quad \Phi_\xi^{(EC)}(\xi, \tau) \to e^{-\hat{r}_f\tau} \text{ when } \xi \to \infty. \tag{4}$$

It is easy to show that $\Phi^{(EC)}$ has the form

$$\Phi^{(EC)} = e^{-\hat{r}_f\tau}\xi N\left[\frac{\ln(\xi) + \upsilon_+\tau}{\sqrt{\tau}}\right] - e^{-\hat{r}_d\tau}N\left[\frac{\ln(\xi) - \upsilon_-\tau}{\sqrt{\tau}}\right], \tag{5}$$

where $\upsilon_\pm = 1/2 \pm (\hat{r}_d - \hat{r}_f)$, and $N(.)$ is the cumulative normal distribution. The corresponding dimensional solution is

$$V^{(EC)} = e^{-r_f(T-t)} MSN \left[\frac{\ln(S/K) + \left(r_d - r_f + \frac{1}{2}\sigma^2\right)(T-t)}{\sigma\sqrt{T-t}} \right]$$

$$- e^{-r_d(T-t)} MKN \left[\frac{\ln(S/K) + \left(r_d - r_f + \frac{1}{2}\sigma^2\right)(T-t)}{\sigma\sqrt{T-t}} \right] . \qquad (6)$$

Expression (5) is much simpler than the dimensional expression (6).

I will show below that self-similar reductions significantly simplify most of the valuation problems for path-dependent options as well.

3. Valuation of Lookback Options Via Self-Similarity

In this section, I show how to value lookback puts (LBP) by virtue of self-similarity. The valuation formula is well-known but is discussed here nonetheless in order to facilitate subsequent developments. Consider the evolution of the exchange rate S between times t (today) and T (maturity) and denote by J the maximum value of S over this period, $J = \max_{t \leq t' \leq T} S(t)$. The LPB has the payoff $M(J - S)$ at maturity which is (almost) always positive, so that, strictly speaking, it is not an option since it is always exercised. The LPB gives the buyer the right to sell the foreign currency at the best exchange rate occurring between inception of the put and its maturity. It is not difficult to obtain the valuation formula for LPBs (and calls) via probabilistic methods which are based on the fact that the joint distribution of (S, J) is readily available, see Goldman *et al.* (1979) and Garman (1989). However, below I need a derivation based on PDE methods.

It is clear that the dimensional price of the put which is denoted by $V^{(LBP)}$ depends on both S and J. In view of the previous discussion, one can treat J as a moving strike and represent this price in the form

$$V^{(LBP)}(S, J, M, T-t, \sigma, r_d, r_f) = MJ\tilde{\Phi}^{(LBP)}(\tilde{\xi}, \tau, \hat{r}_d, \hat{r}_f), \qquad (7)$$

where $\tilde{\xi} = S/J$, $0 < \tilde{\xi} \leq 1$, is the corresponding moneyness. For the purposes of the present paper, it is more convenient to consider the LBP as a call option on J struck at S and write the price in the form

$$V^{(LBP)}(S, J, M, T-t, \sigma, r_d, r_f) = MS\Phi^{(LBP)}(\xi, \tau, \hat{r}_d, \hat{r}_f), \qquad (8)$$

where $\xi = J/S$, $1 \leq \xi < \infty$, is the inverse moneyness. The pricing problem for $\Phi^{(LBP)}$ can be written as

$$\Phi_\tau^{(LBP)} - (\hat{r}_f - \hat{r}_d)\xi\Phi_\xi^{(LBP)} - \frac{1}{2}\xi^2\Phi_{\xi\xi}^{(LBP)} + \hat{r}_f\Phi^{(LBP)} = 0, \quad 1 \le \xi < \infty,$$

$$\Phi^{(LBP)}(\xi,0) = \xi - 1,$$

$$\Phi_\xi^{(LBP)}(1,\tau) = 0, \quad \Phi_\xi^{(LBP)}(\xi,\tau) \to e^{-\hat{r}_d\tau} \text{ when } \xi \to \infty. \tag{9}$$

The above derivation is similar in spirit to the one given by Wilmott *et al.* (1993).

It is clear that problems (4) and (9) are analogous. The biggest distinction between them is in the boundary conditions. One can consider the LPB as a barrier option with a (financially) non-standard Neumann boundary condition at the barrier $\xi = 1$. The solution of the valuation problem can be found by virtue of the generalized method of images due to Sommerfeld (see, e.g., Wilmott *et al.*, 1993, and Lipton, 1999). Here I present an alternative derivation based on the Laplace transform technique which, to my knowledge, is new. (I hope it will help the reader to appreciate the future developments.) The Laplace transform in time $\Phi^{(LBP)}(\xi,\tau) \to \Psi^{(LPB)}(\xi,\lambda) = \int_0^\infty e^{-\lambda\tau}\Phi^{(LBP)}(\xi,\tau)d\tau$ yields

$$(\lambda + \hat{r}_f)\Psi^{(LBP)} - (\hat{r}_f - \hat{r}_d)\xi\Psi_\xi^{(LBP)} - \frac{1}{2}\xi^2\Psi_{\xi\xi}^{(LBP)} = \xi - 1, \quad 1 \le \xi < \infty,$$

$$\Psi_\xi^{(LBP)}(1,\lambda) = 0, \quad \Psi_\xi^{(LBP)}(\xi,\lambda) \to \frac{1}{(\lambda + \hat{r}_d)} \text{ when } \xi \to \infty. \tag{10}$$

I represent Ψ in the form

$$\Psi^{(LPB)}(\xi,\lambda) = \Xi^{(LBP)}(\xi,\lambda) + \frac{\xi}{(\lambda + \hat{r}_d)} - \frac{1}{(\lambda + \hat{r}_f)}, \tag{11}$$

and obtain the following problem for $\Xi^{(LBP)}$

$$(\lambda + \hat{r}_f)\Xi^{(LBP)} - (\hat{r}_f - \hat{r}_d)\xi\Xi_\xi^{(LBP)} - \frac{1}{2}\xi^2\Xi_{\xi\xi}^{(LBP)} = 0, \quad 1 \le \xi < \infty,$$

$$\Xi_\xi^{(LBP)}(1,\lambda) = -\frac{1}{(\lambda + \hat{r}_d)}, \quad \Xi_\xi^{(LBP)}(\xi,\lambda) \to 0 \text{ when } \xi \to \infty. \tag{12}$$

Since Ξ satisfies a homogeneous Euler equation, it can be chosen in the form

$$\Xi^{(LBP)}(\xi,\lambda) = \Delta^{(LBP)}\xi^{-\alpha}, \tag{13}$$

where

$$\alpha = \sqrt{(1/2 + \hat{r}_d - \hat{r}_f)^2 + 2(\lambda + \hat{r}_f)} - (1/2 + \hat{r}_d - \hat{r}_f), \tag{14}$$

$$\Delta^{(LBP)} = \frac{1}{(\lambda + \hat{r}_d)\alpha}. \tag{15}$$

In order to compute the inverse Laplace transform of Ξ I represent this function in the form

$$\Xi^{(LBP)}(\xi, \lambda) = \frac{2\xi^{\upsilon_+} e^{-\ln(\xi)\sqrt{\mu}}}{(\sqrt{\mu} - \upsilon_+)(\mu - \upsilon_-^2)}$$

$$= \frac{2\xi^{\upsilon_+} e^{-\ln(\xi)\sqrt{\mu}}}{(\upsilon_+ - \upsilon_-)(\sqrt{\mu} - \upsilon_+)} + \frac{\xi^{\upsilon_+} e^{-\ln(\xi)\sqrt{\mu}}}{\upsilon_-(\sqrt{\mu} + \upsilon_-)}$$

$$- \frac{\xi^{\upsilon_+} e^{-\ln(\xi)\sqrt{\mu}}}{\upsilon_-(\upsilon_+ - \upsilon_-)(\sqrt{\mu} - \upsilon_-)}, \tag{16}$$

where

$$\mu = 2(\lambda + \hat{r}_f) + \upsilon_+^2. \tag{17}$$

By using the standard inversion formulas for the Laplace transform (see, Abramowitz and Stegun, 1972, Eqs. 29.2.14, 29.3.88), I obtain

$$\mathcal{L}^{-1}[\Xi^{(LBP)}(\xi, \lambda)] = \frac{2\upsilon_+ e^{-\hat{r}_f \tau}}{(\upsilon_+ - \upsilon_-)} N\left(-\frac{\ln(\xi) - \upsilon_+ \tau}{\sqrt{\tau}}\right)$$

$$- \xi e^{-\hat{r}_d \tau} N\left(-\frac{\ln(\xi) + \upsilon_- \tau}{\sqrt{\tau}}\right)$$

$$- \frac{\xi^{(\upsilon_+ - \upsilon_-)} e^{-\hat{r}_d \tau}}{(\upsilon_+ - \upsilon_-)} N\left(-\frac{\ln(\xi) - \upsilon_- \tau}{\sqrt{\tau}}\right). \tag{18}$$

Accordingly,

$$\Phi^{(LBP)}(\xi, \tau) = \mathcal{L}^{-1}\left[\Xi^{(LBP)}(\xi, \lambda) + \frac{\xi}{(\lambda + \hat{r}_d)} - \frac{1}{(\lambda + \hat{r}_f)}\right]$$

$$= \mathcal{L}^{-1}[\Xi^{(LBP)}(\xi, \lambda)] + e^{-\hat{r}_d \tau} \xi - e^{-\hat{r}_f \tau}$$

$$= e^{-\hat{r}_d \tau} \xi \left[N\left(\frac{\ln(\xi) + \upsilon_- \tau}{\sqrt{\tau}}\right) - \frac{\xi^{2(\hat{r}_d - \hat{r}_f)-1}}{2(\hat{r}_d - \hat{r}_f)} N\left(-\frac{\ln(\xi) - \upsilon_- \tau}{\sqrt{\tau}}\right)\right]$$

$$- e^{-\hat{r}_f \tau} \left[N\left(\frac{\ln(\xi) - \upsilon_+ \tau}{\sqrt{\tau}}\right) - \frac{1}{2(\hat{r}_d - \hat{r}_f)} N\left(-\frac{\ln(\xi) - \upsilon_+ \tau}{\sqrt{\tau}}\right)\right]. \tag{19}$$

It is easy to check that this formula is equivalent to the standard one. In the special case when $\hat{r}_d = \hat{r}_f = \hat{r}$ the above formula assumes the form

$$\Phi^{(LBP)}(\xi, \tau) = e^{-\hat{r}\tau} \left\{ \xi N\left(\frac{\ln(\xi) + \tau/2}{\sqrt{\tau}}\right) - N\left(\frac{\ln(\xi) - \tau/2}{\sqrt{\tau}}\right) - [\ln(\xi) - \tau/2] \right.$$

$$\left. \times N\left(-\frac{\ln(\xi) - \tau/2}{\sqrt{\tau}}\right) + \sqrt{\tau} n\left(\frac{\ln(\xi) - \tau/2}{\sqrt{\tau}}\right) \right\}, \tag{20}$$

where $n(.)$ is the standard normal distribution.

4. Valuation of Asian Options Via Self-Similarity and Splitting

In the previous section, I showed how to value LPBs by combining self-similarities and Laplace transforms. It turns out that the same technique can be used in order to value a number of other options. To illustrate this point, in this section I study Asian calls (ACs) while in the next section I consider POs.

The AC gives the buyer the right to purchase the foreign currency at the average rate prevailing between the inception of the option and its maturity. For my purposes, it is convenient to introduce the cumulative FX rate $I(t) = \int_0^t S(\xi)d\xi$, (its dimension is (DC/FC) $\times$ time). In terms of I the payoff of an AC can be written as $M(I/T - K)_+$. The pair $S(t)$, $I(t)$ is governed by the system of SDEs,

$$dS = (r_d - r_f)Sdt + \sigma SdW\,, \quad dI = Sdt\,, \tag{21}$$

so that the corresponding pricing problem is the degenerate two-factor parabolic problem of the form

$$V_t^{(AC)} + (r_d - r_f)SV_S^{(AC)} + SV_I^{(AC)} + \frac{1}{2}\sigma^2 S^2 V_{SS}^{(AC)} - r_d V^{(PO;q)} = 0\,,$$

$$V^{(AC)}(S,T,I) = M(I/T - K)_+\,, \tag{22}$$

where $V^{(AC)}$ is the price of the AC. As before, one can use the self-similarity reduction and represent this price in the form

$$V^{(AC)}(S,I,M,T-t,\sigma,r_d,r_f) = MS\Phi^{(AC)}(\xi,\tau,\hat{r}_d,\hat{r}_f)\,, \tag{23}$$

where $\xi = (I/T - K)/S$ (this non-dimensional variable is used by Rogers and Shi, 1995; slightly different variables have been used by other researchers). The pricing problem for $\Phi^{(AC)}$ has the form

$$\Phi_\tau^{(AC)} + [(\hat{r}_d - \hat{r}_f)\xi - 1/\hat{T}]\Phi_\xi^{(AC)} - \frac{1}{2}\xi^2\Phi_{\xi\xi}^{(AC)} + \hat{r}_f\Phi^{(AC)} = 0\,, \quad -\infty < \xi < \infty\,,$$

$$\Phi^{(AC)}(\xi,0) = \xi_+\,,$$

$$\Phi^{(AC)}(\xi,\tau) \to 0 \text{ when } \xi \to -\infty\,, \quad \Phi_\xi^{(AC)}(\xi,\tau) \to e^{-\hat{r}_d\tau} \text{ when } \xi \to \infty\,, \tag{24}$$

where $\hat{T} = \sigma^2 T$ is the non-dimensional maturity of the option. This problem is difficult to handle directly because it has to be considered on the whole axis including the origin where the diffusion coefficient vanishes. Fortunately, the pricing problem can be split into two weakly dependent problems. Namely, it can be shown that due to a singularity at $\xi = 0$, one can study the pricing problem separately for $\xi \geq 0$ and $\xi \leq 0$ as long as the corresponding solutions $\Phi_\pm^{(AC)}$ match at $\xi = 0$. (Derivatives of these solutions need not match.) For $\xi \geq 0$ the solution can be chosen in the affine form (this ansatz is similar in nature to the one used by Vasicek, Cox *et al.*, and others),

$$\Phi_+^{(AC)}(\xi,\tau) = e^{-\hat{r}_d\tau}\xi + \frac{e^{-\hat{r}_f\tau} - e^{-\hat{r}_d\tau}}{(\hat{r}_d - \hat{r}_f)\hat{T}}\,. \tag{25}$$

Unfortunately, for negative ξ the corresponding $\Phi_-^{(AC)}$ has a much more complicated form. The pricing problem for $\Phi_-^{(AC)}$ has the form[a]

$$\Phi_{-,\tau}^{(AC)} + [(\hat{r}_d - \hat{r}_f)\xi - 1/\hat{T}]\Phi_{-,\xi}^{(AC)} - \frac{1}{2}\xi^2\Phi_{-,\xi\xi}^{(AC)} + \hat{r}_f\Phi_-^{(AC)} = 0, \quad -\infty < \xi < 0,$$

$$\Phi_-^{(AC)}(\xi, 0) = 0,$$

$$\Phi_-^{(AC)}(\xi, \tau) \to 0 \text{ when } \xi \to -\infty, \quad \Phi_-^{(AC)}(0, \tau) = \frac{e^{-\hat{r}_f\tau} - e^{-\hat{r}_d\tau}}{(\hat{r}_d - \hat{r}_f)\hat{T}}. \quad (26)$$

As before, the Laplace transform in time $\Phi_-^{(AC)}(\xi, \tau) \to \Psi_-^{(AC)}(\xi, \lambda)$ makes this problem analytically tractable (up to a point). The transformed problem has the form

$$(\lambda + \hat{r}_f)\Psi_-^{(AC)} + \left[(\hat{r}_d - \hat{r}_f)\xi - \frac{1}{\hat{T}}\right]\Psi_{-,\xi}^{(AC)} - \frac{1}{2}\xi^2\Psi_{-,\xi\xi}^{(AC)} = 0, \quad -\infty < \xi < 0,$$

$$\Psi_-^{(AC)}(\xi, \lambda) \to 0 \text{ when } \xi \to -\infty, \quad \Psi_-^{(AC)}(0, \lambda) = \frac{1}{(\lambda + \hat{r}_d)(\lambda + \hat{r}_f)\hat{T}}. \quad (27)$$

I claim that the pricing equation is a general confluent equation (see, Abramowitz and Stegun, 1972, Eq. 13.1.35). Recall that general confluent equation has the form

$$\Psi_{\xi\xi} + \left[\frac{2A}{\xi} + 2f_\xi + \frac{bh_\xi}{h} - h_\xi - \frac{h_{\xi\xi}}{h_\xi}\right]\Psi_\xi$$
$$+ \left[\left(\frac{bh_\xi}{h} - h_\xi - \frac{h_{\xi\xi}}{h_\xi}\right)\left(\frac{A}{\xi} + f_\xi\right)\right.$$
$$\left. + \frac{A(A-1)}{\xi^2} + \frac{2Af_\xi}{\xi} + f_{\xi\xi} + f_\xi^2 - \frac{ah_\xi^2}{h}\right]\Psi = 0, \quad (28)$$

where f, h are functions of ξ, and A, a, b are constants. In order to represent Eq. (27) in the form (28) one has to rewrite this equation as

$$\Psi_{-,\xi\xi}^{(AC)} + \left[\frac{2}{\hat{T}\xi^2} - \frac{2(\hat{r}_d - \hat{r}_f)}{\xi}\right]\Psi_{-,\xi}^{(AC)} - \frac{2(\lambda + \hat{r}_f)}{\xi^2}\Psi_-^{(AC)} = 0, \quad (29)$$

and choose the corresponding parameters as follows

$$f = 0, \quad h = \frac{2}{\hat{T}\xi},$$

$$A = a = \alpha,$$

$$b = \beta = 2(\alpha + 1 + \hat{r}_d - \hat{r}_f), \quad (30)$$

---

[a]There is an interesting similarity between the above pricing problem and Merton's pricing problem for call options on stocks paying *constant* dividend which will be discussed elsewhere.

where α is given by expression (14). Accordingly, the regular solution of problem (27) has the form

$$\Psi_{-}^{(AC)}(\xi,\lambda) = \Delta^{(AC)}(-\xi)^{-\alpha} M\left(\alpha,\beta,\frac{2}{\hat{T}\xi}\right), \qquad (31)$$

where M is Kummer's function,

$$\Delta^{(AC)} = \frac{2^{\alpha}\Gamma(\beta-\alpha)}{(\lambda+\hat{r}_d)(\lambda+\hat{r}_f)\hat{T}^{\alpha+1}\Gamma(\beta)}, \qquad (32)$$

and Γ is Gamma function. One can check that this expression is equivalent to the one given by Geman and Yor (1993). The complete solution has the form

$$\Phi^{(AC)}(\xi,\tau) = \begin{cases} e^{-\hat{r}_d\tau}\xi + (e^{-\hat{r}_f\tau} - e^{-\hat{r}_d\tau})/(\hat{r}_d - \hat{r}_f)\hat{T} & \text{when} \quad \xi \geq 0 \\ \mathcal{L}^{-1}[\Psi_{-}^{(AC)}(\xi,\lambda)] & \text{when} \quad \xi \leq 0 \end{cases}, \qquad (33)$$

where $\mathcal{L}^{-1}$ is the inverse Laplace transform. The numerical implementation of the inverse Laplace transform is relatively difficult to achieve. Details are given by Geman and Eydeland (1995); the reader should be warned, however, that their important paper contains a typographical error in the key formula (3). (Unfortunately, this error has been reproduced in several textbooks.)

5. Valuation of Passport Options Via Self-Similarity and Splitting

In this section, I consider POs originally introduced by Hyer, Lipton and Pugachevsky (1997). Recall that the PO allows the buyer to select a strategy of her choice (within certain limits) and gives her the right to keep all the gains generated by this strategy while obligating the seller to absorb all the losses. By now the literature on POs is rather extensive; the reader is referred to the papers by Ahn *et al.*, Anderson *et al.*, Delbaen and Yor, Henderson and Hobson, Hyer *et al.*, Nagayama, Shreve and Večeř, and others for details. Let $Q(t)$, $|Q(t)| \leq M$, be the strategy function chosen by the buyer; it shows the amount of foreign currency which she buys or sells at time t, and let $\Pi(t)$ be the value of the corresponding trading account. It is cleat that one can think of a PO as a EC on Π with zero strike.

In principle, $Q(t)$ can depend on all the information available at time t, but without loss of generality one can assume that it is a function of $S(t)$, $\Pi(t)$. The pair $S(t)$, $\Pi(t)$ is governed by the following system of stochastic differential equations (SDEs),

$$dS = (r_d - r_f)Sdt + \sigma SdW, \quad d\Pi = r_d\Pi dt + q\sigma SdW, \qquad (34)$$

the corresponding degenerate two-factor pricing problem is

$$V_t^{(PO;Q)} + (r_d - r_f)SV_S^{(PO;Q)} + r_d\Pi V_\Pi^{(PO;Q)}$$

$$+ \frac{1}{2}\sigma^2 S^2 \left(V_{SS}^{(PO;Q)} + 2qV_{S\Pi}^{(PO;Q)} + q^2 V_{\Pi\Pi}^{(PO;Q)}\right) - r_d V^{(PO;Q)} = 0,$$

$$V^{(PO;Q)}(S,T,\Pi) = \Pi_+, \tag{35}$$

where $V^{(PO;Q)}$ denotes the instantaneous value of the option for a given strategy Q. As before, this value can be represented in the form

$$V^{(PO;Q)}(S,\Pi,M,T-t,\sigma,r_d,r_f) = MS\Phi^{(PO;q)}(\xi,\tau,\hat{r}_d,\hat{r}_f), \tag{36}$$

where $\xi = \Pi/MS$, $-\infty < \xi < \infty$, is the relative value of the trading account, and $q = Q/M$ is the fraction of the notional amount which the option holder buys or sells at any given time.

Problem (35) explicitly depends on the choice of strategy. There are plenty of reasons for the buyer to choose a sub-optimal strategy (see, e.g., Garbade, 1999). However, the seller of the option has to price it assuming that the buyer chooses the optimal strategy. This strategy has the form

$$q = -\text{sgn}(\xi). \tag{37}$$

The corresponding reduced Hamilton-Jacobi-Bellman pricing problem for $\Phi^{(PO)} = \max_q[\Phi^{(PO;q)}]$ can be written as

$$\Phi_\tau^{(PO)} - \hat{r}_f\xi\Phi_\xi^{(PO)} - \frac{1}{2}(|\xi|+1)^2\Phi_{\xi\xi}^{(PO)} + \hat{r}_f\Phi^{(PO)} = 0, \quad -\infty < \xi < \infty,$$

$$\Phi^{(PO)}(\xi,0) = \xi_+,$$

$$\Phi^{(PO)}(\xi,\tau) \to 0 \text{ when } \xi \to -\infty, \quad \Phi_\xi^{(PO)}(\xi,\tau) \to 1 \text{ when } \xi \to \infty. \tag{38}$$

I emphasize that $\Phi^{(PO)}$ is independent of $\hat{r}_d$. Once again, the pricing problem is difficult to solve because the domain of ξ covers the entire axis. The corresponding diffusion coefficient does not vanish so that the naive splitting of the previous section cannot be used. Fortunately, the following observation comes to the rescue. The pricing equation is invariant with respect to the transformation $\xi \to -\xi$, i.e., it preserves parity. Accordingly, the split of the initial data into the odd and even components,

$$\xi_+ = \frac{1}{2}\xi + \frac{1}{2}|\xi|, \tag{39}$$

is preserved for all $\tau > 0$. Thus, the pricing equation supplied with the initial data $\xi/2$ and $|\xi|/2$ can be solved separately. I denote the corresponding solutions by $\Phi_O^{(PO)}$ and $\Phi_E^{(PO)}$, respectively. It is clear that $\Phi_O^{(PO)} = \xi/2$. For $\xi \geq 0$ it is convenient to deal with the adjusted even solution $\bar{\Phi}_E^{(PO)} = \Phi_E^{(PO)} - \xi/2$, rather than with $\Phi_E^{(PO)}$ itself. The function $\bar{\Phi}_E^{(PO)}$ solves the problem

$$\bar{\Phi}_{E,\tau}^{(PO)} - \hat{r}_f \xi \bar{\Phi}_{E,\xi}^{(PO)} - \frac{1}{2}(\xi+1)^2 \bar{\Phi}_{E,\xi\xi}^{(PO)} + \hat{r}_f \bar{\Phi}_E^{(PO)} = 0, \quad 0 \le \xi < \infty,$$

$$\bar{\Phi}_E^{(PO)}(\xi,0) = 0,$$

$$\bar{\Phi}_{E,\xi}^{(PO)}(0,\tau) = -\frac{1}{2}, \quad \bar{\Phi}_E^{(PO)}(\xi,\tau) \to 0 \text{ when } \xi \to \infty. \tag{40}$$

As before, I use the Laplace transform in time $\bar{\Phi}_E^{(PO)}(\xi,\tau) \to \bar{\Psi}_E^{(PO)}(\xi,\lambda)$ and obtain the following problem for $\bar{\Psi}_E^{(PO)}$

$$(\lambda + \hat{r}_f)\bar{\Psi}_E^{(PO)} - \hat{r}_f \xi \bar{\Psi}_{E,\xi}^{(PO)} - \frac{1}{2}(\xi+1)^2 \bar{\Psi}_{E,\xi\xi}^{(PO)} = 0, \quad 0 \le \xi < \infty,$$

$$\bar{\Psi}_{E,\xi}^{(PO)}(0,\lambda) = -\frac{1}{2\lambda}, \quad \bar{\Psi}_E^{(PO)}(\xi,\lambda) \to 0 \text{ when } \xi \to \infty. \tag{41}$$

The shift $\xi \to \zeta = \xi + 1$ yields the pricing problem of the form

$$(\lambda + \hat{r}_f)\bar{\Psi}_E^{(PO)} + \hat{r}_f(1-\zeta)\bar{\Psi}_{E,\zeta}^{(PO)} - \frac{1}{2}\zeta^2 \bar{\Psi}_{E,\zeta\zeta}^{(PO)} = 0, \quad 1 \le \zeta < \infty,$$

$$\bar{\Psi}_{E,\zeta}^{(PO)}(1,\lambda) = -\frac{1}{2\lambda}, \quad \bar{\Psi}_E^{(PO)}(\zeta,\lambda) \to 0 \text{ when } \zeta \to \infty, \tag{42}$$

which is analogous to problem (27). Its regular solution can be written as

$$s\bar{\Psi}_E^{(PO)}(\zeta,\lambda) = \Delta^{(PO)}\zeta^{-\alpha}M\left(\alpha,\beta,-\frac{2\hat{r}_f}{\zeta}\right), \tag{43}$$

where

$$\Delta^{(PO)} = \frac{\beta}{2\lambda\alpha}[\beta M(\alpha,\beta,-2\hat{r}_f) - 2\hat{r}_f M(\alpha+1,\beta+1,-2\hat{r}_f)]^{-1}, \tag{44}$$

α, β are given by expressions (15), (30) with $\hat{r}_d = 0$. Thus,

$$\bar{\Psi}_E^{(PO)}(\xi,\lambda) = \Delta^{(PO)}(\xi+1)^{-\alpha}M\left(\alpha,\beta,-\frac{2\hat{r}_f}{\xi+1}\right), \tag{45}$$

$$\Phi_E^{(PO)}(\xi,\tau) = |\xi|/2 + \mathcal{L}^{-1}[\bar{\Psi}_E^{(PO)}(|\xi|,\lambda)]. \tag{46}$$

Finally, $\Phi^{(PO)}$ can be represented as

$$\Phi^{(PO)}(\xi,\tau) = \xi_+ + \mathcal{L}^{-1}[\bar{\Psi}_E^{(PO)}(|\xi|,\lambda)]. \tag{47}$$

When $\hat{r}_f = 0$ the function $\Phi^{(PO)}$ can be found explicitly,

$$\bar{\Psi}_E^{(PO)}(\xi,\lambda) = \frac{\sqrt{\xi+1}\,e^{-\ln(\xi+1)\sqrt{\mu}}}{(\mu-1/4)(\sqrt{\mu}-1/2)}$$

$$= \frac{\sqrt{\xi+1}\,e^{-\ln(\xi+1)\sqrt{\mu}}}{(\sqrt{\mu}+1/2)} - \frac{\sqrt{\xi+1}\,e^{-\ln(\xi+1)\sqrt{\mu}}}{(\sqrt{\mu}-1/2)}$$

$$+ \frac{\sqrt{\xi+1}\,e^{-\ln(\xi+1)\sqrt{\mu}}}{(\sqrt{\mu}-1/2)^2}, \tag{48}$$

where $\mu = 2\lambda + 1/4$. A tedious calculation yields

$$\Phi^{(PO)}(\xi,\tau) = \frac{1}{2}\left\{ \xi + (|\xi| + 1)N\left(\frac{\ln(|\xi| + 1) + \tau/2}{\sqrt{\tau}}\right)\right.$$

$$- N\left(\frac{\ln(|\xi| + 1) - \tau/2}{\sqrt{\tau}}\right) - [\ln(|\xi| + 1) - \tau/2]$$

$$\left. \times N\left(-\frac{\ln(|\xi| + 1) - \tau/2}{\sqrt{\tau}}\right) + \sqrt{\tau}n\left(\frac{\ln(|\xi| + 1) - \tau/2}{\sqrt{\tau}}\right)\right\}. \quad (49)$$

It can be checked that expression (47) tends to expression (49) when $r_f \to 0$.

6. Similarities Between Lookback and Passport Options

In this section, I discuss relations between LBPs and POs in the special case when $\hat{r}_d = \hat{r}_f = 0$. Specifically, I use the splitting technique to show that there is a close relation between pricing problems (9) and (38) and hence their solutions. Indeed, when $\hat{r}_d = \hat{r}_f = 0$ the corresponding pricing problems are

$$\Phi_\tau^{(LBP)} - \frac{1}{2}\xi^2\Phi_{\xi\xi}^{(LBP)} = 0, \quad 1 \le \xi < \infty,$$

$$\Phi^{(LBP)}(\xi,0) = \xi - 1,$$

$$\Phi_\xi^{(LBP)}(1,\tau) = 0, \quad \Phi_\xi^{(LBP)}(\xi,\tau) \to 1 \text{ when } \xi \to \infty, \quad (50)$$

$$\Phi_\tau^{(PO)} - \frac{1}{2}(|\xi| + 1)^2\Phi_{\xi\xi}^{(PO)} = 0, \quad -\infty < \xi < \infty,$$

$$\Phi^{(PO)}(\xi,0) = \xi_+,$$

$$\Phi^{(PO)}(\xi,\tau) \to 0 \text{ when } \xi \to -\infty, \quad \Phi_\xi^{(PO)}(\xi,\tau) \to 1 \text{ when } \xi \to \infty. \quad (51)$$

Splitting of this problem into even and odd parts yields

$$\Phi_{E,\tau}^{(PO)} - \frac{1}{2}(\xi + 1)^2\Phi_{E,\xi\xi}^{(PO)} = 0, \quad 0 \le \xi < \infty,$$

$$\Phi_E^{(PO)}(\xi,0) = \frac{1}{2}\xi,$$

$$\Phi_{E,\xi}^{(PO)}(0,\tau) = 0, \quad \Phi_{E,\xi}^{(PO)}(\xi,\tau) \to \frac{1}{2} \text{ when } \xi \to \infty. \quad (52)$$

It is clear that problem (52), can be transformed into problem (50) by virtue of a simple shift and scaling, so that for positive ξ one can write

$$\Phi_E^{(PO)}(\xi,\tau) = \frac{1}{2}\Phi^{(LBP)}(\xi + 1,\tau). \quad (53)$$

(The reader of the old school who prefers representation (21) to representation (22) has to use the shifted Kelvin transform to find the relation between $\Phi_E^{(PO)}$ and

$\tilde{\Phi}^{(LBP)}$. Recall that the definition of the Kelvin transform which is frequently used in hydrodynamics and electrodynamics is as follows: $(\mathbf{K}f)(x) = xf(1/x)$.) Thus, the relation between $\Phi^{(LBP)}$ and $\Phi^{(PO)}$ can be written as

$$\Phi^{(PO)}(\xi,\tau) = \frac{1}{2}[\xi + \Phi^{(LBP)}(|\xi| + 1,\tau)]. \tag{54}$$

In particular,

$$\Phi^{(PO)}(0,\tau) = \frac{1}{2}\Phi^{(LBP)}(1,\tau). \tag{55}$$

Relation (54) can also be established via direct comparison of formulas (20), (49).

The corresponding dimensional relations are

$$V^{(PO)}(S,\Pi,M,T-t,\sigma,0,0) = \frac{1}{2}\left[\Pi + V^{(LBP)}\left(S,\frac{|\Pi|}{M} + S,M,T-t,\sigma,0,0\right)\right],\tag{56}$$

$$V^{(PO)}(S,0,M,T-t,\sigma,0,0) = \frac{1}{2}V^{(LBP)}(S,S,M,T-t,\sigma,0,0), \tag{57}$$

so that at inception the price of the PO is equal to one-half the price of the ATM LPB. I emphasize that $V^{(PO)}$ is independent of the value of $\hat{r}_d$, so that relation (56) can be generalized as follows

$$V^{(PO)}(S,\Pi,M,T-t,\sigma,r_d,0)$$
$$= \frac{1}{2}\left[\Pi + V^{(LBP)}\left(S,\frac{|\Pi|}{M} + S,M,T-t,\sigma,0,0\right)\right]. \tag{58}$$

It is useful to keep in mind a simple pictorial explanation of the connection between LBPs and POs. Consider a certain scenario of the FX rate evolution. For example, assume that this rate has a single maximum at $t = t_*$. At time $t = t_0$ the investor has a choice of strategy. For the sake of the argument, assume that she buys the foreign currency at $t = 0$. Since initially the rate increases at a later time $t = t_1$ her trading account becomes positive, so that her optimal strategy is to sell the currency. Sooner or later her trading account becomes negative and she has to go buy the currency. This oscillatory pattern of buying and selling will repeat itself until $t = t_*$. Depending on the value of the trading account (which hovers near zero) two outcomes are possible. If $\Pi(t_*) > 0$ then between $t = t_*$ and $t = t_N$ the investor has to sell the currency thus accumulating profit $\Pi(t_N) \sim M(S_{t_*} - S_{t_N})$. If $\Pi(t_*) < 0$ then she has to go buy the currency which produces a loss $\Pi(t_N) \sim -M(S_{t_*} - S_{t_N})$ (which has to be absorbed by the seller of the option). It is clear the value of $\Pi(t_*)$ depends on the initial choice of strategy and fine details of the rate evolution between t_0 and t_*. Accordingly, it can be both positive and negative with equal probability, so that the price of the PO should be equal to one-half the price of the corresponding LBP.

7. Connections Between Imperfectly Hedged European Options and Passport and Asian Options

Finally, I want to discuss the relation between imperfectly hedged ECs, POs and ACs. I study this relation from the viewpoint of a seller of a EC who chooses a certain imperfect hedging strategy q. For simplicity I assume that both domestic and foreign interest rates are zero. I also assume that the hedging strategy is Markovian and depends only on S, t, although it is not difficult to consider a more general case. There are many reasons for not hedging according to the Black-Scholes strategy. The seller can deviate from this strategy either by choice (for example, because she follows efficient mean-variance or utility maximizing hedging strategies), or by chance (because continuous hedging is impossible, or the actual volatility of S is not known, or for many other reasons). Deviations from the Black-Scholes strategy result in the P&L χ (equal to the difference between the final value of the trading account Π_T and the final value of the European call $M[S_T - K]_+$) which is random. Accordingly, it is necessary to study the probability density function (PDF) for the random variable $\chi = [\Pi_T - M(S_T - K)_+]$. The original discussion of this problem is given in a recent paper by Esipov and Vaysburd (1999), here I present an alternative discussion.

I start with the coupled system of SDEs describing the evolution of S, Π,

$$dS = \mu S dt + \sigma S dW \,, \quad d\Pi = q(S, t)dS \,, \tag{59}$$

which is similar to system (34), except for the fact that the drift μ of S is not necessarily equal to $r_d - r_f$. The transitional PDF $\Theta(S_t, \Pi_t, t; S_T, \Pi_T, T)$ for system (59) is governed by the standard forward Fokker-Plank equation. In terms of Θ one can represent the PDF $\theta(S_t, \Pi_t, t; \chi)$ for the random variable χ as

$$\theta(S_t, \Pi_t, t; \chi) = \int_0^\infty \int_{-\infty}^\infty \Theta(S_t, \Pi_t, t; S_T, \Pi_T, T)$$

$$\times \delta[\Pi_T - M(S_T - K)_+ - \chi]dS_T d\Pi_T \,,$$

$$= V^{(q)}(S_t, \Pi_t - \chi, t) \,, \tag{60}$$

where $V^{(q)}(S, \Pi, t)$ solves the backward Kolmogoroff problem

$$V_t^{(q)} + \mu S V_S^{(q)} + \mu q S V_\Pi^{(q)} + \frac{1}{2}\sigma^2 S^2 (V_{SS}^{(q)} + 2q V_{S\Pi}^{(q)} + q^2 V_{\Pi\Pi}^{(q)}) = 0 \,,$$

$$V^{(q)}(S, \Pi, T) = \delta[\Pi - M(S - K)_+] \,, \tag{61}$$

and δ is the Dirac delta function. Equation (61) is practically identical to the unoptimized pricing problem (35) for the PO, as can be expected.

To see the relation between imperfectly hedged ECs and ACs it is convenient to represent the trading strategy in the form $q(S, t) = R_S(S, t)$, where $R(S, t)$ is the corresponding potential function, introduce the adjusted trading account

$\tilde{\Pi} = \Pi - R(S, t)$ and represent the P&L as $\chi = \tilde{\Pi}_T + Q(S_T, T) - M(S_T - K)_+$. (Esipov and Vaysburd 1999 also use this transformation but for a different purpose.) The evolution of the pair $S, \tilde{\Pi}$ is governed by the system of SDEs of the form

$$dS = \mu S dt + \sigma S dW, \quad d\tilde{\Pi} = \varkappa(S, t) dt, \tag{62}$$

where

$$\varkappa(S, t) = -\left[R_t(S, t) + \frac{1}{2}\sigma^2 S^2 R_{SS}(S, t) \right]. \tag{63}$$

Note that for the standard Black-Scholes trading strategy $\varkappa(S, t) = 0$ and $R(S_T, T) = M(S_T - K)_+$, so that both $\tilde{\Pi}_T$ and χ are non-random and the corresponding P&L is identically equal to zero, or, in other words, the PDF for χ is $\delta(\chi)$.

It is easy to show that $\theta(S_t, \tilde{\Pi}_t, t; \chi) = \tilde{V}^{(q)}(S_t, \tilde{\Pi}_t - \chi, t)$, where $\tilde{V}^{(q)}(S, \tilde{\Pi}, t)$ solves the backward Kolmogoroff problem

$$\tilde{V}_t^{(q)} + \mu S \tilde{V}_S^{(q)} + \varkappa(S, t)\tilde{V}_{\tilde{\Pi}}^{(q)} + \frac{1}{2}\sigma^2 S^2 \tilde{V}_{SS}^{(q)} = 0,$$

$$\tilde{V}^{(q)}(S, \tilde{\Pi}, T) = \delta[\tilde{\Pi} + Q(S, T) - M(S - K)_+]. \tag{64}$$

which strongly resembles the valuation problem (22) for ACs. Unfortunately, for this problem a simple self-similar reduction is not possible, so that it has to be treated as a degenerate two-factor parabolic problem. Fortunately, several established numerical schemes can be used in order to solve it. A viable alternative is to exploit the technique developed by Landau (in order to solve the Vlasov equation) and apply the Fourier transform in $\tilde{\Pi}$,

$$\tilde{V}^{(q)}(S, \tilde{\Pi}, t) \rightarrow \tilde{W}^{(q)}(S, \kappa, t) = \int_{-\infty}^{\infty} e^{-i\kappa\tilde{\Pi}} \tilde{V}^{(q)}(S, \tilde{\Pi}, t) d\tilde{\Pi}, \tag{65}$$

in order to replace the original two-factor problem by a family of one-factor problems

$$\tilde{W}_t^{(q)} + \mu S \tilde{W}_S^{(q)} + i\kappa\varkappa(S, t)\tilde{W}^{(q)} + \frac{1}{2}\sigma^2 S^2 \tilde{W}_{SS}^{(q)} = 0,$$

$$\tilde{W}^{(q)}(S, \kappa, T) = e^{i\kappa[R(S,T)-M(S-K)_+]}, \tag{66}$$

parametrized by κ. Once $\tilde{W}^{(q)}(\kappa)$ is found, one can construct $\tilde{V}^{(q)}(\tilde{\Pi})$ by virtue of the inverse Fourier transform.

In order to illustrate the ideas outlined above, I revisit the stop-loss start-gain (SLSG) strategy studied by Carr and Jarrow, Hull, and others. For this strategy $q(S, t) = M\theta(S - K)$, $R(S, t) = M(S - K)_+$, where θ is the Heaviside function. The hedger who follows this strategy buys M units of the FC every time the FX rate hits the strike level K from below, and sells the same amount of the FC whenever this level is hit from above. Intuitively this looks like a perfect hedge; however, a more careful analysis shows that the SLSG strategy results in a random P&L

with a non-trivial PDF (which nonetheless does have a δ-function component). Broadly speaking, the hedger who follows the SLSG strategy assumes (whether she appreciates it or not) that for hedging purposes the volatility of the FX rate is zero. In the case in question problem (64) has the form

$$\tilde{V}_t^{(SLSG)} + \mu S \tilde{V}_S^{(SLSG)} - \frac{1}{2}\sigma^2 M K^2 \delta(S-K)\tilde{V}_{\tilde{\Pi}}^{(SLSG)} + \frac{1}{2}\sigma^2 S^2 \tilde{V}_{SS}^{(SLSG)} = 0\,,$$

$$\tilde{V}^{(SLSG)}(S,\tilde{\Pi},T) = \delta(\tilde{\Pi})\,. \tag{67}$$

The first two moments of the solution $\tilde{V}^{(SLSG)}$ are numerically evaluated by Esipov and Vaysburd (1999), while the exact analytical expression for $\mu = 0$ is given by the present author in Appendix C of their paper. Here I show how to find the analytical solution for $\mu \neq 0$. By now my methodology should be clear. First, I use the dimensional reduction and rewrite the pricing problem as follows

$$\Phi_\tau^{(SLSG)} - \hat{\mu}\xi\Phi_\xi^{(SLSG)} + \frac{1}{2}\delta(\xi-1)\Phi_\varpi^{(SLSG)} - \frac{1}{2}\xi^2\Phi_{\xi\xi}^{(SLSG)} = 0\,,$$

$$\Phi^{(SLSG)}(\xi,\varpi,\tau) = \delta(\varpi)\,, \tag{68}$$

where

$$\Phi^{(SLSG)} = MK\tilde{V}^{(SLSG)}\,, \quad \xi = S/K\,, \quad \varpi = \tilde{\Pi}/MK\,,$$

$$\tau = \sigma^2(T-t)\,, \quad \hat{\mu} = \mu/\sigma^2\,. \tag{69}$$

Next, I combine the Laplace transform in time and the Fourier transform in space and represent the pricing problem in the form

$$\lambda\Psi^{(SLSG)} - \hat{\mu}\xi\Psi_\xi^{(SLSG)} + \frac{i\kappa}{2}\delta(\xi-1)\Psi^{(SLSG)} - \frac{1}{2}\xi^2\Psi_{\xi\xi}^{(SLSG)} = 1\,, \quad -\infty < \xi < \infty\,, \tag{70}$$

where

$$\Psi^{(SLSG)}(\xi,\kappa,\lambda) = \int_0^\infty \int_{-\infty}^\infty \Phi^{(SLSG)}(\xi,\varpi,\tau)e^{-(\lambda\tau+i\kappa\varpi)}d\tau d\varpi\,. \tag{71}$$

This equation is similar in nature to Eq. (31) of Hyer, Lipton and Pugachevsky (1997) and can be solved by virtue of the technique developed in their paper and extended by the present author is his study of the special case $\mu = 0$. Omitting the corresponding algebra for the sake of brevity, I simply present the final solution of the original problem

$$\tilde{V}^{(SLSG)}(S,\tilde{\Pi},T-t) = \left[1 - \frac{1}{\sqrt{2\pi}}|\ln\xi|\xi^{(1-\hat{\mu})/2}Int_1\right]\frac{\delta(\varpi)}{MK}$$

$$+ \sqrt{\frac{2}{\pi}}\xi^{(1-\hat{\mu})/2}Int_2\frac{\theta(\varpi)}{MK}\,, \tag{72}$$

where

$$Int_1 = \int_0^\tau \exp\left[-\frac{(\ln\xi)^2}{2\eta} - \frac{(1-\hat{\mu})^2\eta}{8}\right]\eta^{-3/2}d\eta\,,$$

$$Int_2 = \int_0^\tau \exp\left[-\frac{(|\ln\xi| + 2\varpi)^2}{2\eta} - \frac{(1-\hat{\mu})^2\eta}{8}\right]$$

$$\times \left[\frac{(|\ln\xi| + 2\varpi)^2}{\eta} - 1\right]\eta^{-3/2}d\eta\,. \tag{73}$$

It is clear that this PDF contains both localized and delocalized components.

8. Conclusions

In this paper, I consider several seemingly unrelated path-dependent options, develop a unified technique for their valuation, and find useful relations between them. My technique combines the self-similarity reduction, splitting and Laplace transform in time (and, if necessary, the Fourier transform is space) in order to make the valuation problem analytically tractable. Even though I restrict myself to the standard Black-Scholes-Garman-Kohlhagen framework, the central idea has a much broader domain of applicability. I will discuss other analytically solvable problems elsewhere.

Acknowledgments

I am grateful to Marsha Lipton, Stewart Inglis, Kenneth Garbade and Alexander Eydeland for useful discussions of the issues considered in the present paper. The invitation of Marco Avellaneda to give a talk at the NYU Mathematical Finance Seminar is much appreciated.

References

M. Abramowitz and I. Stegun, *Handbook of Mathematical Functions*, Dower, New York, 1972.

G. I. Barenblatt, *Scaling, Self-similarity, and Intermediate Asymptotics*, Cambridge University Press, Cambridge, 1996.

F. Black and M. Scholes, "The pricing of options and corporate liabilities", *Journal of Political Economy* **81** (1973) 637–659.

S. Esipov and I. Vaysburd, "On the profit and loss distribution of dynamic hedging strategies", *International Journal of Theoretical and Applied Finance*, to appear (1999).

K. D. Garbade, "Managerial discretion and the contingent valuation of corporate securities", *The Journal of Derivatives*, to appear (1999).

M. Garman, "Recollection in tranquility", *Risk Magazine* **2**(3) (1989).

H. Geman and A. Eydeland, "Asian options revisited: Inverting the Laplace transform", *Risk Magazine* **8**(4) (1995) 65–67.

H. Geman and M. Yor, "Bessel processes, Asian options, and perpetuities", *Mathematical Finance* **3**(4) (1993) 349–375.

M. B. Goldman, H. B. Sosin and M. A. Gatto, "Path dependent options: Buy at the low, sell at the high", *Journal of Finance* **34** (1979) 1111–1128.

V. Henderson and D. G. Hobson, "Local time, coupling and the passport options", *Finance and Stochastics*, to appear (1999).

T. Hyer, A. Lipton and D. Pugachevsky, "Passport to success", *Risk Magazine* **10**(9) (1997) 127–131.

A. Lipton, "Predictability and unpredictability in finance", *Physica D* **133** (1999) 321–347.

R. Merton, "Theory of rational option pricing", *Bell Journal of Economics and Management Science* **4** (1973) 141–183.

L. Rogers and Z. Shi, "The value of an Asian option", *Journal of Applied Probability* **32** (1995) 1077–1088.

G. I. Taylor, "The formation of a blast wave by a very intense explosion. II. The atomic explosion of 1945", *Proceedings of Royal Society* **A201** (1950) 175–186.

P. Wilmott, S. Howison and J. Dewynne, *The Mathematics of Financial Derivatives*, Cambridge University Press, Cambridge, 1995.

Reprinted from Risk Magazine
12(2) (1999) 55–59

MONTE CARLO WITHIN A DAY

JUAN D. CÁRDENAS, EMMANUEL FRUCHARD, JEAN-FRANÇOIS PICRON,
CECILIA REYES, KRISTEN WALTERS and WEIMING YANG

This article presents an innovative approach to measuring intra-day VaR that combines
the use of a robust parametric technique, *Gamma VaR*,[a] with Monte Carlo simulation
to capitalize on the respective strengths of these models. The simulation is optimized by
using parametric VaR results to limit the required number of portfolio revaluations to
those random scenarios that are statistically relevant given the greek-estimated profit and
loss distribution, and as a variance reduction tool to minimize the standard Monte Carlo
error term. As the results presented here will show, these techniques, combined with
portfolio and market risk factor compression, significantly enhance the performance and
precision of the Monte Carlo engine. Although VaR alone, no matter how sophisticated
the model, is not sufficient to effectively capture all possible market moves, it is an
invaluable intra-day tool for risk managers.

1. Estimating the Loss Tail of Monte Carlo

A naive or "brute force" Monte Carlo simulation for VaR performs full trade
revaluations of an entire portfolio under thousands of random scenarios to approximate the profit and loss distribution. The trade revaluation process is computationally intensive, thereby preventing Monte Carlo from being a viable tool to measure
risk during the trading day. Given that the goal in VaR is to obtain the "loss tail"
(e.g., 5% loss tail for a 95% confidence level) of the distribution, it is reasonable
to only perform full trade revaluations for those scenarios that will result in large
losses for a portfolio. Our approach uses the Gamma VaR distribution as a tool for
determining which scenarios will result in losses in the tail of the Monte Carlo distribution and discarding irrelevant scenarios using the following step-by-step process:

- **Parametric VaR Calculation.** First, our model calculates the first- and
  second-order derivatives with respect to all relevant portfolio risk factors which
  will result in a vector of deltas and vegas and a matrix of gammas. Next, instead
  of approximating the profit and loss function (e.g., by using partial simulations or
  its first few moments), the characteristic function is determined analytically and
  the full profit and loss distribution is recovered using a fast Fourier Transform
  (FFT).

---

[a]Co-authored by two authors of this article and detailed in the following Risk article: VAR: One
Step Beyond (Juan Cárdenas, Emmanuel Fruchard, *et al.*), Volume 10, No. 10, October 1997.

- **Random Rate Path Generation.** The next step involves generating pseudo-random scenarios for the portfolio risk factors (interest rates, bond & equity prices, FX rates, volatilities, etc.). Risk factors may either represent individual market rates or statistically based shift scenarios[b] calculated using principal component analysis (PCA). Employing PCA to compress the universe of market risk factors into their salient principal components enhances performance. The definition of the risk factors is a critical input to the analysis. Recognizing that during crises historical measures of volatilities and correlations break down, it is very important to supplement VaR with event risk analysis.

- **Modelling of Error Distribution.** For each random scenario, the change in portfolio mark-to-market (MTM) is approximated using the greeks. The true profit and loss based on full revaluation is calculated for those scenarios where the greek-based profit and loss falls below a user-specified upper bound (see graph below). The specified upper bound will logically reside to the right of the parametric loss tail to reflect possible error in the parametric VaR model. Next, the distribution of the error between the greek-based and full revaluation for the scenarios whose greek-based profit and loss falls below the upper bound is modelled.

- **Upper Bound Adjustment.** Based on the calculated error distribution, the upper bound may be adjusted further to prevent any relevant scenarios from being inappropriately discarded in the Monte Carlo process. The graph below depicts the error distribution and the adjusted upper bound used to determine relevant random scenarios. In addition, a minimum number of tail scenarios may be specified to ensure sufficient sampling in the tail region.

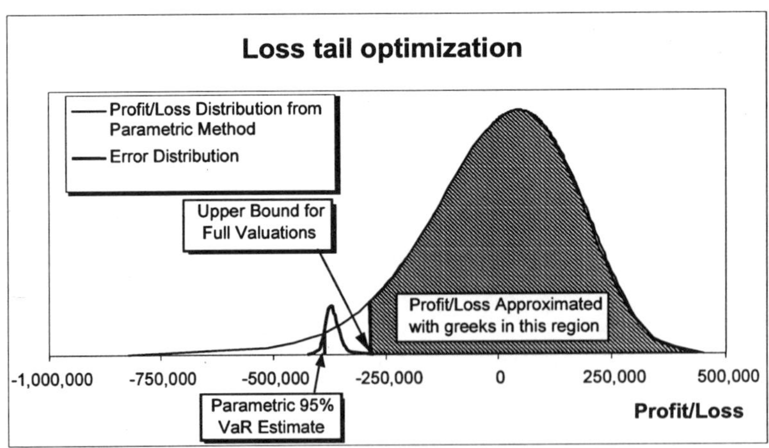

- **Monte Carlo VaR Calculation.** Finally, the full valuation profits and losses for the scenarios that fall below the adjusted upper bound are ordered and the

[b]For example, rather than modelling the processes of a set of zero-coupon rates that comprise a yield curve, it is possible to concisely represent its process using statistically relevant shift scenarios without significantly losing information, e.g., parallel, steepening and curvature shifts.

Monte Carlo VaR is obtained by assessing the α-percentile based on the original number of scenarios generated, where $(1 - \alpha)$ is the confidence level.

2. Variance Reduction

Monte Carlo results explicitly include a calculated error term. Besides determining which randomly generated scenarios should be valued, the greek approximation of the profit and loss distribution can also be used to reduce the variance of the estimators. Our method employs two variance reduction techniques, "control variate" and "stratified sampling", both of which are based on the Gamma VaR distribution, to reduce the Monte Carlo error term. By reducing the error term, risk managers may obtain a higher degree of precision with Monte Carlo based on the same number of random scenarios or improved performance by reducing the number of required scenarios and obtaining the same specified error.

3. Control Variate Case

Given that, in most cases, the Gamma VaR distribution will be highly correlated with the true distribution being estimated via Monte Carlo simulation, it may be used as a control for the true profit and loss distribution as shown in the diagram below.

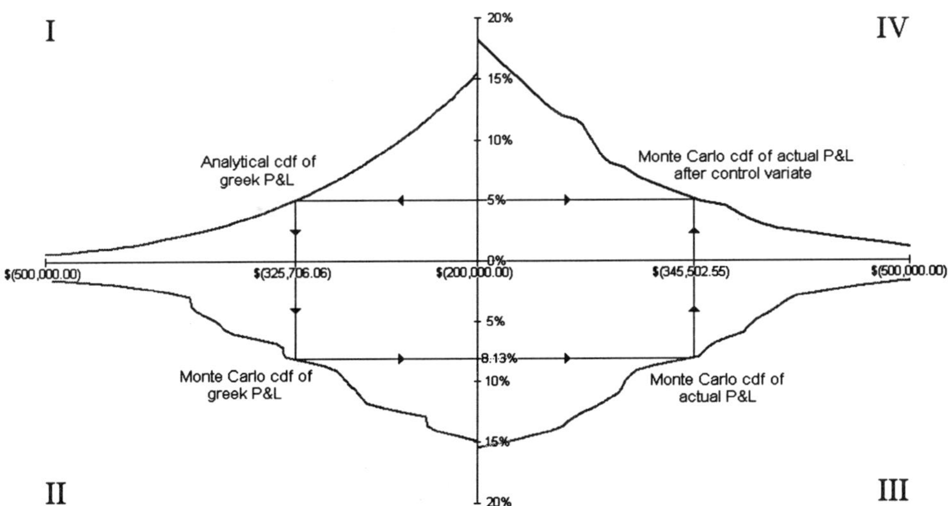

1. In the first quadrant (I), the greek VaR is obtained using the analytical approximation (i.e., Gamma VaR) of the profit and loss distribution.
2. In the second quadrant (II), the Monte Carlo estimate of the greek VaR's percentile is obtained by performing a partial simulation (i.e., computing the greek profit and loss for randomly-generated scenarios).

3. In the third quadrant (III), the true profit and loss corresponding to the percentile found above is calculated.
4. The actual profit and loss results are subsequently mapped in the fourth quadrant (IV) as corresponding to the VaR percentile.

By using a control, the variance of the estimator is reduced by a factor, $2(1 - \rho)$ where ρ is the correlation. For most test cases, using the full Gamma matrix for the computation of the parametric VaR results in a correlation above 90%, making the simulation more than five times faster.

One of the most attractive features of the control variate is that it can be easily applied to the full distribution, yielding a much smoother curve than brute force (i.e., full) Monte Carlo. The figure below shows a comparison of brute force Monte Carlo and control variate for 100 scenarios and a correlation of 95%. The curve labeled "exact" was obtained by running 50,000 scenarios and using the control variate.

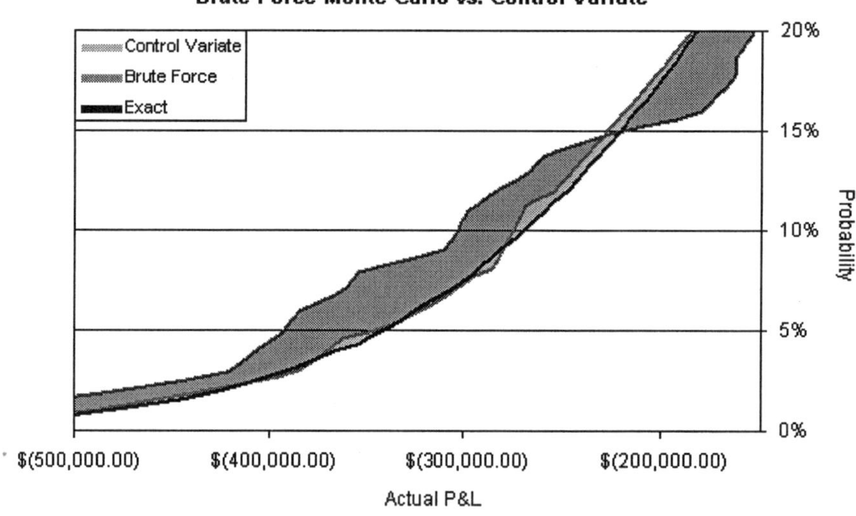

4. Stratified Sampling

Stratified sampling can be viewed as a "divide and conquer" variance reduction technique in which the following steps are performed (as shown in the figure below):

1. The pseudo-randomly generated scenarios are divided into two independent sets: one containing the scenarios that result in greek-approximated profit and losss falling below the parametric VaR, and the other set containing those scenarios that yield greek-approximated profits and losses falling above the parametric VaR.

2. Each subset is sampled proportionally to its probability measure. For example, if a 95% VaR is desired, 5% of the scenarios should be below the parametric VaR and 95% should be above it. The true profit and loss distribution is computed for each set separately using full valuation.

3. Since the two sets are independent by construction, the overall Monte Carlo distribution is simply a weighted sum of the two sub-distributions.

Cumulative distribution function (cdf) using stratified sampling

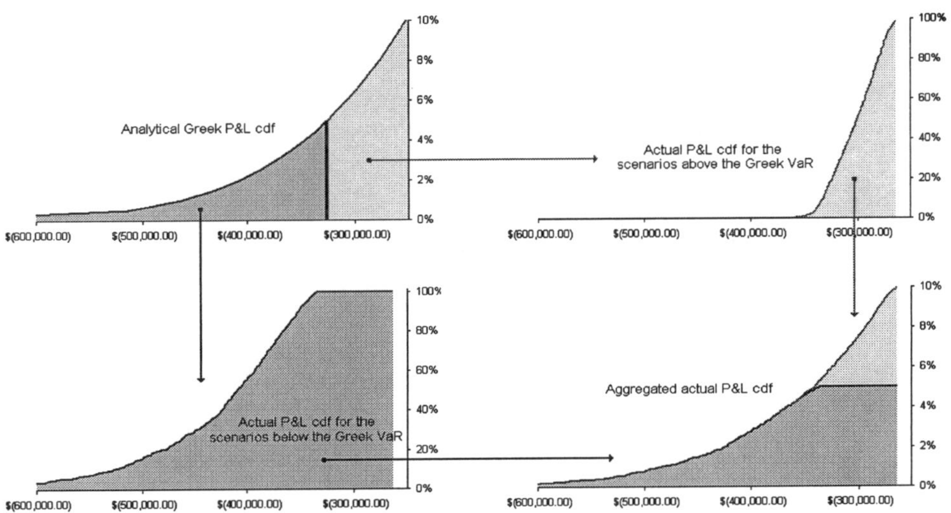

5. Portfolio Mapping

Mapping a portfolio into fewer equivalent trades is an additional feature incorporated in the design of our Monte Carlo engine. Equivalence, in this sense, means that the sensitivities of the mapped trades and the original portfolio to all the risk factors are equal. This technique can yield significant improvements in performance and may be employed for linear (e.g., interest rate swaps) as well as non-linear products (e.g., plain vanilla caps and floors). A large swap portfolio of 20,000 trades, for example, can be mapped to less than 100 trades.

For option portfolios, the logic is similar with the addition of a second dimension to the mapping process. In any given cap and floor portfolio, for example, there are usually a few major currencies in which many plain vanilla caplets share the same tenor and have similar expiry dates and/or strikes. These trades can be readily mapped to a smaller subset of caplets that depict similar two-dimensional expiry/strike and risk behavior. The mapping method calculates the vega by expiry and strike and determines the equivalent quantity of bucketed caplets. Once the vega hedge has been determined, the differential between the real portfolio and the vega hedge is delta hedged and the resulting additional linear trades are included in the mapping portfolio. This method is effective as long as the number of plain vanilla caplets in the portfolio significantly exceeds the number of expiry dates multiplied by the representative strikes. While a similar approach may be used for swaptions and bond options, the underlying structures to be populated for these instruments are three-dimensional and there are generally no concentrations on any dominant underlying swap tenor or bond maturity. Thus, our experience has shown that mapping these products is only useful for extremely large portfolios.

6. Parametric VaR in Monte Carlo

Market practitioners frequently cite the need for a full Monte Carlo simulation based on the assumption that parametric VaR models give inaccurate results for short-dated at-the-money options whose greeks are rather unstable and hence unreliable to use to predict profit and loss. While this is true when using analytic greeks, the error is often of a significantly smaller magnitude when the greeks are calculated empirically using reasonably sized perturbations of the risk factors. This key observation formed the basis of our development of a parametric Monte Carlo approach and is illustrated below by a "worst case" example of a bond option with one day to expiry date. For this example, the VaR is calculated for various strikes and with greeks obtained locally (i.e., equivalent to the analytic greeks) and globally with step sizes equaling 1/2, 1, and 2 standard deviations (i.e., 3, 6, and 12 basis points shifts, respectively).[c] The results show that the parametric VaR is inaccurate

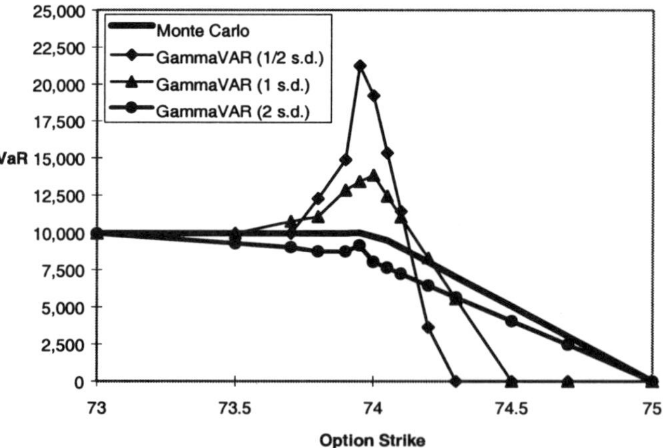

[c]For this example, greeks were calculated using finite difference methods. Local and global greek calculations refer to the respective amplitude of the market shift employed to calculate the greeks. It is well known that using minute "local" shifts in market rates can yield gammas that approach infinity for at-the-money options at expiry. Thus, it is reasonable to increase the magnitude of the shift to reflect expected market moves which will yield more reliable gammas. In this case, the option MTM was recalculated based on the specified # of standard deviation change in market rates - i.e., a 1/2 standard deviation move represents a 3 basis point shift in market rates.

when the shift is small, because the gamma surges to an artificially high level. On the other hand when the shift size is set to a number of standard deviations consistent with the confidence interval used in the VaR calculation, the gap narrows to a small fraction of the correct result. Therefore the few close-to-the-money options that are in every large portfolio will not reduce the effectiveness of the parametric VaR optimization.

7. Monte Carlo Results

To illustrate the beneficial effects on performance and precision of VaR estimate provided by our Monte Carlo approach, VaR results for a 97.5% confidence level and one day time horizon are generated for a representative portfolio of a multi-national dealer with 500 trades in 24 currencies (see results in tables below). Including numerous currencies and risk factors is relevant for any performance benchmark given that the size of the variance/covariance matrix and corresponding processing time for Monte Carlo expand with the number of relevant risk factors. This portfolio contains 100 risk factors with options comprising 40% of the trades (including FX barriers, Asian options and 100 American bond options).

The fact that this sample portfolio contains exotic options with nonmonotonic payoffs and exhibits significant convexity is key to demonstrating the power of the parametric Monte Carlo method given that, for these instrument types, parametric methods used in isolation are not sufficient VaR estimators. As the tables show, our Monte Carlo method handles these trades (as well as instruments priced with lattices) remarkably well with intermediate events (e.g., barrier crossings) incorporated into the modeling process.[d] The same performance is observed after mapping for portfolios of any size (e.g. 50,000 swaps) with the same number of options.

VaR was generated on a single Sparc Ultra-10 workstation with a 512-megabyte memory. Performance times could be substantially reduced in a distributed processing environment with multiple servers, which is the desired configuration for generating global VaR. Our Monte Carlo engine was designed based on an object model to support this technology structure.

Table 1 shows the processing time of our Monte Carlo engine for the sample portfolio given differing VaR precisions (and associated required scenarios). Our experience indicates that 5% precision, which is considered more than adequate, will generally result in less than 1,000 random scenarios. As shown, the effect on performance of employing the parametric Monte Carlo VaR technique instead of brute force Monte Carlo is quite dramatic.

Table 2 shows the results of VaR and estimated error before and after applying variance reduction techniques. The computation time for variance reduction is

---

[d]As an alternative to aging barrier trades on several dates from the initial date to the horizon date, a one step simulation is performed where the simulated MtM value of the barrier option on the final date takes into account the probability that the underlying asset has crossed the barrier over the simulation period for each scenario. Thus correlations across barrier crossing events are taken into consideration.

Table 1. Performance run time of Monte Carlo engine.

VaR Precision	5%	3%
Total Number of Scenarios	700	1900
Brute Force Monte Carlo Engine - No Optimization	1 hr, 44 min	4 hrs, 31 min
Optimized Monte Carlo VaR	**13 min.**	**24 min.**

Table 2. Monte Carlo VaR results with variance reduction.

Total Number of Scenarios	700		1900	
Results	VaR	Standard Error	VaR	Standard Error
No Variance Reduction	65,122,890	1,543,517	64,931,509	1,221,515
Stratified Sampling	66,200,240	695,671	65,754,475	484,224
Gamma VaR Control Variate	66,371,881	628,933	66,085,661	362,389

negligible compared with total time of Monte Carlo engine and, as shown, the variance reduction is substantial. For most cases, the reductions are more than 55% which may be interpreted as a more than five times performance advantage after applying the variance reduction technique.

The *exact* VaR for the portfolio shown in the example above was calculated by generating 50,000 pseudo-random rate paths with variance reduction; the resulting VaR was 66,075,770. The figure below compares the exact VaR results with the Gamma VaR estimate as well as our optimized Monte Carlo VaR, with and without employing variance reduction. These results readily demonstrate the improvement in precision of results obtained by generating VaR using our Monte Carlo method.

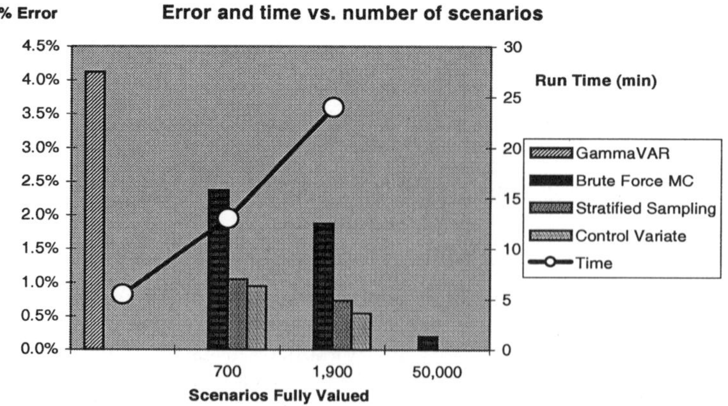

8. Conclusion and Implementation Issues

The results displayed above clearly validate the use of parametric VaR with Monte Carlo simulation and demonstrate the performance gains obtained associated with the combined approach. The parametric Monte Carlo VaR alone provides a tenfold increase in performance over brute force Monte Carlo. With the variance reduction, an additional (five) to 20 times boost in performance is obtained. There will be instances (i.e., when the greeks are highly unstable) where the parametric approximation is not reasonable to use together with the Monte Carlo simulation. Key to the design and practical implementation of the parametric Monte Carlo method is the ability to configure the VaR calculation process based on portfolio risk profiles and current market conditions. With our methodology, risk managers have complete flexibility to determine the extent to which performance optimizations are used when generating VaR results.

Intra-day, risk managers may take advantage of all applicable optimizations to generate VaR in real time and/or determine the "marginal" VaR of certain trades. At close of business day, if the full distribution of profit and loss is required (i.e., not just the tail region) the Monte Carlo engine may be run in "brute force" mode involving full revaluations of the portfolio for all randomly generated rate paths. If the full distribution is not required, then the combined parametric method and Monte Carlo provides a very efficient way to estimate VaR. VaR obtained using the optimized Monte Carlo may be back-tested against brute force Monte Carlo, providing risk managers with an empirical view on the accuracy of intra-day VaR measures and the resulting ability to allocate capital and risk limits confidently and optimally during the trading day.

Appendix. Variance Reduction

A.1 *Brute force Monte Carlo*

For a given level x_a of actual profit and loss, the percentile is estimated using a sum of indicator function:

$$\overline{F_a}(x_a) = \frac{\sum_{i=1}^{N} I(a_i \leq x_a)}{N} \tag{A.1}$$

where a_i is the actual profit and loss of scenario i. For a sufficiently large number of scenarios, this estimator will be approximately normally distributed with a mean equal to the true percentile and a variance approximately equal to:

$$\overline{\sigma^2}(\overline{F_a}(x_a)) = \frac{\overline{F_a}(x_a)(1 - \overline{F_a}(x_a))}{N-1}. \tag{A.2}$$

Note that in order to find a confidence interval for the VaR, we need to map the confidence interval on the percentile back to the profit and loss space using the cdf.

A.2 *Control variate*

For given levels x_g of greek profit and loss and x_a of actual profit and loss, the following is an unbiased estimator of the percentile of x_a:

$$\overline{Fcv}(x_a) = \overline{F_a}(x_a) - (\overline{F_g}(x_g) - F_g(x_g)) \tag{A.3}$$

where a bar denotes a Monte Carlo approximation, as opposed to an analytical result, and the index indicates whether the percentiles and profit and loss refer to the greek approximation (g) or the actual valuation (a). Since we are looking for the νth percentile, we can replace x_g by VaR_g in Eq. (A.3) to obtain:

$$\nu\% = \overline{F_a}(\overline{VaR_a}) - (\overline{F_g}(VaR_g) - \nu\%) \Rightarrow \overline{VaR_a} = \overline{F_a}^{-1}(\overline{F_g}(VaR_g)).$$

The standard error on this estimator is simply:

$$\sigma_{cv}^2 = \sigma_a^2 + \sigma_g^2 - 2\rho\sigma_a\sigma_g \Rightarrow \sigma_{cv}^2 = 2\sigma_a^2(1-\rho).$$

In order to compute the above variance, one can simply realize that the estimator (3) is a difference of indicator functions that has an average of zero:

$$\overline{F_{cv}}(\overline{VaR_a}) = \nu\% + \frac{\displaystyle\sum_{i=1}^{N}(I(a_i \leq \overline{VaR_a}) - I(g_i \leq VaR_g))}{N}$$

$$\overline{\sigma_{cv}^2} = \frac{\displaystyle\sum_{i=1}^{N}(I(a_i \leq \overline{VaR_a}) - I(g_i \leq VaR_g))^2}{N-1}$$

$$= \frac{\displaystyle\sum_{i=1}^{N}\left|I(a_i \leq \overline{VaR_a}) - I(g_i \leq VaR_g)\right|}{N-1}. \tag{A.4}$$

Equation (A.4) indicates that the variance of the estimator is asymptotically equal to the "disagreement probability": the probability that the greek profit and loss of a random scenario will be above the greek VaR and its actual profit and loss below the actual VaR or vice-versa.

A.3 *Stratified sampling*

For stratified sampling, we have a weighted sum of percentile estimates:

$$\overline{F_{ss}}(x_a) = \nu\overline{F_1}(x_a) + (1-\nu)\overline{F_2}(x_a). \tag{A.5}$$

The indices refer to the sub-space in which the sampling is done: 1 for the scenarios that fall below the greek VaR and 2 for the scenarios above the greek VaR. The standard error of the estimator is equal to:

$$\sigma_{ss}^2 = \nu^2 \sigma_1^2 + (1-\nu)^2 \sigma_2^2$$

$$= \nu^2 \frac{\overline{F_1}(x_a)(1-\overline{F_1}(x_a))}{N_1-1} + (1-\nu)^2 \frac{\overline{F_2}(x_a)(1-\overline{F_2}(x_a))}{N_2-1} \, .$$

If we use biased estimates (i.e., dividing by N instead of $N-1$) then the proportional sampling allows use to rewrite the above equation as:

$$\sigma_{ss}^2 = \nu \frac{\overline{F_1}(x_a)(1-\overline{F_1}(x_a))}{N} + (1-\nu) \frac{\overline{F_2}(x_a)(1-\overline{F_2}(x_a))}{N} \, . \qquad (A.6)$$

Comparing Eqs. (A.5) and (A.6) to Eqs. (A.1) and (A.2), we see that the variance of the stratified sampling estimator is a convex combination of the variance of two brute force Monte Carlo estimators, as illustrated below.

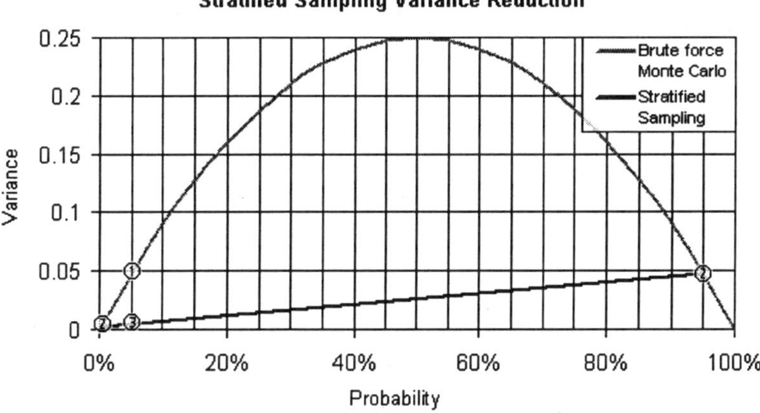

In the figure above, the parabola represents the variance of brute force Monte Carlo as a function of the target percentile, as defined by Eq. (A.2). Point 1 represents the variance for a 95% confidence level VaR, while the two points labeled are the two individual terms of the stratified sampling variance. According to Eqs. (A.5) and (A.6), the variance will be located on the line joining those two points. The exact stratified sampling variance 3 will be located on that line, right below the brute force Monte Carlo variance, as indicated by Eq. (A.5).

DECOMPOSITION AND SEARCH TECHNIQUES IN DISJUNCTIVE PROGRAMS FOR PORTFOLIO SELECTION

KATHERINE WYATT

Logic Based Systems Lab, Brooklyn College, City University of New York

Received 30 August 1999

1. Introduction

There are many problems in financial portfolio selection in which the primary objective of maximizing return and minimizing risk must be pursued subject to various constraints. Some examples of these portfolio selection problems are analysis of different types of portfoios *vis á vis* benchmarks in historical simulations or stress tests; designing portfolios with particular characteristics that track an index across different scenarios; and making an optimal hedging designation for items and derivatives under *SFAS 133* requirements.[1,2]

These problems can be formulated as disjunctive programs, where the desired portfolio is described by means of logical disjunctions. For example, some portfolio requirements that can be modeled with logical disjunctions are that the number of securities is an integer, or bounded by an integer; securities have to be purchased in set denominations; securities have to have certain credit ratings; or there is a minimum investment requirement for a portfolio. In many portfolio selection problems, the function to be optimized can be expressed in terms of the absolute value of the difference between a portfolio's return and the return of some known collection of securities. Using the L_1 distance instead of variance produces disjunctive linear programs. Finally, because analysis takes place across many scenarios, these disjunctive linear programs can be very large.

This paper presents a disjunctive formulation for a portfolio selection problem and an algorithm for decomposing the large linear programs that result with this model.[20] A main feature of the algorithm is that variable decomposition of the linear programs is combined with branch and bound search among the disjunctive sets that describe the desired portfolio.

2. Background

Markowitz[3,17] pioneered portfolio analysis using variance of returns as the definition of risk. In mean variance portfolio analysis, the ratio of expected return to variance is maximized, usually as a quadratic programming problem. Recently, Konno[4,5] advocated using mean absolute deviation of returns in portfolio selection, since mean absolute deviation can be represented by a piecewise linear risk function. One method of portfolio optimization involves picking the best "tracking" portfolio, i.e., one whose returns are closest to those of a benchmark portfolio or index. Worzel *et al.* minimized the mean absolute deviation of portfolio and benchmark returns.[6] The *absolute deviation trade-off model* we present uses mean absolute deviation from a benchmark on interest rate scenarios as the measure of risk, and has as objective maximizing the difference between expected return and expected risk. Optimizing mean absolute deviation can be modeled with linear constraints and a linear objective function, thus selecting the optimal portfolio is now a large linear problem.

The choice of an optimal portfolio is often subject to structural constraints. There may be integral trading amount requirements for certain instruments, e.g., Treasury bills have to be purchased in increments of \$5,000, after an initial investment of \$10,000. Investors may demand a certain level of diversification or a certain credit grade for their portfolios. Dedicated bond portfolios often have sector specifications that have to be satisfied. All these are logical requirements; they can be modeled with disjunctions that describe allowable choices. The complexity of quadratic programming makes adding disjunctive requirements to a traditional mean-variance model computationally prohibitive. However, if absolute deviation is used to measure dispersion of returns instead of variance,[18,19] then structural requirements can be added to a portfolio selection model with linear constraints and a linear objective function. The portfolio selection problem can then be represented as a disjunctive linear program. An additional advantage of this formulation is that optimal portfolios for LP-based models have fewer nonzero holdings than do portfolios found optimal for quadratic models,[4,7] thereby reducing transaction costs.

3. Model Formulation

The absolute deviation trade-off model is an extension of earlier mean absolute deviation models.[4,5,8] We use the L_1 distance from a given benchmark as the measure of risk and maximize the trade-off between the expected present value of reinvested cash flows and this L_1 distance. The objective function is similar in spirit to one utilized by Hiller and Eckstein;[9] they used downside deviation from a liability stream as the measure of risk. In this context, let R_{ij} be the sample returns for asset j on scenario i; let $bench_i$ be the known returns for the benchmark on scenario i. Then minimizing the risk for the absolute deviation trade-off model is equivalent to minimizing the function

$$f(x) = \sum_{i=1}^{S} \left| bench_i - \sum_{j=1}^{N} R_{ij} x_j \right|.$$

As an example of this technique, we present a solution to the problem of selecting a portfolio from a universe of N securities that maximizes the trade-off between expected return over a set of scenarios and expected L_1 distance from a benchmark's return over the scenarios, with the additional requirement that investment be at least ν.

The objective function for this problem is to maximize *Expected portfolio returns - | Expected difference of portfolio and benchmark returns |*.

This can be expressed as a linear function if we introduce variables for the difference between benchmark and portfolio returns on each scenario, and add constraints to force these variables to the absolute value of the difference. We can then decompose the set of variables into the variables for securities, which may have disjunctive requirements, and the variables for the L_1 distance on each scenario.

3.1. *The absolute deviation trade-off model for fixed-income portfolio selection*

We first present the linear model without logical requirements. For our model we assume that we have a fixed universe of available instruments and that there is a cashflow model available to calculate the present value of each security's cash payments and embedded options. Further, we have fixed a benchmark with known returns on the set of scenarios. We assume that the only constraints are that all asset holdings are non-negative, i.e., there is no short selling, and there is a set budget for the investment. If there are N available assets and the holding period for the portfolio comprises M dates on which cash flows are collected, then there are 2^M scenarios. If the number of scenarios becomes too large, sampling techniques or Monte Carlo simulation methods can be used in conjunction with the basic model. In this research, we have used a discrete model of the future evolution of spot rates.[10]

For every asset, the present value of investing the cash flows at the current spot rate over the holding period of the portfolio is calculated. A feasible portfolio is a linear combination of the assets for which the price does not exceed the budget. A variable y_i is introduced for every scenario i to model the absolute value of the difference between a portfolio and the benchmark yielded by this scenario. Two constraints are added to the constraint set for every scenario:

$$\forall i \left(\sum_{j=1}^{N} r_{ij} x_j \right) - y_i \leq benchmark_i$$

$$\forall i \left(-\sum_{j=1}^{N} r_{ij} x_j \right) - y_i \leq -benchmark_i .$$

The objective function then is the difference between expected return and expected risk:

$$\max \sum_{i=1}^{S} \rho_i \sum_{j=1}^{N} r_{ij} x_j - \rho_i \sum_{i=1}^{S} y_i$$

where ρ_i is the probability associated with scenario i.

At optimality, since the y_i are non-negative, subtracting the expected y_i is equivalent to minimizing the y_i, or forcing y_i to the absolute value of the difference between benchmark and portfolio on each scenario. We now have a linear problem with linear constraints instead of a piecewise linear problem. The definitions and constraints for the model are as follows:

Definitions.

S = number of scenarios
N = number of securities available
M = maximum to be invested
B_i = expected return of benchmark on scenario i
r_{ij} = expected return from security j on scenario i
p_j = current price of security j
x_j = amount of security j held in portfolio
y_i = absolute value of the difference between expected portfolio and benchmark
      returns on scenario i
ν = minimum trade amount
w = minimum investment amount
ρ_i = probability associated with scenario i. $\square$

The linear program that models the problem of selecting the best tracking portfolio is:

$$\max \left\{ \sum_{i=1}^{S} \rho_i \sum_{j=1}^{N} r_{ij} x_j - \sum_{i=1}^{S} \rho_i y_i \right\} = z_P$$

$$\text{s.t.} \qquad p_1 x_1 + \cdots + p_N x_N \le M$$
$$-x_1 - \cdots - x_N \le -w$$
$$\forall j \qquad \qquad -x_j \le 0$$
$$\forall i \qquad r_{i1} x_1 + \cdots + r_{iN} x_N - y_i \le B_i$$
$$\forall i \qquad -r_{i1} x_1 - \cdots - r_{iN} x_N - y_i \le -B_i$$
$$\forall i \qquad \qquad -y_i \le 0$$

(P) can be written formally as:

$$(P) \quad \max \quad \{cx + dy\} = z_P$$
$$\text{s.t.} \quad A_1^1 x + A_1^2 y \le t$$
$$A_2^1 x + A_2^2 y \le b$$
$$A_3^1 x + A_3^2 y \le 0$$

or, in full detail,

$$
\begin{array}{llllllll}
A_1^1 = & p_1 & \cdots & p_N & A_1^2 = 0 & \cdots & 0 & t = M \\
& -1 & -1 & -1 & 0 & \cdots & 0 & -w \\
& -1 & \cdots & 0 & 0 & \cdots & 0 & 0 \\
& \vdots & -1 & \vdots & \vdots & 0 & \vdots & \vdots \\
& 0 & \cdots & -1 & 0 & \cdots & 0 & 0 \\
A_2^1 = & r_{11} & \cdots & r_{1N} & A_2^2 = -1 & \cdots & 0 & b = B_1 \\
& \vdots & \vdots & \vdots & \vdots & -1 & \vdots & \vdots \\
& r_{S1} & \cdots & r_{SN} & 0 & \cdots & -1 & B_S \\
& -r_{11} & \cdots & r_{1N} & -1 & \cdots & \cdots & -B_1 \\
& \vdots & \vdots & \vdots & 0 & -1 & 0 & \vdots \\
& -r_{S1} & \cdots & -r_{SN} & 0 & \cdots & -1 & -B_S \\
A_3^1 = & 0 & \cdots & 0 & A_3^2 = -1 & \cdots & 0 & 0 \\
& \vdots & 0 & \vdots & \vdots & -1 & \vdots & \vdots \\
& 0 & \cdots & 0 & 0 & \cdots & -1 & 0.
\end{array}
$$

Among standard structural requirements for a fixed-income portfolio are that asset holdings satisfy minimum trade requirements and that specified levels of diversification be maintained. We can model these requirements with the following constraints:

$$x \in \cap_i \cup_k F_{ik}$$
$$\forall j \qquad x_j \geq d_j \nu$$
$$\forall j \qquad 0 \leq d_j \leq 1$$
$$\sum_{j=1}^{N} d_j \geq m$$
$$d \in \cap_r \cup_s D_{rs}$$

where m is the desired level of diversification and we have $x \in \cap_i \cup_k F_{ik}$ if for all j, $x_j = 0 \vee x_j \geq \nu$ and $d \in \cap_r \cup_s D_{rs}$ if for all j, $d_j = 0 \vee d_j = 1$.

3.2. *Step-shaped programs*

The programs that correspond to the absolute deviation models share several family characteristics. First, auxiliary variables are introduced to represent the absolute value of the difference between portfolio and benchmark returns; second, there are constraints on the primary variables in which the auxiliary variables don't appear; third, the constraints in which both the primary and auxiliary variables appear do not further constrain the primary variables; and, finally, the only constraints on the auxiliary variables alone are bounding constraints restricting their values to the non-negative orthant. We will call the primary variables *portfolio variables* and the auxiliary variables *risk variables*.

These characteristics motivate the following definitions.

Definition. Let $R = \{x \in \Re^n : Ax \leq b\}$ and let $Dx + Ey \leq f$ be a system of linear inequalities. Then $Dx + Ey \leq f$ is *free* for $x \in R$ if for every $x \in R\ \exists y \geq 0$ such that (x, y) satisfies $Dx + Ey \leq f$. $\Box$

Let $x \in \Re_+^N$ and $y \in \Re_+^S$. Consider the following linear program (P):

$$
\begin{aligned}
(P) \quad \max \quad & \{cx + dy\} = z_P \\
\text{s.t.} \quad & A_1^1 x + A_1^2 y \leq t \\
& A_2^1 x + A_2^2 y \leq b \\
& A_3^1 x + A_3^2 y \leq 0.
\end{aligned}
$$

Definition. Let $R = \{x \in \Re_+^N : A_1^1 x \leq t\}$. A program (P) is a *step-shaped program* if the coefficient matrix for (P) can be written as above and if, in this formulation for (P), (1) A_1^2 and A_3^1 are zero submatrices, and (2) the constraints $A_2^1 x + A_2^2 y \leq b$ are free for $x \in R$. $\Box$

3.3. *Decomposition for step-shaped programs*

The absolute deviation trade-off model produces large programs with two kinds of variables, portfolio variables and risk variables, so it seems useful to explore variable decomposition methods for these programs. Benders[11] used a partition of the variable set to solve mixed integer problems. Van Slyke and Wets[12] reported on a method called L-shaped decomposition, which is akin to Benders decomposition, for large-scale and stochastic linear problems.

The form of the coefficient matrix for step-shaped programs is close to the Van Slyke and Wets description of L-shaped programs. The shape of the non-zero submatrices we are considering differs from the L-shaped form, however, in that we need to explicitly consider the bounding constraints on y since we will be interested in dual multipliers for the constraints. In fact, our definition of *free* makes precise an informal property.[12] We will first outline variable decomposition of linear step-shaped programs.

3.4. *Variable decomposition for linear step-shaped programs*

Let (P) be a linear step-shaped program:

$$
\begin{aligned}
(P) \quad \max \quad & \{cx + dy\} = z_P \\
\text{s.t.} \quad & A_1^1 x + A_1^2 y \leq t \\
& A_2^1 x + A_2^2 y \leq b \\
& A_3^1 x + A_3^2 y \leq 0.
\end{aligned}
$$

Let $x \in \Re_+^N$ and $y \in \Re_+^S$. Let m_1 denote the number of rows in $\left(A_1^1 \; A_1^2 \right)$, let m_2 denote the number of rows in $\left(A_2^1 \; A_2^2 \right)$, and let m_3 denote the number of rows in $\left(A_3^1 \; A_3^2 \right)$. Then A_1^2 is an $m_1 \times S$ zero submatrix, and A_3^1 is an $m_3 \times N$ zero submatrix.

Let $M = m_1 + m_2 + m_3$, let $u \in \Re_+^M$ and let π be a projection operator. Then $\pi_1(u) = \mathbf{u}_1 \in \Re_+^{m_1}$, $\pi_2(u) = \mathbf{u}_2 \in \Re_+^{m_2}$, and $\pi_3(u) = \mathbf{u}_3 \in \Re_+^{m_3}$.

To carry out variable decomposition, we rearrange the constraints for (P), and write the program as

$$\max_{\{x \geq 0\}} \{ cx + \max_{y \in G} dy \}$$

$$G = \{ y \in \Re_+^S : A_1^2 y \leq t - A_1^1 x, \, A_2^2 y \leq b - A_2^1 x, \, A_3^2 y \leq 0 \}.$$

Then by LP-duality, we can write (P) as

$$\max_{\{x \geq 0\}} \{ cx + \min_{u \in J} \{ \mathbf{u}_1(t - A_1^1 x) + \mathbf{u}_2(b - A_2^1 x) + \mathbf{u}_3 0 \} \}$$

$$J = \{ u \in \Re_+^M : \mathbf{u}_1 A_1^2 + \mathbf{u}_2 A_2^2 + \mathbf{u}_3 A_3^2 = d \}.$$

Notice that since A_1^2 is a zero matrix, $\mathbf{u}_1 = \pi_1(u)$ is unconstrained in the description of J. Since the constraints $A_2^1 x + A_2^2 y \leq t$ are free for $x \in \Re^N$, then if x is feasible for the constraints $A_1^1 x \leq t$, we can work in J', which is the projection of J, where $J' = \{ \mathbf{u}_2, \mathbf{u}_3 : \mathbf{u}_2 A_2^2 + \mathbf{u}_3 A_3^2 = d \}$.

Therefore, since by weak duality $dy \leq \mathbf{u}_2(b - A_2^1 x)$ for all feasible x, y, and $\mathbf{u}_2$, a program equivalent to (P) is the following one, where one continuous variable λ is substituted for dy and is bound by the dual objective function value at extreme points of J'. Let H be an *a priori* bound for y.

$$
\begin{aligned}
(MP) \quad & \max \quad \{ cx + \lambda \} = z_{MP} \\
& \text{s.t.} \quad\quad A_1^1 x \leq t \\
& \quad\quad\quad\quad\;\; |\lambda| \leq dH \\
& \quad\quad\quad\quad\;\; \lambda \leq \mathbf{u}_2^t (b - A_2^1 x) \\
& \quad\quad\quad\quad\;\; \forall u^t = \langle \mathbf{u}_2^t, \mathbf{u}_3^t \rangle.
\end{aligned}
$$

Here u^t is an extreme point of $\{ u \in \Re_+^{m_2 + m_3} : \mathbf{u}_2 A_2^2 + \mathbf{u}_3 A_3^2 = d \}$.

(MP) is the master problem for the variable decomposition reformulation of (P). The variable decomposition method for solving (MP) is (briefly) as follows. Assume $R = \{ x \in \Re_+^N : A_1^1 x \leq t \} \neq \emptyset$. At the first iteration, solve the relaxation (MP^0) of (MP) by dropping the constraints $\lambda \leq \mathbf{u}_2^t (b - A_2^1 x)$ for all extreme points of J' from (MP). Let (x^0, λ^0) be the optimal solution. If we have

$$(1) \quad \min \{ \mathbf{u}_2(b - A_2^1 x) : \mathbf{u}_2 A_2^2 + \mathbf{u}_3 A_3^2 = d, \langle \mathbf{u}_2, \mathbf{u}_3 \rangle \geq 0 \} \geq \lambda^0$$

then $\lambda^0 = dy^0$ is the optimal solution of the program

$$(2) \quad \max\{dy : A_2^2 y \geq b - A_2^1 x^0, A_3^2 y \geq 0\}$$

and (x^0, y^0) is optimal for (P). If

$$\lambda^0 > \min\{\mathbf{u}_2(b - A_2^1 x^0) : \mathbf{u}_2 A_2^2 + \mathbf{u}_3 A_3^2 = d, \langle \mathbf{u}_2, \mathbf{u}_3 \rangle \geq 0\}$$

then (x^0, λ^0) is not feasible for (MP) and we have a valid inequality (or cut) that can be added to (MP^0)

$$\lambda \leq \min\{\mathbf{u}_2(b - A_2^1 x) : \mathbf{u}_2 A_2^2 + \mathbf{u}_3 A_3^2 = d\}$$

to produce the relaxed master problem at stage 1, (MP^1).

At each iteration, this method either proves a point optimal or generates a cut to be added to the current relaxed master problem. (1) is called the dual subproblem $(DSP(x))$ and (2) is called the primal subproblem $(SP(x))$.

Using this method, we replace (P) with a series of relaxations of (MP) and easier subproblems. Solving (P) is carried out by first solving a relaxation of the master problem (MP^k) at stage k and then solving a subproblem $(DSP(x^k))$ to obtain the extreme point that will figure in the test for optimality, and if this fails, in an inequality that is added for stage $k + 1$. In this way the relaxations of (MP) are progressively tightened.

3.5. *Decomposition for disjunctive step-shaped programs*

3.5.1. *Disjunctive linear programs*

We give a precise definition of programs with *disjunctive constraints*.

Definitions. A *polyhedral set* is the solution set for a finite system of linear inequalities, and a finite union of polyhedral sets is called a *disjunctive set*. A disjunctive set is defined by a *disjunctive constraint* and a *disjunctive (linear) program* is the problem of maximizing (or minimizing) a linear functional over the intersection of a family of disjunctive sets. $\square$

Let $\cup_k F_{ik}$ be a disjunctive set for each $i, i = 1, \ldots, n$. Then the intersection of the disjunctive sets $\cup_k F_{ik}$ can be described by a logical formula in disjunctive normal form (DNF): $\vee_{h \in Q}(x \in K_h)$, where $|Q| = \prod_{i=1}^n M_i$. If $M_i = M_j = M$ for $i, j \leq n$, then $|Q| = M^n$. We shall call $K_h, \forall h \in Q$, a *disjunctive feasible region*, since $K_h \subseteq \cap_i \cup_k F_{ik} \, \forall h \in Q$. The set of vectors $\cup_{h \in Q} K_h$ is a disjunctive set. See Refs. 13 and 14 for a discussion of disjunctive linear programming. Benders decomposition[11] was developed for mixed integer programs and involves decomposition with respect to a set of "complicating" variables, that is, ones required to take on integer values. Linear subproblems are solved for a fixed integer-valued variable, and the optimal solution over all feasible integer values is found. An account of Benders decomposition can be found in Ref. 15. Hooker[16] used Benders decomposition after expressing a problem with logical requirements as a 0–1 program. Here, we will generalize Benders decomposition to disjunctive programs; i.e., we will solve linear problems with respect to variables required to take on values in certain disjunctive sets. Since each disjunctive set is defined by a "large" linear step-shaped program, we can then apply variable decomposition to solving each of these linear programs. Note that step-shaped programs are *hereditary* in the sense that adding constraints on x yields another step-shaped program.

Assume we have a problem with the constraint set of the step-shaped linear problem (P), but with additional logical requirements for x. Assume also that these requirements can be represented by the condition $x \in \cap_i \cup_j F_{ij}$ where F_{ij} is a polyhedral set for all i and j. Then the program below is a step-shaped disjunctive program.

$$
\begin{aligned}
(L) \quad \max \quad & cx + dy \\
\text{s.t.} \quad & A_1^1 x \le t \\
& A_2^1 x + A_2^2 y \le b \\
& A_3^2 y \le 0 \\
& x \in \cap_i \cup_j F_{ij} \,.
\end{aligned}
$$

Proposition 1. *Let $\{(x,y) : (x,y) \ is feasible for \ (L)\} \neq \emptyset$. Then the program (L) is equivalent to the following formulation for Benders decomposition:*

$$
(1) \quad \max_{x \in \cap_i \cup_j F_{ij}} \{cx + \min_{u \in J}\{\mathbf{u}_1(r - A_1^1 x) + \mathbf{u}_2(b - A_2^1 x) + \mathbf{u}_3 0\}\}
$$

where

$$
J = \{u \in \Re^{3S+N+2} : \mathbf{u}_1 A_1^2 + \mathbf{u}_2 A_2^2 + \mathbf{u}_3 A_3^2 = d\} \,.
$$

Proof. We can write (1) as

$$
\max_{x \in \cap_i \cup_j F_{ij}} G(x)
$$

where

$$
G(x) = cx + \min_{u \in J}\{\mathbf{u}_1(t - A_1^1 x) + \mathbf{u}_2(b - A_2^1 x) + \mathbf{u}_3 0\} \,.
$$

For each fixed $\bar{x}$, the linear program with $G(\bar{x})$ as its objective function and constraint set $\{u \in J\}$ is dual to

$$
\begin{aligned}
\max \quad & c\bar{x} + dy \\
\text{s.t.} \quad & A_1^2 y \le t - A_1^1 \bar{x} \\
& A_2^2 y \le b - A_2^1 \bar{x} \\
& A_3^2 y \le 0 \,.
\end{aligned}
$$

The result follows from LP duality. $\qquad\square$

We shall express the linear program corresponding to each *disjunctive region* of (L) separately. For each K_h, we have a linear step-shaped program (L_h).

$$\begin{aligned}
(L_h) \quad \max \quad & cx + dy \\
\text{s.t.} \quad & A_1^1 x \le t \\
& A_2^1 x + A_2^2 y \le b \\
& A_3^2 y \le 0 \\
& x \in K_h\,.
\end{aligned}$$

We can combine the constraints $A_1^1 x \le t$ with the requirement $x \in K_h$ and rewrite the constraint set as $E^h x \le e^h$. Then we have

$$\begin{aligned}
(L_h) \quad \max \quad & cx + dy \\
\text{s.t.} \quad & E^h x \le e^h \\
& A_2^1 x + A_2^2 y \le b \\
& A_3^2 y \le 0\,.
\end{aligned}$$

Since this is a linear step-shaped program, the previous discussion applies, and the full master problem for Benders decomposition of (L_h) is:

$$\begin{aligned}
(ML_h) \quad \max \quad & cx + \lambda \\
\text{s.t.} \quad & E^h x \le e^h \\
& \lambda \le \mathbf{u}_2^t (b - A_2^1 x)
\end{aligned}$$

for $\mathbf{u}_2^t = \pi(u^t)$, where $u^t = \langle \mathbf{u}_2^t, \mathbf{u}_3^t \rangle$ is an extreme point of the set $\{u \in \Re^{m_2 + m_3} : \mathbf{u}_2 A_2^2 + \mathbf{u}_3 A_3^2 = d\}$. The dual and primal subproblems are

$$\begin{aligned}
(DSP_h \bar{x}) \quad \min \quad & \mathbf{u}_2 (b - A_2^1 \bar{x}) + c\bar{x} \\
\text{s.t.} \quad & \mathbf{u}_2 A_2^2 + \mathbf{u}_3 A_3^2 = d \\
& u \ge 0
\end{aligned}$$

$$\begin{aligned}
(SP_h \bar{x}) \quad \max \quad & c\bar{x} + dy \\
\text{s.t.} \quad & A_2^1 \bar{x} + A_2^2 y \le b \\
& A_3^2 y \le 0\,.
\end{aligned}$$

Notice that the feasible region for the dual subproblem $(DSP_h \bar{x})$ is the same for every (L_h) and is in fact identical to the feasible region for the dual subproblem for the linear problem (P) which is a relaxation of (L). Let U be the collection of all inequalities of the form $\lambda \le \mathbf{u}_2^t (b - A_2^1 x)$ where $(\mathbf{u}_2, \mathbf{u}_3)$ is an extreme point of (DSP). Relaxations of the full master problem (MP) contain a subset of the inequalities from U and relaxations of (ML_h) contain a subset of inequalities from U in addition to a subset of the inequalities that define the requirement $x \in K_h$.

So the master problem at node n, where n is an ancestor of leaf h, is a relaxation of (ML_h) and of (MP). We can test for optimality for any $x \in \cup_{h \in Q} K_h$ both for some L_h and for (P). We also can use the fact that if a point x satisfies $x \in \cup_{h \in Q} K_h$ and tests optimal for a master problem without disjunctive constraints at a node and therefore for (P), then this point must be optimal for (L).

3.5.2. *Combined depth-first search and decomposition*

The constraint set for each polyhedral set K_h has a linear step-shaped coefficient matrix and if $i \in \{1, \ldots, N\}$ and $j \in \{1, \ldots, M\}$ there are M^N linear step-shaped programs to solve in any exhaustive search of this problem. Since the number of programs is exponential in the number of variables with disjunctive requirements, it is necessary computationally to shorten this search. We will use a depth-first search among the disjunctive regions and interleave the steps of the search with the cut generation steps of Benders (variable) decomposition. The point of this interleaving is to use the decomposition algorithm to direct the search for the maximal objective function among the disjunctive sets, and to use the maximum objective function value of any disjunctive feasible set as a lower bound to eliminate branches of the search tree.

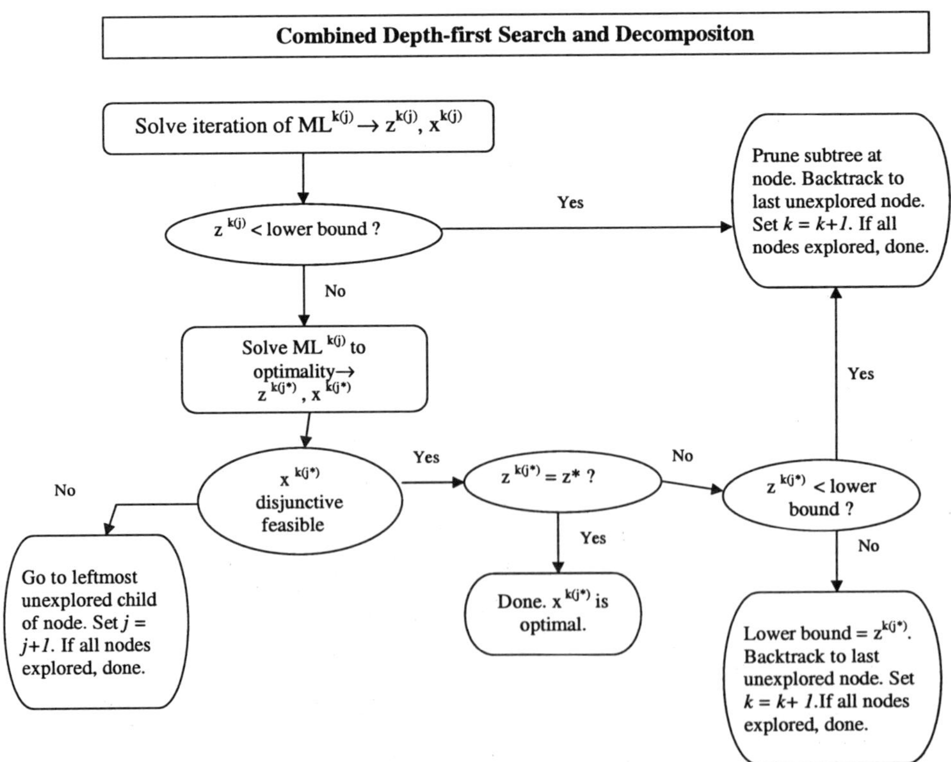

In combined depth-first-search and variable decomposition, a solution is found at a node and then tested for disjunctive feasibility. If the solution is in one of the disjunctive regions, then it is tested for optimality at a node. If it does not satisfy the disjunctive constraints, then a new disjunctive set is branched on. If a point x satisfies the disjunctive constraints, there is some K_h such that $x \in K_h$; and whether this point is optimal for (ML_h) or not, its objective function value can be used as a lower bound. Further, a problem at node n, (ML_n^k) at stage k is a relaxation of (ML_n^{k+1}) and all nodes descendent from it. So if a disjunctive feasible point is optimal for (ML_n^k) then all children of (ML_n^k) can be pruned; if the initial objective function value for (ML_n^k) is less than the current lower bound, (ML_n^k) and its children can be pruned. Further, if the candidate solution satisfies the disjunctive constraints, we can test it for local and global optimality at the same time, by checking if it is a solution to the global master problem without the disjunctive constraints.

In carrying out combined decomposition and depth-first search, all of the disjunctive requirements are satisfied at a leaf; only some of them may be satisfied at a node. The algorithm is as follows:

Stage 0:

Let $n = 0, \ldots, T$ index the nodes of the search tree. Set $k := 0$. Select the root node as the first node to examine. Set lower bound equal to $-\infty$. Let (ML_0^0) be the full master problem at the root node.

$$
\begin{aligned}
(ML_0^0) \quad \max \quad & cx + \lambda \\
\text{s.t.} \quad & A_1^1 x \leq r \\
& \lambda \leq \mathbf{u}_2(b - A_2^1 x).
\end{aligned}
$$

Solve $ML_0^0 (\equiv P)$ to optimality, obtaining x^* and z^*. If x^* is disjunctive feasible, then x^* is optimal solution to disjunctive problem L. If x^* is not disjunctive feasible, continue.

Stage k $(k > 0)$:

(1) Solve the relaxed master problem (ML_n^k) at node n to get $(\bar{x}_n^k, \bar{\lambda}_n^k)$ and the relaxed objective function value $\bar{z}_n^k$. If $\bar{z}_n^k \geq$ lower bound, go to (2). If $\bar{z}_n^k <$ lower bound, go to (3). If (ML_n^k) is infeasible, go to (3).
(2) Solve ML_n^k to optimality, obtaining z_n^{k*}, x_n^{k*}.

    (a) If $z_n^{k*} <$ lower bound, then go to (3).
    (b) If $z_n^{k*} \geq$ lower bound, then test if x_n^{k*} is disjunctive feasible.

        (i) If x_n^{k*} is disjunctive feasible, then if $z_n^{k*} = z^*$, the optimal solution to ML_0^0, then x_n^{k*} is optimal and we're done. If $z_n^{k*} < z^*$, then z_n^{k*} is new lower bound. Go to (3).

   (ii) If x_n^{k*} is not disjunctive feasible then go to leftmost unexplored child of node. Set $n = n + 1$. If all nodes have been explored, stop; current lower bound is optimal. Else, go to (1).

(3) Prune subtree at node m. Set $k := k + 1$. Backtrack to last unexplored node m. If all nodes have been explored, stop. The lower bound is optimal. Else, go to (1).

Proposition 2. *Assume that the linear relaxation* (P) *of the disjunctive program* (L) *is feasible. If* (L) *is feasible, then the algorithm combining variable decomposition and depth-first search will terminate with the optimal solution. If* (L) *is infeasible, then the algorithm will end with no solution.*

Proof. If no nodes are pruned in the search for the optimal solution to (L), then all M^N leaves in the search tree will be explored and the optimal objective function value will be found for each disjunctive feasible set. In this case, the algorithm clearly terminates and the optimal solution is the best objective function value found at a leaf. Since any nodes pruned by this algorithm achieved relaxed objective function values strictly less than the current lower bound, and since any parent node is a relaxation of its descendents, leaves which contain optimal solutions are not removed. Therefore, the algorithm will terminate with the optimal value.

However, if there is no feasible solution to (L), then every master problem at a leaf is infeasible. This infeasibility may be detected at a node higher up in the search tree, in which case the descendents of this node will not be explored. But since every disjunctive region is considered, the algorithm will terminate with no solution. $\qquad\square$

References

[1] Katherine Wyatt, "Maximizing hedge effectiveness under FASB 133 accounting standards", talk at The Fourth International Congress on Industrial and Applied Mathematics, Edinburgh, Scotland, 5–9 July 1999.

[2] Katherine Wyatt, "Optimal hedging relationships", FAS 133 Special Report, *Energy & Power Risk Management*, May 2000, p. 17.

[3] Harry M. Markowitz, *Portfolio Selection: Efficient Diversification of Investments*, Cowles Foundation for Research in Economics at Yale University, New Haven, 1959.

[4] Hiroshi Konno, "Piecewise linear risk function and portfolio optimization", *Journal of the Operations Research Society of Japan* **33**(2) (1990) 139–156.

[5] Hiroshi Konno and Hiroaki Yamazaki, "Mean-absolute deviation portfolio optimization and its applications to Tokyo stock market", *Management Science* **37**(5) (1991) 519–531.

[6] J. Kenneth Worzel, Christiana Vassidou-Zeniou and Stavros A. Zenios, "Integrated simulation and optimization models for tracking indices of fixed-income securities", *Operations Research* **42**(2) (1994) 223–233.

[7] Daniel Bienstock, "Computational study of a family of mixed-integer quadratic programming problems", *Mathematical Programming* **74** (1996) 121–140.

[8] Stavros A. Zenios and Pan Kang, "Mean-absolute deviation portfolio optimization for mortgage-backed securities", *Annals of Operations Research* **45** (1993) 433–450.

[9] S. Randall Hiller and Jonathan Eckstein, "Stochastic dedication: Designing fixed income portfolios using massively parallel Benders decomposition", *Management Science* **39**(11) (1993) 1422–1437.

[10] Robert A. Jarrow, *Modelling Fixed Income Securities and Interest Rate Options*, The McGraw-Hill Companies, Inc., New York, 1996.

[11] J. F. Benders, "Partitioning procedures for solving mixed variables programming problems", *Numerische Mathematik* **4** (1962) 238–252.

[12] R. M. Van Slyke and Roger Wets, "L-Shaped linear programs with applications to optimal control and stochastic programming", *SIAM J. Appl. Math.* **17**(4) (1969) 638–663.

[13] Ken McAloon and Carol Tretkoff, *Optimization and Computational Logic*, Wiley, New York, 1996.

[14] Ken McAloon and Carol Tretkoff, "Logic, modeling and programming", *Annals of Operations Research*, to appear.

[15] Alexander Schrijver, *Theory of Linear and Integer Programming*, Wiley, Chichester, 1986.

[16] J. N. Hooker and H. Yan, "Verifying logic circuits by Benders decomposition", in *Principles and Practice of Constraint Programming*, Vijay Saraswat and Pascal Van Hentenryck, eds., (MIT Press, 1995).

[17] Harry M. Markowitz, *Mean-Variance Analysis in Portfolio Choice and Capital Markets*, Basil Blackwell, Oxford, 1987.

[18] Sharpe and F. William, "A linear programming algorithm for mutual fund portfolio selection", *Management Science* **13**(7) (1967) 499–510.

[19] Sharpe and F. William, "Mean-absolute-deviation characteristic lines for securities and portfolios", *Management Science* **18**(2) (1971) B1–B13.

[20] Katherine Wyatt, "Decomposition Techniques and Disjunctive Linear Programming for Fixed-Income Portfolio Selection", doctoral dissertation, Mathematics, City University of New York, 1997.